Optimal Power Flow Using Evolutionary Algorithms

Provas Kumar Roy
Kalyani Government Engineering College, India

Susanta Dutta
Dr. B. C. Roy Engineering College, India

A volume in the Advances in Computer and Electrical Engineering (ACEE) Book Series

IGI Global
DISSEMINATOR OF KNOWLEDGE

Published in the United States of America by
IGI Global
Engineering Science Reference (an imprint of IGI Global)
701 E. Chocolate Avenue
Hershey PA, USA 17033
Tel: 717-533-8845
Fax: 717-533-8661
E-mail: cust@igi-global.com
Web site: http://www.igi-global.com

Library of Congress Cataloging-in-Publication Data

Names: Roy, Provas Kumar, 1973- author. | Dutta, Susanta, 1976- author.
Title: Optimal power flow using evolutionary algorithms / by Provas Kumar Roy and Susanta Dutta.
Description: Hershey, PA : Engineering Science Reference, an imprint of IGI Global, [2019] | Includes bibliographical references and index.
Identifiers: LCCN 2018010795| ISBN 9781522569718 (hardcover) | ISBN 9781522569725 (ebook)
Subjects: LCSH: Electric power systems--Control. | Electric power systems--Load dispatching--Mathematics. | Electric power transmission--Mathematical models. | Evolutionary computation. | Genetic algorithms. | Mathematical optimization.
Classification: LCC TK1007 .R69 2019 | DDC 621.310285/63823--dc23 LC record available at https://lccn.loc.gov/2018010795

This book is published in the IGI Global book series Advances in Computer and Electrical Engineering (ACEE) (ISSN: 2327-039X; eISSN: 2327-0403)

British Cataloguing in Publication Data
A Cataloguing in Publication record for this book is available from the British Library.

Advances in Computer and Electrical Engineering (ACEE) Book Series

ISSN:2327-039X
EISSN:2327-0403

Editor-in-Chief: Srikanta Patnaik, SOA University, India

MISSION

The fields of computer engineering and electrical engineering encompass a broad range of interdisciplinary topics allowing for expansive research developments across multiple fields. Research in these areas continues to develop and become increasingly important as computer and electrical systems have become an integral part of everyday life.

The **Advances in Computer and Electrical Engineering (ACEE) Book Series** aims to publish research on diverse topics pertaining to computer engineering and electrical engineering. **ACEE** encourages scholarly discourse on the latest applications, tools, and methodologies being implemented in the field for the design and development of computer and electrical systems.

COVERAGE

- Applied Electromagnetics
- Chip Design
- Sensor Technologies
- Computer Architecture
- VLSI Fabrication
- Analog Electronics
- Digital Electronics
- Optical Electronics
- Programming
- Electrical Power Conversion

IGI Global is currently accepting manuscripts for publication within this series. To submit a proposal for a volume in this series, please contact our Acquisition Editors at Acquisitions@igi-global.com or visit: http://www.igi-global.com/publish/.

The Advances in Computer and Electrical Engineering (ACEE) Book Series (ISSN 2327-039X) is published by IGI Global, 701 E. Chocolate Avenue, Hershey, PA 17033-1240, USA, www.igi-global.com. This series is composed of titles available for purchase individually; each title is edited to be contextually exclusive from any other title within the series. For pricing and ordering information please visit http://www.igi-global.com/book-series/advances-computer-electrical-engineering/73675. Postmaster: Send all address changes to above address.

Titles in this Series

For a list of additional titles in this series, please visit:
https://www.igi-global.com/book-series/advances-computer-electrical-engineering/73675

Multidisciplinary Approaches to Service-Oriented Engineering
Mehdi Khosrow-Pour, D.B.A. (Information Resources Management Association, USA)
Engineering Science Reference • ©2018 • 389pp • H/C (ISBN: 9781522559511) • US $245.00

Cyber-Physical Systems for Next-Generation Networks
Joel J. P. C. Rodrigues (National Institute of Telecommunications (Inatel), Brazil & Instituto de Telecomunicações, Portugal & University of Fortaleza (UNIFOR), Brazil) and Amjad Gawanmeh (Khalifa University, UAE)
Engineering Science Reference • ©2018 • 293pp • H/C (ISBN: 9781522555100) • US $210.00

Augmented Reality for Enhanced Learning Environments
Gerardo Reyes Ruiz (Universidad Autónoma del Estado de México, Mexico) and Marisol Hernández Hernández (Universidad Autónoma del Estado de México, Mexico)
Information Science Reference • ©2018 • 322pp • H/C (ISBN: 9781522552437) • US $195.00

Handbook of Research on Cloud and Fog Computing Infrastructures for Data Science
Pethuru Raj (Reliance Jio Infocomm. Ltd., India) and Anupama Raman (Societe Generale Global Solution Center, India)
Engineering Science Reference • ©2018 • 440pp • H/C (ISBN: 9781522559726) • US $245.00

Advanced Synchronization Control and Bifurcation of Chaotic Fractional-Order Systems
Abdesselem Boulkroune (Jijel University, Algeria) and Samir Ladaci (National Polytechnic School of Constantine, Algeria)
Engineering Science Reference • ©2018 • 539pp • H/C (ISBN: 9781522554189) • US $215.00

Innovative Methods, User-Friendly Tools, Coding, and Design Approaches in People-Oriented ...
Steve Goschnick (Swinburne University of Technology, Australia)
Engineering Science Reference • ©2018 • 488pp • H/C (ISBN: 9781522559696) • US $215.00

For an entire list of titles in this series, please visit:
https://www.igi-global.com/book-series/advances-computer-electrical-engineering/73675

IGI Global
DISSEMINATOR OF KNOWLEDGE

701 East Chocolate Avenue, Hershey, PA 17033, USA
Tel: 717-533-8845 x100 • Fax: 717-533-8661
E-Mail: cust@igi-global.com • www.igi-global.com

Table of Contents

Preface

Electric energy is an essential need of today's society. Past blackout experiences have shown that an extensive failure of the electric power system may paralyze the entire country. The reasons for such blackouts are manifold but often they occur as a consequence of cascading events. When a component fails, subsequent failures become more probable. Today's power grids are driven closer to their transfer capacities due to the increased consumption and power transfers, endangering the security of the system.

Transmission lines in congested areas are often driven close to or even beyond their limits in order to satisfy the increased electric power consumption and trades. Thus, secure operation and reliable supply is endangered by the higher risks for faulted lines. However, the construction of additional power lines is often difficult for environmental, economic, and political reasons. This is where the technology of Flexible AC Transmission Systems (FACTS) provides a significant opportunity.

The optimal power flow (OPF) and optimal reactive power dispatch (ORPD) are considered as the emerging areas for getting quality power under normal operating condition. The OPF and ORPD utilize all control variables to minimize the costs, transmission loss, voltage deviation and L-index of the power system operation. It also yields valuable economic information and insight into the power system.

At present, another important issue in power system operation is that, power systems are forced to operate at almost full capacity. More and more power generation patterns result in heavy line flows that tend to incur greater losses as well as threatening stability and security of the system. This ultimately creates undesirable increased risk of power outages of different levels of severity. For this reason, there is a general consensus that the power grid has to be reinforced, to make it smart and aware, fault tolerant and self-healing, and dynamically and statically controllable. A traditional alternative to reinforce the power network consists of upgrading the electrical transmission system infrastructure through the addition of new transmission lines, substations and associated equipment.

The utilization of the existing power system may be improved through the application of advanced power electronics devices. FACTS devices provide technical solutions to address the new operating challenges. Popular FACTS devices, such as thyristor controlled series compensator (TCSC), thyristor controlled phase shifter (TCPS), static compensator (STATCOM), static synchronous series compensator (SSSC) and unified power flow controller (UPFC) can be connected in series, shunt or a combination of two to achieve numerous control functions, including voltage regulation and power flow control. In this way, the system performance can considerably be improved by controlling the power flows without changing the generation rescheduling. Furthermore, the thermal limits may not be violated, losses may be minimized and the stability margin may also be increased. The potential benefits of FACTS equipment are nowadays widely recognized by the power systems engineering community. However, the current challenge is now to obtain the maximum benefit from these devices at minimum cost. Thus, the problem becomes an optimization problem which involves a very complex formulation and it is certainly difficult to solve in an efficient manner.

On the one hand, the power system itself is a highly non-linear and non-stationary system, subject to noise and uncertainties. On the other hand, the operation of the FACTS devices is quite difficult to optimize because of the complexity in the control schemes. From the optimization perspective, the inclusion of discontinuous and discrete domain variables, the existence of a non-convex feasible region and multiple local minima, all make it difficult to find a suitable technique with the ability to pursue the global optimum with reasonable computational effort.

Classical mathematical optimization algorithms such as Newton-based method (Liu et al., 1992), interior point method (Yan et al., 2006), quadratic programming (Quintana & Santos-Nieto, 1989), gradient based methods (Burchett, 2013) suffer from certain limitations and are not able to solve the above-mentioned problems efficiently. This is because the conventional methods depend upon the derivative of the objective function. Due to the non-differentiable objective functions of the different problems discussed in this thesis, the conventional methods fail to give global optimal results. But with the advent of evolutionary intelligent computational techniques, near global optimal solutions may be obtained. To overcome this problem some intelligent optimization algorithms known as heuristic techniques are applied to solve OPF and ORPD problem.

The basic idea of intelligent optimization algorithm is starting from an initial solution population, according to the probability principle of transfer, searching for the most adaptive optimal solution by using some means. Particle swarm optimization (PSO; Yoshida et al., 2000), differential evolution (DE; Liang et al., 2007), evolutionary programming (EP; Liang et al., 2006), bacteria foraging optimization (BFO; Edward et al., 2013), gravitational search algorithm (GSA; Roy et al., 2012), artificial bee colony (ABC; Karaboga & Basturk, 2008), harmony search algorithm (has; Sirjani et al., 2012), firefly algorithm (FA; Manoranjitham & Shunmugalatha, 2015), etc. are known as efficient population based optimization techniques to solve large scale non-linear optimization problems. It is not largely affected by the size and nonlinearity of the problem, and may converge to the optimal solution in many problems where most of the classical methods fail to converge. The above-mentioned techniques can therefore be effectively applied to power systems optimization problem.

However, the problem associated with the aforesaid population-based techniques is, most of them suffer from poor convergence rate and low exploitation ability. Therefore, proposing and implementing new evolutionary methods to real world problem are always welcome. This encourages the present researcher to pay more attention to the development of some effective algorithms. Three relatively new modern heuristic optimization algorithms, named, teaching-learning-based optimization algorithm (TLBO; Mandal & Roy, 2013), chemical reaction optimization (CRO; Xu et al., 2011), grey wolf optimization (GWO; Pradhan et al., 2016) and biogeography based optimization (BBO; Roy et al., 2010) are introduced in this book to solve FACTS based OPF and ORPD problems. These approaches have some advantages over other population based optimization techniques such as (1) they are easier to implement and there are few parameters to adjust, therefore they are able to perform an efficient search of the problem space, and (2) they maintain the diversity of the particles, thus it is capable to avoid getting trapped in local minima. However, the above stated algorithms have got some disadvantages such as they sometimes require large iteration cycles and computational time to reach to the optimal solution. To enhance the convergence speed of these basic algorithms and to provide the global optimum solution of the same, the present researcher has proposed three modified techniques, namely hybridization of DE with BBO (HDE-BBO), opposition-based teaching-learning-based optimization (OTLBO), and hybrid chemical reaction optimization with differential evolution (HCRO). These algorithms have hardly been used to solve power system optimization problem so far. They are implemented successfully to some selected power system optimization problems in this research work to check their capability of achieving global optimality. It is observed through rigorous simulation that HDE-BBO, TLBO, and HCRO are more efficient than the previously stated BBO, TLBO, and CRO algorithms.

This book is composed of nine different chapters.

Chapter 1 deals with different optimization methods, including BBO, ABC, TLBO, CRO, KHA, GWA, and some modified versions like OTLBO, HDE-BBO, and HCRO. Steps of these evolutionary optimization techniques are presented in this chapter.

Chapter 2 represents the limitation of economic load dispatch and concept of optimal power flow and optimal reactive power dispatch problems.

Chapter 3 describes BBO, TLBO, CRO, KHA, GWA, and HCRO algorithms applied to ELD problems. To check the effectiveness, the aforesaid techniques are implemented on 10-unit, 15-unit, 40-unit, 80-unit and 140-unit test systems. In order to judge the superiority, the proposed algorithms are compared with other techniques available in the literature.

Chapter 4 deals with OPF and ORPD problems without FACTS devices and the aforesaid techniques are implemented on IEEE 30-bus, 57-bus, systems. Comparative simulation study is made to judge the performances of DE, BBO, HDE-BBO, TLBO, OTLBO, CRO, KHA, GWA, and HCRO.

Chapter 5 presents the different types of Flexible AC transmission systems (FACTS) controller in power transmission. Moreover, their objectives, connection method and advantages and drawbacks are also focus in this chapter.

Chapter 6 deals with power flow modeling, active and reactive power flow after connecting either thyristor-based or voltage-source-based FACTS devices in power transmission line.

Chapter 7 deals with OPF and ORPD problems with the introduction of two FACTS devices namely, thyristor-controlled series compensator (TCSC) and thyristor-controlled phase shifter (TCPS). Moreover, the performance of the aforesaid techniques for minimization of generation cost, minimization of transmission loss and minimization of voltage deviation are analyzed. To test the effectiveness and optimization performance of above technique, they are applied on IEEE 30-bus systems and the test results are compared to the same of other optimization techniques published in the literature.

Chapter 8 presents OPF and ORPD problems with the optimal position of various FACTS devices namely, TCSC, SVC, and TCPS. The proposed techniques are implemented on IEEE 30-bus system to find the optimal sizing and positions of the aforesaid FACTS devices. The comparative results are thoroughly analyzed to judge the performances of BBO, HDE-BBO, ABC, TLBO, OTLBO, CRO, and HCRO approaches.

Chapter 9 finds optimal location of STATCOM, SSSC, and UPFC in order to minimize the fuel cost, installation cost of UPFC devices, transmission loss, voltage deviation of stressed in IEEE 30-bus and IEEE 57-bus power system systems. To assess the superiority and effectiveness, the simulation results of BBO, HDE-BBO, ABC, TLBO, OTLBO, CRO, and HCRO are compared with those obtained with other optimization techniques available in the literature.

REFERENCES

Burchett, B. T. (2013). Aerodynamic parameter identification for symmetric projectiles: An improved gradient based method. *Aerospace Science and Technology*, *30*(1), 119–127. doi:10.1016/j.ast.2013.07.010

Edward, J. B., Rajasekar, N., Sathiyasekar, K., Senthilnathan, N., & Sarjila, R. (2013). An enhanced bacterial foraging algorithm approach for optimal power flow problem including FACTS devices considering system load ability. *ISA Transactions*, *52*(5), 622–628. doi:10.1016/j.isatra.2013.04.002 PMID:23759251

Karaboga, D., & Basturk, B. (2008). On the performance of artificial bee colony (ABC) algorithm. *Applied Soft Computing*, *8*(1), 687–697. doi:10.1016/j.asoc.2007.05.007

Liang, C. H., Chung, C. Y., Wong, K. P., & Duan, X. Z. (2006). Comparison and improvement of evolutionary techniques for power system optimal reactive power flow. *IEE Proceedings. Generation, Transmission and Distribution*, *153*(2), 228–236. doi:10.1049/ip-gtd:20050081

Liang, C. H., Chung, C. Y., Wong, K. P., Duzn, X. Z., & Tse, C. T. (2007). Study of differential evolution for optimal reactive power flow. *IET Generation, Transmission & Distribution*, *1*(2), 253–260. doi:10.1049/iet-gtd:20060123

Liu, W. H. E., Papalexopoulos, A. D., & Tinney, W. F. (1992). Discrete shunt controls in a Newton optimal power flow. *IEEE Transactions on Power Systems*, *7*(4), 1509–1518. doi:10.1109/59.207375

Mandal, B., & Roy, P. K. (2013). Optimal reactive power dispatch using quasi-oppositional teaching learning based optimization. *International Journal of Electrical Power & Energy Systems*, *53*, 123–134. doi:10.1016/j.ijepes.2013.04.011

Manoranjitham, G. E., & Shunmugalatha, A. (2015). Application of firefly algorithm on optimal power flow control incorporating simplified impedance UPFC model. *International Journal of Electrical Power & Energy Systems*, *71*, 358–363. doi:10.1016/j.ijepes.2015.03.009

Pradhan, M., Roy, P. K., & Pal, T. (2016). Grey wolf optimization applied to economic load dispatch problems. *International Journal of Electrical Power & Energy Systems*, *83*, 325–334. doi:10.1016/j.ijepes.2016.04.034

Quintana, V. H., & Santos-Nieto, M. (1989). Reactive-power dispatch by successive quadratic programming. *IEEE Transactions on Energy Conversion*, *4*(3), 425–435. doi:10.1109/60.43245

Rizk-Allah, R. M., El-Sehiemy, R. A., & Wang, G.-G. (2018). A novel parallel hurricane optimization algorithm for secure emission/economic load dispatch solution. *Applied Soft Computing*, *63*, 206–222. doi:10.1016/j.asoc.2017.12.002

Roy, P. K., Ghoshal, S. P., & Thakur, S. S. (2010). Multi-objective Optimal power flow using biogeography based optimization. *Electric Power Components and Systems*, *38*(12), 1406–1426. doi:10.1080/15325001003735176

Roy, P. K., Mandal, B., & Bhattacharya, K. (2012). Gravitational search algorithm based optimal reactive power dispatch for voltage stability enhancement. *Electric Power Components and Systems*, *40*(9), 956–976. doi:10.1080/15325008.2012.675405

Sirjani, R., Mohamed, A., & Shareef, H. (2012). Optimal allocation of shunt Var compensators in power systems using a novel global harmony search algorithm. *International Journal of Electrical Power & Energy Systems*, *43*(1), 562–572. doi:10.1016/j.ijepes.2012.05.068

Xu, J., Lam, A. Y. S., & Li, V. O. K. (2011). Chemical reaction optimization for task scheduling in grid computing. *IEEE Transactions on Parallel and Distributed Systems*, *22*(10), 1624–1631. doi:10.1109/TPDS.2011.35

Yan, W., Lu, S., & Yu, D. C. (2006). A hybrid genetic algorithm–interior point method for optimal reactive power flow. *IEEE Transactions on Power Systems*, *21*(3), 1163–1169. doi:10.1109/TPWRS.2006.879262

Yoshida, H., Fukuyama, Y., Kawata, K., Takayama, S., & Nakanishi, Y. (2000). A particle swarm optimization for reactive power and voltage control considering voltage security assessment. *IEEE Transactions on Power Systems*, *15*(4), 1232–1239. doi:10.1109/59.898095

Chapter 1
Description of Various Evolutionary Optimization Techniques

ABSTRACT

Despite the success of various classical optimization techniques, there remains a large class of problems where either these methods are unable to find the satisfactory results or the computational times are sufficiently large. Several heuristic methods have emerged in the recent years as complementary tools to various mathematical approaches. These methods include genetic algorithm (GA), evolutionary programming (EP), particle swarm optimization (PSO), differential evolution (DE), and so on. Researchers are constantly trying to learn from the behavioral pattern of organisms and implementing those ideas and philosophies in solving optimizing problems. In this chapter, a few efficient optimization algorithms, namely grey wolf optimization (GWO), teaching-learning-based optimization (TLBO), biogeography-based optimization (BBO), krill herd algorithm (KHA), chemical reaction optimization (CRO) algorithms, and hybrid CRO (HCRO) are discussed, and in the subsequent chapters, the performance of the aforesaid algorithms are investigated by applying them in a few areas of power systems.

DOI: 10.4018/978-1-5225-6971-8.ch001

INTRODUCTION

We often encounter optimization problems in scientific and technological research and development. The word "optimum" is latin, and means "the ultimate ideal" similarly, "optimus" means "the best." Therefore, to optimize refers to try to bring whatever optimize we are dealing with towards its ultimate state. Let us take a closer look at what that means in terms of an example, and at the same time bring the definition of the term optimization forward, as the scientific field understands and uses it. Optimization is the process of making something better. In other words, optimization is the process of adjusting the inputs to or characteristics of a device, mathematical process, or experiment to find the minimum or maximum output or result.

Optimized operation of the power system is a challenging problem due to its complex and non-linear nature. There are also many uncertainties in power system, which add its complexity further. Recently, with the introduction of deregulation in power industry (Roy, 2018) many new complexities are added to the existing problems. The application of optimization techniques to various power system problems has been an area of research for many years.

The different techniques are applied to power system optimization problems over the years. These algorithms are generally divided into two categories: (i) classical mathematical optimization algorithms and (ii) intelligent optimization algorithms. The basic idea of the classical algorithm is starting from an initial point, continuously improving the current solution through a certain orbit, and ultimately converging to the optimal solution. These algorithms include linear programming (LP) (Aoki et al., 1988; Deeb, & Shahidehpour, 1988; Momoh et al., 1999; Urdaneta et al., 1999), Lagrangian decomposition method (Granada et al., 2012), quadratic programming (QP) (Grudinin, 1998; Lo & Zhu, 1991), non-linear programming (NLP) (Habiabollahzadeh et al., 1989; Lai & Ma, 1997), interior point method (IPM) (Capitanescu & Wehenkel, 2013; Granville, 1995; Yan. & Quintana, 1999; Zhu & Xiong, 2003) Newton method (Santos & Costa, 1995; Sun et al., 1984), mixed integer linear programming (MILP) (Kim et al., 1998), gradient search (Burchett, 2013; Yu et al., 1986), neural network (NN) (Dhillon, 1991; Park et al., 1993), Lagrange Hopfield network (Vo & Ongsakul, 2012), etc. Some of these techniques, however, have a good convergence, but most of them suffer from reaching the global optimal solution because the most of the power system optimization problems are a multimodal and non-linear problem and severely depends on the initial guess.

Recently, with the advent of computer technology, various evolutionary algorithms (Li et al., 2018) are evolved and successfully applied to solve the various optimization problems. Some of the well popular evolutionary techniques are genetic algorithm (GA) (Wei et al., 2006) simulated annealing (SA) (Keyan et al., 2006; Roa-Sepulveda & Pavez-Lazo, 2003), tabu search (TS) (Wennan et al.,2008). ant swarm optimization (ASO) (Cai et al., 2012 ; Jiejin et al., 2010), differential evolution (DE) (Cai et al., 2008; Shouzheng et al., 2010; Sivasubramani & Swarup, 2012), evolutionary algorithm (EA) (Abido, 2004; Das & Patvardhan, 2003), particle swarm optimization (PSO) (Abido, 2002; AlRashidi & El-Hawary, 2007; Gaing, 2005), artificial bee colony optimization (ABC) (Karaboga & Basturk, 2007; Ayan & Kılıc, 2012; Kursat & Ulas, 2013), biogeography based optimization (BBO) (Roy et al., 2011), imperialist competitive algorithm (ICA) (Mohammadi-ivatloo et al., 2012), ant colony optimization (ACO) (Huang, 2001), seeker optimization algorithm (SOA) (Chaohua et al., 2009), honey bee mating optimization (HBMO) (Bozorg et al., 2006; Javidan & Ghasemi, 2012; Niknam et al., 2011), bacteria foraging optimization (BFO) (Tang et al., 2006; Tripathy & Mishra, 2007; Nanda et al., 2009), gravitational search algorithm (GSA) (Hadi et al., 2012; Roy et al., 2012), ongroup search optimization (OSO), (Guo et al., 2012), harmony search algorithm (HSA) (Khazali & Kalantar, 2011), cuckoo optimization algorithm (COA) (Rajabioum, 2011), charge system search algorithm (CSSA) (Ozyon et al., 2012), krill herd algorithm (KHA) (Gandomi & Alavi, 2012), multi verse optimization (MVO) (Mirjalili, 2015; Sivasubramani & Swarup, 2011), parallel hurricane optimization algorithm (PHO) (Rizk-Allah et al., 2018) etc. Modern heuristic optimization techniques are paid much more attention with respect to conventional techniques by many researchers due to their ability to find an almost global optimal solution with satisfying all operating constraints.

LITERATURE SURVEY

Classical optimization techniques were recognized as early as 1930. In power systems, the Lagrangian relaxation (LR) approach (Wood, 1982) is conventionally used for economic dispatch. Linear programming (LP) is also used in various power system applications including ELD, OPF and reactive power flow. Wells (Wells, 1968) presented a LP approach to determine an economic schedule for loading plants in a power system. The objective function and constraints were linearized and solved using the simplex method. Shen and Laughton (1970) (Shen & Laughton, 1970) developed a dual linear programming technique for load scheduling. Aoki *et al.* (Aoki et al., 1987) presented recursive quadratic programming technique for solving constrained load flow problems. Petcharaks *et al.* (Petcharaks & Ongsakul, 2007) reported a hybrid method based on enhanced Lagrangian relaxation and quadratic programming technique for solving hydrothermal scheduling problems. Dynamic programming is a powerful mathematical optimization technique developed by Richard Ernest Bellman (Bellman, and Dreyfus, 1962). Lu and Hsu (1995) (Lu & Hsu, 1995) reported dynamic programming method to solve reactive power control. Hobbs *et al.* (Hobbs et al., 1988) applied a method based on enhanced dynamic programming approach in unit commitment problem. An algorithm based on dynamic programming technique for hydro thermal scheduling was developed by Engles *et al.* (Engles et al., 1976). Ramirez *et al.* (Ramirez & Onate, 2003) applied GA in the short term hydrothermal generation scheduling problem. Ghoshal (Ghoshal, 2004) developed GA/GA-SA based fuzzy automatic generation control of a multi-area thermal generating system in 2004. Wong and Wong (1994) (Wong & Wong, 1994) proposed a short term hydrothermal scheduling technique based on the simulated annealing. Ghoshal *et al.* (Ghoshal et al., 2009) proposed a bio-inspired fuzzy logic based tuning of power system stabilizer. Huang (Huang, 2001) presented an ant colony optimization technique for the enhancement of hydroelectric generation scheduling. Taher and Amooshahi (2012) (Taher & Amooshahi, 2012) developed a hybrid immune algorithm based optimal power flow using FACTS device. Shaheen *et al.* (Shaheen et al.,2011) reported differential evolution (DE) algorithm for optimal placement of UPFC for OPF problem. Rajasomashekar and Aravindhababu (2012) (Rajasomashekar & Aravindhababu, 2012) developed an elegant biogeography based optimization (BBO) algorithm for obtaining the best compromise solution of EED problem. Mondal *et al.* (Mondalet et al., 2013) presented gravitational search algorithm (GSA) to solve combined economic emission dispatch problem.

The literature of different heuristic techniques applied for OPF incorporating FACTS devices, ORPD incorporating FACTS devices and optimal location of FACTS device for solving OPF and ORPD problem are presented in details in the section "Literature Review" of the corresponding chapters.

EVOLUTIONARY ALGORITHMS UNDER STUDY

In this chapter, variants of evolutionary techniques which are investigated and applied to some selected power system problems are discussed in the following subsections. These include biogeography based optimization (BBO), artificial bee colony (ABC) optimization, differential evolution (DE), teaching learning based optimization (TLBO), chemical reaction based optimization (CRO), krill herd algorithm (KHA), grey wolf optimization (GWO). Moreover, to improve the convergence speed and the simulation results some modifications are made on BBO, TLBO and CRO. The modified versions are termed as follows:

1. Hybrid biogeography-based optimization with differential evolution (HDE-BBO)
2. Hybrid chemical reaction optimization with differential evolution (HCRO)
3. Opposition teaching learning-based optimization (OTLBO).

Biogeography Based Optimization (BBO)

Biogeography

Biogeography is nature's way of distributing species. Figure 1 describes a model of species abundance in a habitat (MacArthur & Wilson, 1997). The immigration curve of Figure 1 shows that immigration rate (I_r) to the habitat is maximum (I_{max}), when there are zero species in the habitat. As the number of species increases, the habitat becomes more crowded, fewer species are able to successfully survive immigration to the habitat, and the immigration rate decreases. The largest possible number of species that the habitat can support is $N_{S\max}$, at which the immigration rate becomes zero. The emigration curve of Figure 1 depicts that if there is no species in the habitat, the emigration rate will be zero. As the number of species increases, the habitat becomes more crowded, more species are able to leave the habitat to explore other possible residences and the emigration rate increases. The maximum emigration rate is E_{max}, which occurs when the habitat contains the largest number of species ($N_{S\max}$). The equilibrium number of species is N_{S0}, at which the immigration rate (I_r) and emigration rates (E_r) are equal. The immigration and emigration curves in Figure 1 are shown as straight lines but, in general, they might be more complicated curves. Nevertheless, this simple model gives us a general description of the process of immigration and emigration.

From Figure 1, I_r & E_r for N_S number of species may be formulated as follows:

$$I_r = I_{max} \cdot \left(1 - \frac{N_S}{N_{S\max}}\right) \quad (1)$$

$$E_r = \frac{E_{max} \cdot N_S}{N_{S\max}} \quad (2)$$

Mathematically the concept of emigration and immigration can be represented by a probabilistic model. P_{N_S} is the probability of the habitat that contains exactly N_S species. P_{N_S} changes from time t to time $t+\Delta t$ as follows (Simon, 2008).

$$P_{N_S}\left(t+\Delta t\right) = P_{N_S}\left(1 - I_r\Delta t - E_r\Delta t\right) + P_{N_{s-1}} I_{r_{S-1}} \Delta t + N_{P_{S+1}} I_{r_{S+1}} \Delta t \quad (3)$$

Figure 1. Species model of a single habitat

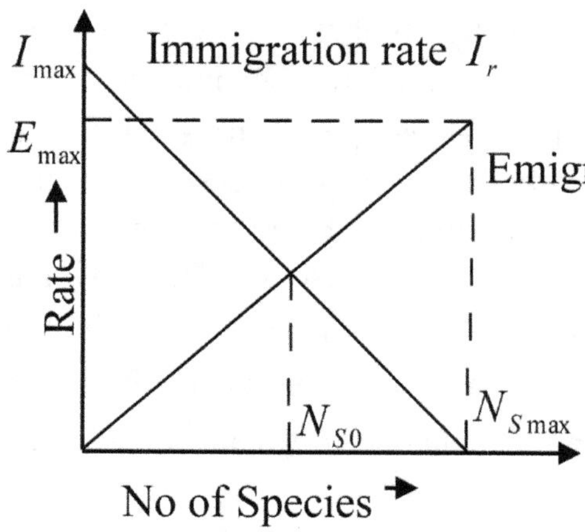

where $I_{r_{S+1}}, I_{r_{S-1}}$ are the immigration rate when there are N_{S+1} and N_{S-1} species in the habitat; $E_{r_{S+1}}, E_{r_{S-1}}$ are the emigration rate when there are N_{S+1} and N_{S-1} species in the habitat.

This equation holds because in order to have N_S species at time $t+\Delta t$, one of the following conditions must hold:

1. There are N_S species at time t, and no immigration or emigration occurs between t and $t+\Delta t$;
2. There are N_{S-1} species at time t, and one species immigrates;
3. There are N_{S+1} species at time t, and one species emigrates.

If time t is small enough so that the probability of more than one immigration or emigration can be ignored then taking the limit of (3) as $\Delta t \to 0$ gives the following equation:

$$\overset{*}{P}_{N_S} = \begin{cases} -(I_r + E_r)P_{N_S} + E_{rs+1}P_{Ns+1} & N_s = 0 \\ -(I_r + E_r)P_{N_S} + I_{rs-1}P_{Ns-1} + E_{rs+1}P_{Ns+1} & 1 \le N_s \le N_{s\max} - 1 \\ -(I_r + E_r)P_{N_S} + I_{rs+1}P_{Ns-1} & N_s = N_{s\max} \end{cases} \tag{4}$$

Overview of BBO Technique

BBO (Simon, 2008) is developed based on the theory of biogeography. BBO technique is similar to other population-based optimization techniques. In BBO, each possible solution is termed as habitat and their features that characterize habitability are called suitability index variables (SIV). The fitness of each solution is called its habitat suitability index (HSI). In BBO, each habitat H is a vector of N (number of control variables) integers. Each element of each vector is considered as one suitability index variable (SIV). The fitness of each candidate solution is evaluated using its $SIVs$. HSI represents the quality of each candidate solution. High HSI solution represents a better quality solution and low HSI solution represents an inferior solution in the optimization problem. The concept is mainly based on migration and mutation. The basic idea and the mathematical formulation of migration and mutation steps are given below:

Migration

This is the initial but the most important part of BBO. In migration, the information is shared between habitats that depend on emigration rates and immigration rates of each solution. Immigration rate (I_r) and emigration rate (E_r) can be evaluated by the equations (1) and (2). Each solution is modified based on other solutions using habitat modification probability. Immigration rate, I_r of each solution is used to probabilistically decide whether or not to modify each SIV in that solution. If a SIV in a given solution is selected for modification, emigration rates, E_r of other solutions are used to probabilistically select which solution's SIV should migrate to that solution. Poor solutions accept more useful information from the good solution, which improve the exploitation ability of the technique. The main difference between recombination approach of evolutionary strategies (ES) and migration process of BBO is that in ES, global recombination process is used to create a completely new solution, while in BBO; migration is used to bring changes within the existing solutions. In order to prevent the best solutions from being corrupted by the immigration process, few elite solutions are kept in BBO algorithm.

Mutation

Due to some natural calamities or other events, HSI of a natural habitat can change suddenly and it may deviate from its equilibrium value. In BBO, this event is represented by the mutation of SIV and species count probabilities are used to determine the mutation rates. The species count probability can be calculated using the differential equation (4). Each population member has an associated probability, which indicates the likelihood that it exists as a solution for a given problem. Mutation rate of each set of solution can be calculated in terms of species count probability using the equation (5):

$$M_r = m_{\max}\left(\frac{1 - P_{N_S}}{P_{\max}}\right) \tag{5}$$

where, M_r is the mutation rate for a habitat that contains N_s species; $m_{\max}$ is the maximum mutation rate; $P_{\max}$ is the maximum probability.

From (5), it can be inferred that if the probability (P_{N_S}) of a given solution is very low, the solution is likely to mutate to some other solution. Similarly, if the probability (P_{N_S}) of a solution is high, the solution has very little chance to mutate. Now, both habitats having very high HSI values and very low HSI values have low species count probability (P_{NS}) and therefore are equally probable for mutation i.e. they have high chances to produce much improved $SIVs$ in the later stage. But, medium HSI solutions have higher values of species count probability (P_{NS}) and less chance to create better solutions after mutation operation. This mutation scheme tends to increase diversity among the populations. Without this modification, the highly probable solutions will tend to be more dominant in the population. This mutation approach makes both low and high HSI solutions likely to mutate, which gives a chance of improving both types of solutions in comparison to their earlier values. Few elite solutions are kept in the mutation process to save the best solutions, so if a solution becomes inferior after mutation process then, previous solution (solution of that set before mutation) can revert back to that place again, if needed. So, mutation operation is a high-risk process. It is normally applied to both poor and better solutions. Since medium quality solutions are in improving stage, so it is better not to apply mutation on medium quality solutions.

Here, mutation of a selected solution is performed simply by replacing it with a randomly generated new solution. Other than this any other mutation scheme that has been implemented for GA can also be implemented for BBO.

BBO Algorithm

The algorithm of BBO technique may be briefly described by the following steps:

- **Step 1:** Initialize the parameters of BBO like habitat modification probability P_{mod}, mutation probability, maximum mutation rate $m_{\max}$, maximum immigration rate $I_{\max}$, maximum emigration rate $E_{\max}$, the step size for numerical integration dt etc. Also set maximum generation, maximum species count $S_{\max}$ and an elitism parameter. Generate the *SIVs* of the given problem within their feasible region. This means deriving a method of mapping problem solutions to *SIVs*. A complete solution consisting of *SIVs* is known as habitat H.

Figure 2. The flowchart of BBO

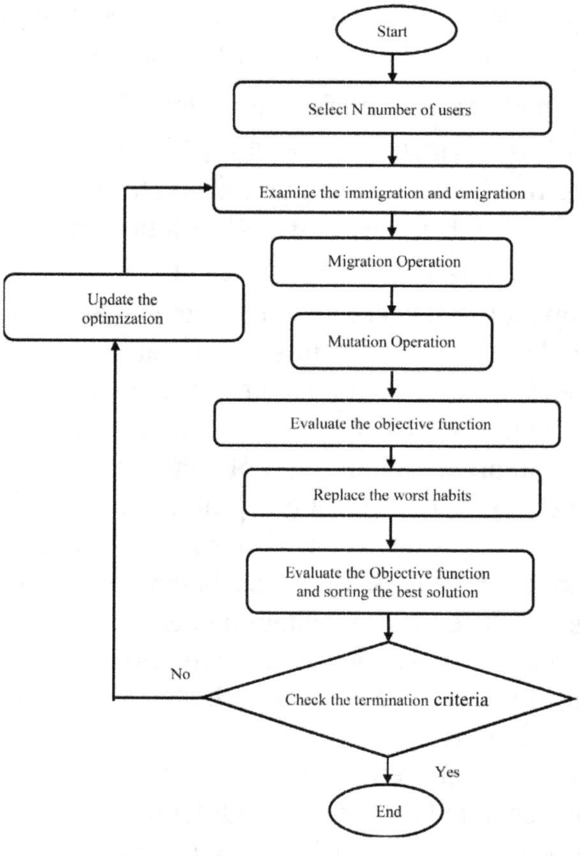

- **Step 2:** Initialize several numbers of habitats depending upon the population size. Each habitat represents a potential solution to the given problem.
- **Step 3:** Calculate the *HSI* for each habitat of the population set for given emigration rate E_r, immigration rate I_r and species N_S.
- **Step 4:** Based on the *HSI* value, elite habitats are identified.
- **Step 5:** Probabilistically immigration rate and emigration rate are used to modify each non-elite habitat using migration operation. The probability that a habitat H_i is modified is proportional to its immigration rate I_{ri} and the probability that the source of the modification comes from a habitat H_j is proportional to the emigration rate E_{rj}. After

modification of each non-elite habitat using migration operation, each *HSI* is recomputed.

- **Step 6:** For each habitat, the species count probability is updated using (4). Afterward, mutation operation is performed on each non-elite habitat to compute each *HSI* value again.
- **Step 7:** Go to step 3 for the next generation. This loop can be terminated after a predefined number of generations or after an acceptable solution is found.

Artificial Bee Colony (ABC) Optimization

Overview of ABC Technique

Artificial bee colony (ABC) (Dutta et al., 2014) is another efficient population based optimization technique developed by Karaboga and Basturk in 2006. ABC is based on the adoptive process of the natural system. In this method, an artificial system is designed that retains the robustness of natural systems. It is a swarm intelligence approach inspired by the collective intelligent behavior of bees colony. The intelligence behavior of honey bees consists of three essential components.

1. Food Sources which are evaluated by each bee's distances from the hive and amount of nectar in it. The nectar of the food source represents the quality or fitness of the solution.
2. Employed foragers, which are currently associated with the food source, exploit it and carry the information about the source. They abandon worse sources after discovering better sources.
3. Unemployed foragers, which look for the new food sources. The unemployed foragers are of two types: onlooker and scout.

Onlooker bees wait in the nest and establish a food source through the information shared by employed foragers and they employed themselves to the most profitable food source. Scouts bees search the new food source. They fly and choose the food sources randomly without using experience. Each food source chosen represents a possible solution to the problem under consideration. The number of scout bees is generally taken between 5% -10% of the population.

Figure 3. The flow chart of ABC

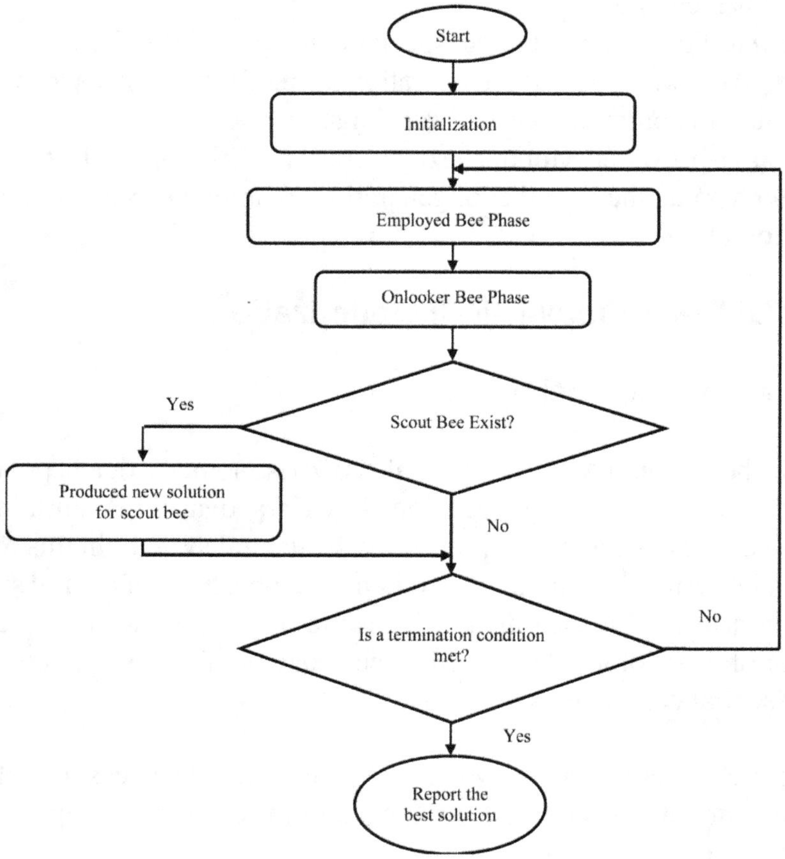

ABC Algorithm

Main steps of ABC technique is given below:

- **Step 1:** Generate the initial population (P) of the solutions (X_i) for the employed bees and calculate their fitness values where $X_i^e = [x_{i1}^e, x_{i2}^e, ..., x_{id}^e]$ is d- dimensional optimization parameters
- **Step 2:** Modify each solution of the employed bees using following equation:

$$v_{ij}^e(t+1) = x_{ij}^e(t) + \varphi_{ij}[x_{ij}^e(t) - x_{kj}^e(t)] \quad (6)$$

where k is the randomly chosen index from population (P), t is the iteration number, ϕ_{ij} is a random number between -1 to 1, $x^e_{ij}(t)$ is the jth dimension (variable) of the ith employed bee at t^{th} iteration and $v^e_{ij}(t+1)$ is the new position of the jth dimension of the ith employed bee.

- **Step 3:** Calculate the objective function values using the modified positions of the employed bees and compare these values with the objective function values using the old position of the employed bees. If the new objective function values are better than the old values then the new positions are retained otherwise the old positions are retained in its memory. This may be described briefly as follows:

If $f[v^e_{i1}(t+1), v^e_{i2}(t+1), ..., v^e_{id}(t+1)] > f[x^e_{i1}(t), x^e_{i2}(t), ..., x^e_{id}(t)]$

$$[x^e_{i1}(t+1), x^e_{i2}(t+1), ..., x^e_{id}(t+1)] = [v^e_{i1}(t+1), v^e_{i2}(t+1), ..., v^e_{id}(t+1)]$$

else

$$[x^e_{i1}(t+1), x^e_{i2}(t+1), ..., x^e_{id}(t+1)] = [x^e_{i1}(t), x^e_{i2}(t), ..., x^e_{id}(t)];$$

endwhere, $f[v^e_{i1}(t+1), v^e_{i2}(t+1), ..., v^e_{id}(t+1)]$ is the modified objective function value of the ith employed bee, $f[x^e_{i1}(t), x^e_{i2}(t), ..., x^e_{id}(t)]$ is the current objective function value of ith employed bee, $[x^e_{i1}(t+1), x^e_{i2}(t+1), ..., x^e_{id}(t+1)]$ is the modified solution parameters of the ith employed bee

- **Step 4:** The new solutions of the employed bees share the information with onlooker bees (i.e. the solution of employed bee becomes the solution of onlooker bee).
- **Step 5:** Calculate the probability of each solution by the following equation:

$$P_i = \frac{f_i}{\sum f_i} \quad (7)$$

where p_i, f_i are the probability and objective function value of ith onlooker bee.

- **Step 6:** Depending on the probability, modify the solution of onlooker bees using following equation:

$$v_{ij}^{o}(t+1) = x_{ij}^{o}(t) + \varphi_{ij}[x_{ij}^{o}(t) - x_{kj}^{o}(t)] \tag{8}$$

where $x_{ij}^{o}(t)$ is the jth dimension of the ith onlooker bee at t^{th} iteration and $v_{ij}^{o}(t+1)$ is the new position of the jth dimension of the ith onlooker bee.

- **Step 7:** Calculate the nectar amount (objective function value) of the onlooker bee. If the new food has equal or better nectar than the old source, it is replaced with the old one in the memory. Otherwise, the old one is retained in the memory. This may be described briefly as follows:

If $f[v_{i1}^{o}(t+1), v_{i2}^{o}(t+1), ..., v_{id}^{o}(t+1)] > f[x_{i1}^{o}(t), x_{i2}^{o}(t), ..., x_{id}^{o}(t)]$

$$[x_{i1}^{o}(t+1), x_{i2}^{o}(t+1), ..., x_{id}^{o}(t+1)] = [v_{i1}^{o}(t+1), v_{i2}^{o}(t+1), ..., v_{id}^{o}(t+1)]$$

else

$$[x_{i1}^{o}(t+1), x_{i2}^{o}(t+1), ..., x_{id}^{o}(t+1)] = [x_{i1}^{o}(t), x_{i2}^{o}(t), ..., x_{id}^{o}(t)];$$

endwhere $f[v_{i1}^{o}(t+1), v_{i2}^{o}(t+1), ..., v_{id}^{o}(t+1)]$ is the modified objective function value of the ith onlooker bee; $f[x_{i1}^{o}(t), x_{i2}^{o}(t), ..., x_{id}^{o}(t)]$ is the current objective function value of ith employed bee, $[x_{i1}^{o}(t+1), x_{i2}^{o}(t+1), ..., x_{id}^{o}(t+1)]$ is the modified solution parameters of the ith employed bee.

- **Step 8:** Replace the worst solutions by the new solutions. Numbers of replacement solutions are controlled by scout limit (SL). New Solutions are generated using the following equation:

$$x_{ij} = x_j^{\min} + r_1[x_j^{\max} - x_j^{\min}] \tag{9}$$

where $x_j^{\max}, x_j^{\min}$ are the maximum and minimum value of the jth independent variable, r_1 is a random number between 0 to 1.

- **Step 9:** Calculate the objective function value of new solutions.
- **Step 10:** Go to step 2 for the next iteration.
- **Step 11:** Stop iterations after a predefined number of iterations is reached.

Differential Evolution (DE)

Overview of DE Technique

Differential evolution (DE) (Shaheen et al., 2011) is one of the most popular and widely used population based evolutionary algorithm, capable of handling non-differentiable, non-linear and multi-modal objective functions. A brief description of different steps of DE algorithm is given below:

Initialization

The population is initialized by randomly generating individuals within the boundary constraints, The jth independent variable of the ith individual is as given below:

$$x_i^j = x_{\min}^j + rand * [x_{\max}^j - x_{\min}^j] \tag{10}$$

where $x_{\max}^j, x_{\min}^j$ are the upper and lower bound, respectively, of the jth variable of the given problem; x_i^j is the jth independent variable of the ith target vector; *rand* is a uniformly distributed random number between 0 and 1.

Mutation

As a step of generating off spring, the operations of 'mutation' are applied. 'Mutation' occupies quite an important role in the reproduction cycle. The mutation operation creates mutant vectors by perturbing a randomly selected vector with the difference of two other randomly selected vectors as per following equation.

$$v_i^j = x_i^j + F * (x_m^j - x_n^j) \tag{11}$$

where, x_i^j, x_m^j, x_n^j are three target vectors chosen randomly from the current population; F is a positive control parameter used for scaling the difference vector.

Crossover

Crossover represents a typical case of a 'genes' exchange. The parent vector is mixed with the mutated vector to create a trial vector, according to the following equation:

$$u_i^j = \begin{cases} v_i^j & if \quad rand \le CR \\ x_i^j & otherwise \end{cases} \tag{12}$$

where CR is the predefined crossover probability and $rand$ is a random number within (0, 1) for the ith vector of the jth variable.

Selection

Selection procedure is used among the set of trial vector and the updated target vector to choose the best. Each solution in the population has the same chance of being selected as parents. Selection is realized by comparing the objective function values of target vector and trial vector. For minimization problem, if the trial vector has a better value of the objective function, then it replaces the updated one as per. The selection operation may mathematically be expressed as follows:

$$X_i(t+1) = \begin{cases} U_i(t) & if \quad (U_i(t)) \le f(X_i(t)) \\ & X_i(t) otherwise \end{cases} \tag{13}$$

Where, $U_i = (u_i^1, u_i^2, u_i^3, \ldots, u_i^j, \ldots, u_i^d)$; $X_i = (x_i^1, x_i^2, \ldots, x_i^j, \ldots, x_i^d)$;

$f(x)$ is the objective function to be minimized and d is the number of control variables. This optimization process is repeated for several iterations allowing individuals to improve their fitness as they explore the solution space in search for optimal solutions.

Figure 4. The flowchart of DE

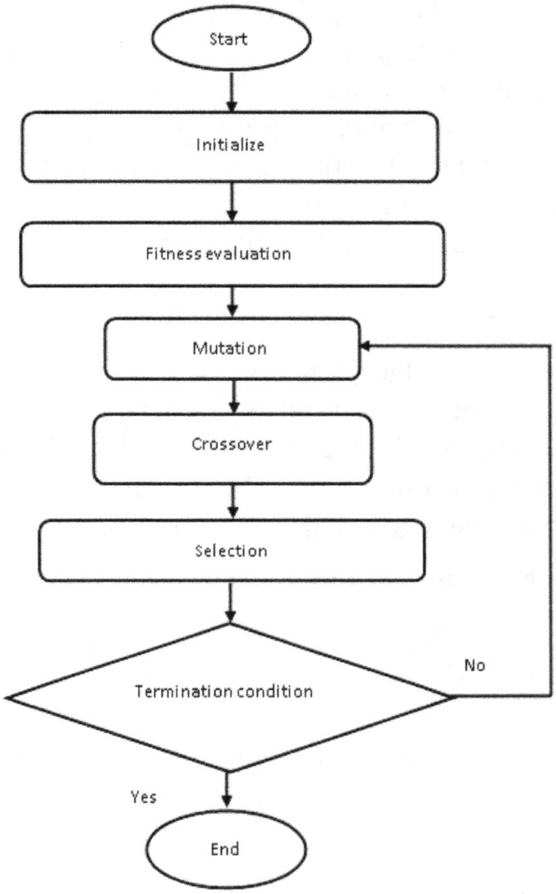

Teaching-Learning-Based Optimization

Teaching-learning-based optimization algorithm (TLBO) is a nature-inspired population-based meta-heuristic search method. TLBO proposed by Rao *et al.* (2012) is based on the effects of the influence of a teacher on the output of the learners in a class. For TLBO, the population is considered as a group of learners or a class of learners. In TLBO, the different control variables (i.e. dependent and independent variables of problem specific objective function) are considered as different subjects offered to the students and the result of a student is analogous to fitness function of the other population-based techniques which represents the quality of each solution set. The algorithm mimics teaching learning ability of teacher and learners in a class room. The teacher is a highly educated person who imparts knowledge to his students by always trying to improve the average knowledge level of the students. Moreover, the students also learn through mutual interaction between themselves which in turn helps them to improve their results in terms of their marks or grades. The result of the students in the class is the basis of evaluation of the level of their knowledge. The TLBO algorithm is based on the following two steps: first is the teaching phase and second is the learning phase. The concept and mathematical formulation of teacher and learning phases are given below:

Teacher Phase

This is the first part of the algorithm where students improve their knowledge level with the help of their teacher and the teacher tries to improve the average result of the class from initial value of z to his own level. However, in practice, only the mean grade of the class can be improved to some extent by the teacher. If the new mean grade of the kth subject at the pth iteration is $z^{p}_{new_k}$, the difference between the existing and new mean $z^{p}_{diff_k}$ may be defined as

$$z^{p}_{diff_k} = rand \times (z^{p}_{new_k} - t_f z^{p}_{k}) \tag{14}$$

where $rand$ is a random number between [0,1]; t_f is the teaching factor and is given by:

$$t_f = round \times (1 + rand \times (0,1)) \tag{15}$$

where *round* is the nearest integer value.

The updated grade of the kth subject of the nth student may be modified as follows:

$$g^{p+1}_{n,k} = g^{p}_{n,k} + z^{p}_{diff_k} \tag{16}$$

where, $g^{p}_{n,k}$, $g^{p+1}_{n,k}$ are the grades of the kth subject of the nth student at the pth and the $(p+1)^{th}$ iterations; $z^{p}_{diff_k}$ is the difference between the mean grade of the kth subject at the pth and the $(p+1)^{th}$ iterations.

Learner Phase

It is the second part of the algorithm where students enhance their knowledge through mutual interaction among themselves. A student can randomly interact with another student, and can gather knowledge whether the other student has more knowledge than him on not. This interactive learning phenomenon may be mathematically expressed as follows:

$$g_{n,k}^{p+1} = g_{nk}^{p} + rand \times (g_{n,k}^{p} - g_{o,k}^{p}) \quad if \quad f(X_n) < f(X_o) \tag{17}$$

$$g_{n,k}^{p+1} = g_{nk}^{p} + rand \times (g_{o,k}^{p} - g_{n,k}^{p}) \quad if \quad f(X_n) > f(X_o) \tag{18}$$

where $g_{n,k}^{p}$, $g_{n,k}^{p+1}$ are the grade points of the kth subject of the nth student at p^{th} and $(p+1)^{th}$ iterations; $g_{o,k}^{p}$ is the grade points of the kth subject of the oth student (randomly selected) at p^{th} iteration; $f(X_o)$ is the overall grade point of the oth student.

Chemical Reaction Based Optimization

In the year of 2011, Xu *et al.* (Xu et al., 2011). proposed chemical-reaction optimization (CRO) algorithm, based on chemical reaction process where molecules undergo a sequence of reactions with each other. The CRO has a good searching ability that shows the excellent operation of intensification and diversification which are two important features of the evolutionary algorithm. In CRO, the atomic structure of molecule represents a solution of the optimization problem. Potential energy (PE) and kinetic energy (KE) are two key factors for a molecule. The fitness of a solution is judged by the PE energy of the molecule, while KE is used to control the acceptance of new solutions with worse fitness.

Chemical reaction process may be classified into four different categories namely (i) reactions are on-wall ineffective collision, (ii) decomposition, (iii) inter-molecular ineffective collision and (iv) synthesis. On-wall ineffective collision and decomposition reactions are of single molecular reactions, where as inter-molecular ineffective collision and synthesis reaction are of the later category.

Figure 5. The flowchart of TLBO

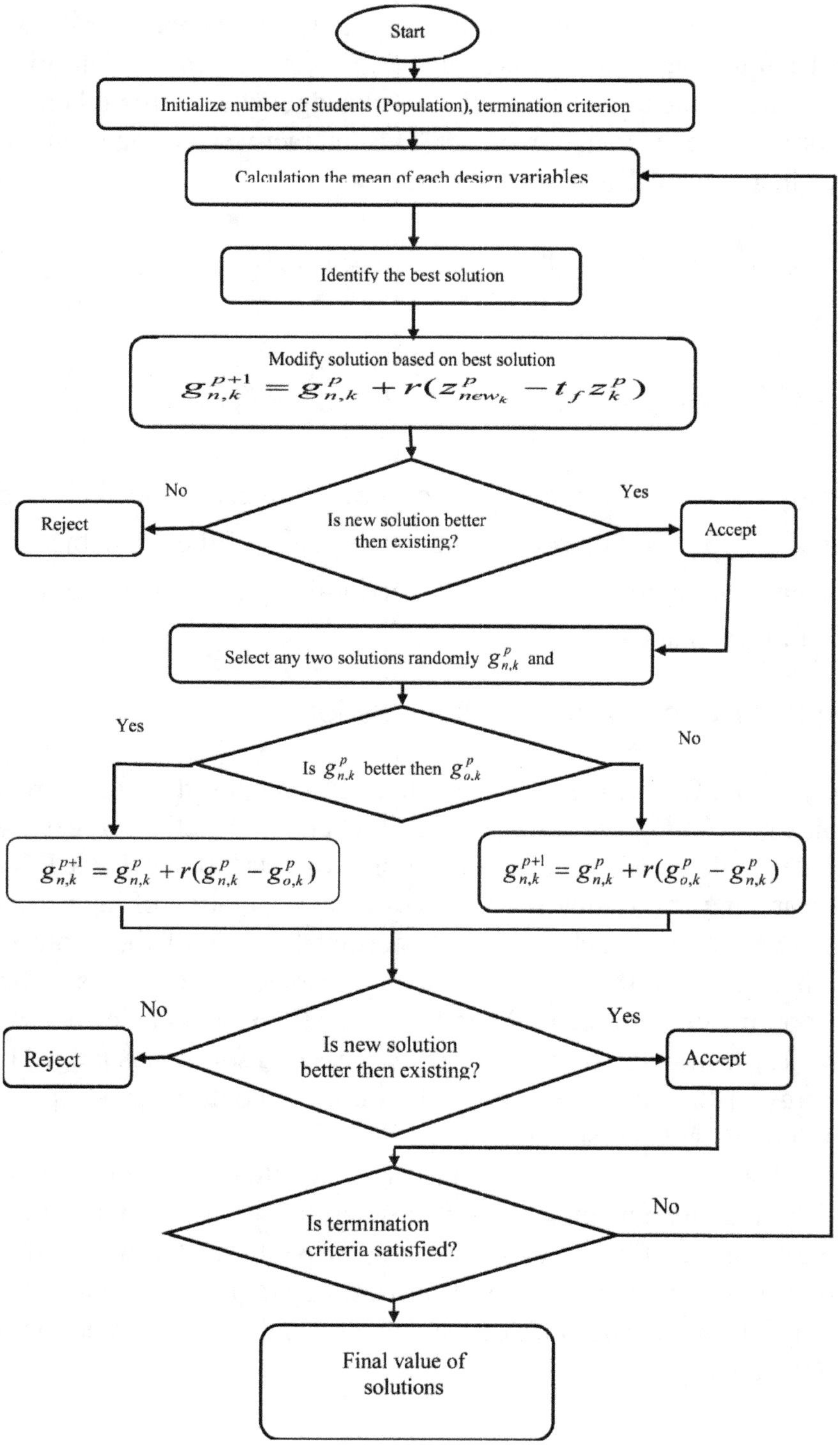

On-Wall Ineffective Collision

The on-wall ineffective collision reaction occurs when a molecule hits the wall and then bounces back. In this reaction, a molecule ms is allowed to change to another molecule ms_1 if the following condition is satisfied:

$$E_k^{ms} + E_p^{ms} \geq E_p^{ms_1} \tag{19}$$

Decomposition

The decomposition reaction is used to mimic the process of hitting the wall and then decomposing into two or more pieces. Because of the severe collision, the resultant molecular structure of the two newly formed molecules ms_1, ms_2 are entirely different from the original molecule ms and is also different from the neighbourhood molecules. In decomposition process takes place if

$$E_k^{ms} + E_p^{ms} \geq E_p^{ms_1} + E_p^{ms_2} \tag{20}$$

Inter-Molecular Ineffective Collision

The inter-molecular ineffective collision mimics the process that two molecules ms_1, ms_2 collide with each other and generate two new molecules $ms_1^{'}, ms_2^{'}$. As the collision is not severe, the molecular structures of the generated molecules are closer to the original molecules. Inter-molecular ineffective collision occurs if

$$E_k^{ms_1} + E_p^{ms_1} + E_k^{ms_2} + E_p^{ms_2} \geq E_p^{ms_1^{'}} + E_p^{ms_2^{'}} \tag{21}$$

Synthesis Collision

In this process, two molecules ms_1, ms_2 collides and combines together to form a single molecule $ms_1^{'}$. The molecular structure of the newly formed molecule is entirely different from the molecular structure of original molecules. The condition for synthesis collision is as follows.

Figure 6. The Flowchart of CRO algorithm

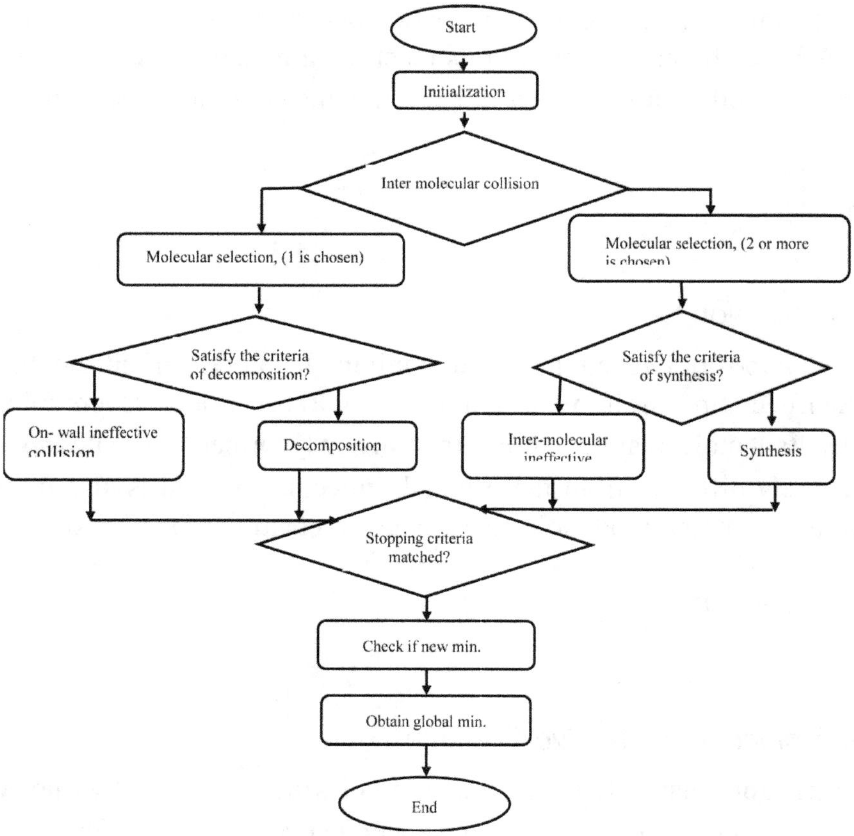

$$E_k^{ms_1} + E_p^{ms_1} + E_k^{ms_2} + E_p^{ms_2} \geq E_p^{ms'} \tag{22}$$

Krill Herd Algorithm

Overview of KHA Technique

Krill herd algorithm (KHA) (Mukherjee & Mukherjee, 2016) is a novel meta-heuristic swarm intelligence optimization method for solving optimization problems, which is based on the simulation of the herding of the krill swarms in response to specific biological and environmental processes. The time-dependent position of an individual krill in 2D surface is governed by the following three main actions:

1. Movement induced by other krill individuals;
2. Foraging action; and
3. Random diffusion.

These actions are briefly explain and mathematically expressed as follows:

Motion Induced by Other Krill Individuals

The direction of motion induced, α_i is approximately estimated by the target swarm density (target effect), a local swarm density (local effect), and a repulsive swarm density (repulsive effect). For a krill individual, this movement can be defined as follows:

$$M_i^k = \alpha_i M_i^{\max} + \omega_n M_i^{k-1} \tag{23}$$

where,

$$\alpha_i = \alpha_i^{new} + \alpha_i^{target} \tag{24}$$

$$\alpha_i^{new} = \sum_{j=1}^{s} \Gamma_{i,j} \Psi_{i,j} \tag{25}$$

where,

$$\Psi_{i,j} = \frac{x_i - x_j}{\left|x_i - x_j\right| + rand(0,1)} \tag{26}$$

$$\Gamma_{i,j} = \frac{F_i - F_j}{F_w - F_b} \tag{27}$$

$$\alpha_i^{terget} = 2\left(rand(0,1) + \frac{i}{i_{\max}}\right)\Gamma_i^{best}\Psi_i^{best} \tag{28}$$

where $\mathrm{M}_{\mathrm{i}}^{\max}$. $\mathrm{M}_{\mathrm{i}}^{\max}$ s the maximum induced motion; M_i^k $\mathrm{M}_{\mathrm{i}}^{\mathrm{prevoius}}$. $\mathrm{M}_{\mathrm{i}}^{\mathrm{prevoius}}$ M_i^{k-1} are the induced motion of the ith krill at the kth and $(k-1)^{th}$ movement; ω_n . ω_n is the inertia weight of the motion induced; $\alpha_i^{new}, \alpha_i^{target}$ are the local and the target effect, respectively; F^{w} . F^{w} are the worst and the best position respectively, among all krill individuals, of the population; $\mathrm{F}^{\mathrm{i}}, \mathrm{F}^{\mathrm{k}}$. $\mathrm{F}^{\mathrm{i}}, \mathrm{F}^{\mathrm{k}}$ F_i, F_jre the fitness value of ith and j^{th} individuals, respectively; i is the current iteration number and $i_{\max}$ $\mathrm{I}_{\max}$. $\mathrm{I}_{\max}$ ithe maximum iteration number.

To identify the neighbouring members of each krill individual, a sensing distance (S_{d_i} S_{d_i} .)arameter is used. If the distance between the two individual krill is less than the sensing distance, that particular krill is considered as neighbour of the other krill. The sensing distance may be defined by:

$$sd_i = \frac{1}{5s}\sum_{j=1}^{s} \left|X_i - X_j\right| \tag{29}$$

where S is the number of krill individuals surrounding the particular krill; Z_{i} . Z_{i} X_i, X_jre the position of the ith and j^{th} krill, respectively.

Foraging Action

The foraging motion $M_{f_i}^k$ is covered in terms of two main effective parameters. The first one is the current food location and the second one is the previous experience about the food location. This motion can be expressed for the ith krill individual as follows:

$$M_{f_i}^k = 0.02\beta_i + \omega_x M_{f_i}^{k-1} \tag{30}$$

$$\beta_i = 2\left(1 - \frac{i}{i_{\max}}\right) F_i \frac{\sum_{j=1}^{s} \frac{x_j}{F_j}}{\sum_{j=1}^{N_S} \frac{1}{F_j}} + \Gamma_i^{best} \Psi_i^{best} \tag{31}$$

where ω_x is the inertia weight of the foraging motion; $\mathrm{F}_{\mathrm{i}}^{\mathrm{previous}}$. $\mathrm{F}_{\mathrm{i}}^{\mathrm{previous}}$ is the foraging motion of the ith krill at $(k\text{-}1)^{th}$ and k^{th} movement.

Random Diffusion

The diffusion process of the krill individuals is considered as a random phenomenon. It may be represented in terms of a maximum diffusion speed and a random directional factor and may mathematically be expressed by:

$$M_{d_i}^{k} = \lambda M_{d}^{\max} \tag{32}$$

where, $M_{d}^{\max}$ is the maximum diffusion motion; $\sigma\lambda$ is the directional vector uniformly distributed between (-1,1).

Position Update

Finally, the position of the ith krill during the time interval t . to Δt may be expressed as:

$$x_i(t + \Delta t) = x_i(t) + \Delta t(M_{i}^{k} + M_{f_i}^{k} + M_{d_i}^{k}) \tag{33}$$

where

$$\Delta t = c_p \sum_{i=1}^{N} (u_i - l_i) \tag{34}$$

$$\Delta \mathrm{t} = \mathrm{K_t} \sum_{\mathrm{j}=1}^{\mathrm{V}} (\mathrm{UL_j} - \mathrm{LL_j})$$

$$\Delta \mathrm{t} = \mathrm{K_t} \sum_{\mathrm{j}=1}^{\mathrm{V}} (\mathrm{UL_j} - \mathrm{LL_j})$$

where V V N s the total number of control variables; u_i, l_i are the upper and lower limits of the ith control variable; c_p is the position constant factor.

Moreover, to improve the behaviour of the individual Krill, two adaptive genetic operators of DE are added to the proposed algorithm. These two operators are briefly described below:

Genetic Operators

To improve the performance of the algorithm, genetic reproduction mechanisms are incorporated into the algorithm. The introduced adaptive genetic reproduction mechanisms are crossover and mutation which are inspired from the classical DE algorithm. These two operations are described as follow:

Crossover

In this process, depending on crossover probability, each krill individual interacts with others to update its position. The jth components of the ith krill may be updated by

$$\psi_{i,j} = \begin{cases} \psi_{r,j} \quad if \quad rand < c_r \quad where \quad r = 1,2,3...,i-1,i+1,...N_p \\ \quad \psi_{i,j} \quad else \end{cases} \tag{35}$$

where,

$$c_r = 0.2\Gamma_i^{best}$$

Mutation

The mutation operation creates mutant vectors Z_{ij}^{M} . Z_{ij}^{M} by perturbing the vector Z_{ibest} . Z_{ibest} $\Psi_{best,j}$ the difference of two other randomly selected vectors $\Psi_{o,j}$ and Z_{mj} . Z_{mj} s per following equation.

$$Z_{ij}^{M} = Z_{ibest} + \mu\left(Z_{lj} - Z_{mj}\right)$$

$$Z_{ij}^{M} = Z_{ibest} + \mu\left(Z_{lj} - Z_{mj}\right)$$

$$\Psi_{i,j}^{m} = \Psi_{best,j} + \mu(\Psi_{o,j} - \Psi_{p,j}) \tag{36}$$

The updated position of Z_{ij}^{U} . Z_{ij}^{U} s selected from Z_{ij}^{M} . Z_{ij}^{M} nd Z_{ij} . Z_{ij} sing mutation probability μ_p as follows:

$$Z_{ij}^{U} = \begin{cases} Z_{ij}^{M} \; if\, rand < M_{\mu} \\ Z_{ij} \; if\, rand > M_{\mu} \end{cases}$$

$$Z_{ij}^{U} = \begin{cases} Z_{ij}^{M} \; if\, rand < M_{\mu} \\ Z_{ij} \; if\, rand > M_{\mu} \end{cases}$$

$$\Psi_{i,j}^{\text{mod}} = \begin{cases} \Psi_{i,j}^{m} & if\ rand \leq \mu_p \\ \Psi_{i,j} & if\ rand > \mu_p \end{cases} \quad (37)$$

Grey Wolf Algorithm

Overview of GWA Technique

Grey wolf optimization (Pradhan et al., 2016) is a novel meta heuristic optimization technique proposed by Mirjalili et al, in 2014, grey wolves are generally considered at the top of the food chain and usually living in packs of (5 to 12). This optimization method mainly focuses on the natural behaviour of the grey wolves their hunting mechanism and social leadership hierarchy, in the pack, the wolves are categorised as alpha, beta, delta and omega. In the social hierarchy of the grey wolves, alpha is the fittest solution followed by beta and delta and rest are categorised under omega, which always follow the alpha, beta and delta. the whole group is guided by the alphas, For the functionality of GWO it only requires to define the population size and maximum number of iterations, where number of grey wolves are considered as population size, and in GWO the fitness value is defined as the distance of the wolf from the prey. the main constraint is on attaining a population size target. the GWO algorithm does not depend on the gradient of the objective function while determining the global optimum solution. The basic functions that the wolves perform are

Searching for Prey

In the exploration phase the whole group diverges from each other and explores the search area as broadly as possible and when the prey is identified the whole group converges to that location, when the random value $\vec{A}$ is greater than 1 and if less than -1 the wolves diverge to prey. Another variable useful for searching the prey is $\vec{C}$

Searching for the prey is the most important objective, in GWO algorithm. Vector $\vec{C}$ gives a random weight for prey searching, this vector varies between [0 2] and it also affects in defining the distance from the prey, hence these two vectors helps to search the area globally.

Encircling the Prey

In GWO algorithm for encircling the prey by the wolves following equations are considered, using (Mirjalili, et al. 2014) .

$$\vec{D} = \left|\vec{C}\,\vec{X}_P(t) - \vec{X}(t)\right| \quad (38)$$

$$\vec{X}(t+1) = \vec{X}_P(t) - \vec{A}\,\vec{D} \quad (39)$$

where t is the current iteration and $\vec{X}_P(t)$ denotes the current position of the victim

A and C are constants, using (Mirjalili et al., 2014) .

$$\vec{A} = 2\,\vec{a} * \vec{r_1} - \vec{a} \quad (40)$$

$$\vec{C} = 2\,\vec{r_2} \quad (41)$$

where r_1 and r_2 are two random vectors between [0,1] and Over each course of iteration $\vec{a}$ decreases linearly from 2 to 0

Hunting

The alphas guide the whole group to hunt the prey, although beta and deltas also participate in the hunting mechanism. The position of the wolves is always updated since the final position of the prey is still not known three best solutions in terms of alpha, beta and delta are stored and again the wolves update their position to find the best possible location of the prey although the final position is also random where alpha beta and delta forms of a circle around the prey. Below are the mathematical equations that follows to describe the hunting phase to update the positions of the wolves (Mirjalili et al., 2014)

$$\vec{D}_\alpha = \left| \vec{C}_1 \vec{X}_\alpha - \vec{X} \right| \tag{42}$$

$$\vec{D}_\beta = \left| \vec{C}_2 \vec{X}_\beta - \vec{X} \right| \tag{43}$$

$$\vec{D}_\delta = \left| \vec{C}_3 \vec{X}_\delta - \vec{X} \right| \tag{44}$$

Finally, position modified as using (Mirjalili et al., 2014)

$$\vec{X}_1 = \vec{X}_\alpha - \vec{A}_1 (\vec{D}_\alpha) \tag{45}$$

$$\vec{X}_2 = \vec{X}_\beta - \vec{A}_2 (\vec{D}_\beta) \tag{46}$$

$$\vec{X}_3 = \vec{X}_\delta - \vec{A}_3 (\vec{D}_\delta) \tag{47}$$

$$\vec{X}(t+1) = \frac{\vec{X}_1 + \vec{X}_2 + \vec{X}_3}{3} \tag{48}$$

Attacking Prey (Exploitation)

In the GWO algorithm the location of the alpha, beta and delta wolves helps for the up gradation of the position of the search agents following the Equations given in the hunting phase (48), the wolves will stop hunting only when the prey stops moving, which is mainly dependent on the value of $\vec{a}$ And $\vec{A}$ is a random value that lies in the range of [-2a, 2a]

Hybridization Algorithm

Hybridization of DE With BBO

In this Section, a hybridization of the DE with the BBO is proposed to take advantage of the exploration capabilities of DE and exploitation capabilities of BBO. Exploration capability of DE is used to prevent the solutions from being trapped in suboptimal points during the search process since BBO suffers from the lack of the diversity, which can lead to stagnation. But at the later stages when fine tuning operation is required, DE fails to perform satisfactorily. On the other hand, in BBO, the solutions improve gradually as the process goes on through migration operation. The main operator of HDE-BBO is the hybrid migration operator, which hybridizes the DE operator with the migration operator of BBO. Moreover, HDE-BBO is able to explore the new search space with the mutation operator of DE and to exploit the population information with the migration operator of BBO.

The proposed HDE-BBO approach is described as follows:

- **Step 1:** Using uniform distribution, iinitialize the population (*P*) randomly on the search space.
- **Step 2:** Evaluate the fitness value (*HSI*) of each individual of the current population.
- **Step 3:** Sort the population from the best to the worst and select few elite individuals to prevent the solutions.
- **Step 4:** Assign number of species to all individual habitats based on the *HSI* value.
- **Step 5:** Evaluate the immigration and emigration rates of each individual.
- **Step 6:** Modify habitats using hybrid migration operation as described in Box 1.

Box 1.

```
for i=1: N_P
    l=randint (1, n_d)
    for j =1: n_d
        if rand < λ_i
            if rand < CR  or  j == l
                v_i^j = x_l^j + F * (x_m^j - x_n^j)
            else
                v_i^j = x_k^j
            end
        else
            v_i^j = x_l^j
        end
    end
end
```

Where N_P is the population size; n_d is the number of dependent variables; x_k^j is the *j*th *SIV* of the *k*th habitat which is probabilistically selected as emigrating island.

- **Step 7:** Calculate mutation rate of each habitat.
- **Step 8:** Modify non-elite individuals of the population based on probabilistic mutation operation of BBO.
- **Step 9:** Go to Step 2 if the termination criterion is not met.

Hybridization of DE With CRO Algorithm (HCRO)

In this part, hybrid CRO (HCRO) (Dutta et al., 2015) algorithm based on CRO and DE is proposed for solving the optimal UPFC allocation problem of the transmission network. The CRO algorithm that emphasizes on exploring the entire search space and a local version of DE algorithm that emphasizes on exploiting the local search space are combined together to make an impact on the performance of the algorithm in terms of the solution quality and convergence speed.

The general steps of the HCRO algorithm are summarized as follows:

- **Step 1:** Initialize several numbers of molecules by randomly generated molecular structure depending upon the population size. The molecular structure of each molecule represents a potential solution to the given problem.

- **Step 2:** The fitness values of the specific problem of the population are assigned as the potential energy (PE) of the individual molecule. A random kinetic energy (KE) is set to the different molecules.
- **Step 3:** Based on the PE values (fitness values) best solutions are retained by elite molecules.
- **Step 4:** The non-elite molecules are modify using on-wall ineffective collision operations as described below:
    - ***Step 4.1:*** One molecule ms is selected randomly.
    - ***Step 4.2:*** Using the mutation operation of DE, thenew molecule ms_1 is generated which may mathematically be expressed as:

$$ms_1 = ms_k + f * (ms_m - ms_n) \tag{49}$$

where ms_k, ms_m, ms_n are the three different molecules chosen randomly from the current population.

- ***Step 4.3:*** The potential energy $E_p^{ms_1}$ of the molecule ms_1 is evaluate and the old molecule ms is replaced by new one ms_1 if the condition $E_k^{ms} + E_p^{ms} \geq E_p^{ms_1}$ is satisfied and the KE of the molecule ms_1 is evaluated by

$$E_k^{ms_1} = rand(0,1) * (E_k^{ms} + E_p^{ms} - E_p^{ms_1}) \tag{50}$$

- **Step 5:** Decomposition operation is performed to modify the molecular structure of the molecules by the following steps:
    - ***Step 5.1:*** Two molecules, one molecule ms from the population and another randomly generated molecule ms_1 are selected for decomposition operation.
    - ***Step 5.2:*** Crossover operation of DE is applied on ms and ms_1 to generate two new molecules ms' and ms_2''
    - ***Step 5.3:*** Potential energy $E_p^{ms'}$ and $E_p^{ms_1'}$ of molecules ms' and ms_1' are evaluated. If $E_k^{ms} + E_p^{ms} \geq E_p^{ms'} + E_p^{ms_1'}$ molecule ms is deleted and the molecules ms' and ms_1' are pushed into the population. Modify the KE of ms' and ms_1' as below:

$$E_k^{ms'} = rand(0.1) * [E_k^{ms} + E_p^{ms} - (E_p^{ms'} + E_p^{ms_1'})] \tag{51}$$

$$E_k^{ms_1'} = [1 - rand(0.1)] * [E_k^{ms} + E_p^{ms} - (E_p^{ms'} + E_p^{ms_1'})] \tag{52}$$

- **Step 6:** Inter-molecular ineffective collisions are made to modify the molecular structure each molecule using the following steps:
    - ***Step 6.1:*** Randomly select two molecules ms^1 and ms^2 from the population.
    - ***Step 6.2:*** Two new molecules ms_1^1 and ms_1^2 are generated using crossover operations of DE.
    - ***Step 6.3:*** Evaluate potential energy $E_p^{ms_1^1}$ and $E_p^{ms_1^2}$ of molecules ms_1^1 and ms_1^2 and replace molecules ms^1 and ms^2 by molecules ms_1^1 and ms_1^2, respectively, if $E_p^{ms^1} + E_k^{ms^1} + E_p^{ms^2} + E_k^{ms^2} \geq E_p^{ms_1^1} + E_p^{ms_1^2}$. The KE of the molecules ms^1 and ms^2 are evaluated using

$$E_k^{ms_1^1} = rand(0,1) * [E_p^{ms^1} + E_k^{ms^1} + E_p^{ms^2} + E_k^{ms^2} - (E_p^{ms_1^1} + E_p^{ms_1^2})] \tag{53}$$

$$E_k^{ms_1^2} = [1 - rand(0,1)] * [E_p^{ms^1} + E_k^{ms^1} + E_p^{ms^2} + E_k^{ms^2} - (E_p^{ms_1^1} + E_p^{ms_1^2})] \tag{54}$$

- **Step 7:** Synthesis collision operations are performed to update the molecular structure of the molecules using the following steps:
    - ***Step 7.1:*** Two molecules ms^1 and ms^2 are randomly selected from the population set.
    - ***Step 7.2:*** Apply the conventional cross over operation of GA on ms^1 and ms^2 by considering them as parents chromosomes and generate a child chromosome ms_1^1.
    - ***Step 7.3:*** Evaluate $E_p^{ms_1^1}$ of the molecule ms_1^1. The molecules ms^1 and ms^2 are omitted and the molecule ms_1^1 is push into the population if $E_k^{ms^1} + E_p^{ms^1} + E_k^{ms^2} + E_p^{ms^2} \geq E_p^{ms_1^1}$. The KE of the new molecule ms_1^1 is calculated using

$$E_k^{ms_1^1} = rand(0,1) * (E_k^{ms^1} + E_p^{ms^1} + E_k^{ms^2} + E_p^{ms^2} - E_p^{ms_1^1}) \tag{55}$$

- **Step 8:** The feasibility of a problem solution is verified, and the infeasible solutions are replaced by feasible solution set.
- **Step 9:** The updated molecules are sorted.

- **Step 10:** The best elite molecules are replaced by the worst molecule.
- **Step 11:** The processes of generating new molecules and selecting those with better function values are continued until the given stopping conditions are satisfied. The iteration process can be stopped after a fixed number of generations or when any significant improve improvement in the solution does not occur.

Opposition Based Optimization

Opposition-Based Learning

Opposition-based learning (OBL) developed by Tizhoosh (Tizhoosh, 2005) is a new concept in computational intelligence. OBL is one of the most successful concepts, which enhances the search abilities of the conventional population based optimization techniques in solving the nonlinear optimization problem. The main idea behind OBL is to consider the opposite of an assumption or a guess and comparing it with the original assumption, thereby improving the chances to find a solution faster. The OBL concept is based on the opposite point and opposite number which are defined as the following:

Opposite number: Let $x \in [\alpha, \beta]$ be a real number. Its opposite number x^o is defined by:

$$x^o = \alpha + \beta - x \tag{56}$$

Opposite point: Let $P(x_{1,} x_2, ..., x_n)$ be a point in n-dimensional space, where $x \in [\alpha, \beta]$; $r = \{1, 2, ..., n\}$. The opposite point $OP(x_1^0, x_2^0, x_3^0, ..., x_n^0)$ is defined by:

$$x_r^o = \alpha_r + \beta_r - x_r \tag{57}$$

Opposition-Based Optimization

Opposition based optimization is based on opposition-based initialization and opposition-based generation jumping which are briefly described as follows:

Box 2.

```
for r = 1 : n_p (n_p = poopulation size)
    for s = 1 : n_d (n_d = no. of control variables)
        OP_{r,s} = a_s + b_s − P_{r,s}
    end
end
```

Opposition-Based Initialization

In absence of priori knowledge, the only choice to create an initial population is random number generation. Oppositional learning can be utilized to obtain a better starting candidate solution without any priori knowledge about the solutions.

Initialization of opposite population (*OP*) is described in Box 2.

Opposition-Based Generation Jumping

A similar approach can be applied to the current population by which the evolutionary process can be forced to jump to a new solution candidate, which may be more fit than the current one. After generating a new population by teaching and learning process of TLBO, the opposite population is generated based on a jumping rate j_r. Opposite population jumping based on jumping rate as described in Box 3.

Box 3.

```
if rand(0,1) < j_r (j_r = jumping rate)
        for r = 1 : n_p (n_p = population size)
            for s = 1 : n_d (n_d = no. of control variables)
                OP_{r,s} = a_s + b_s − P_{r,s}
            end
        end
end
```

CONCLUSION

This chapter describes BBO, DE, ABC, TLBO, CRO and KHA algorithms elaborately. In addition to that, to improve the convergence speed, the modified versions of BBO, DE, TLBO, and CRO are adopted. These modified algorithms viz. hybrid BBO, hybrid CRO and opposition TLBO techniques are successfully applied for further improvement of the simulation results. All the optimization techniques are tested for different power system problems like OPF and ORPD without and with different FACTS device. The performances of different techniques are compared and are shown in details in the subsequent chapters with the help of curves and tables. Moreover, in order to judge the superiority, the results obtained using the proposed methods are compared with those of other algorithms reported in the literature.

REFERENCES

Abido, M. A. (2002). Optimal power flow using particle swarm optimization. *International Journal of Electrical Power & Energy Systems*, *24*(7), 563–571. doi:10.1016/S0142-0615(01)00067-9

Abido, M. A. (2004). Multi-objective optimal power flow using strength pareto evolutionary algorithm. *The 39th International Universities Power Engineering Conference*, 457–461.

AlRashidi, M. R., & El-Hawary, M. E. (2007). Hybrid particle swarm optimization approach for solving the discrete OPF problem considering the valve point effect. *IEEE Transactions on Power Systems*, *22*(4), 2030–2038. doi:10.1109/TPWRS.2007.907375

Aoki, K., Fan, M., & Nishikori, A. (1988). Optimal Var planning by approximation method forrecursive mixed integer linear programming||. *IEEE Transactions on Power Systems*, *3*(4), 1741–1747. doi:10.1109/59.192990

Aoki, K., Nishikory, A., & Yokoyana, R. T. (1987). Constrained load flow using recursive quadratic programming. *IEEE Transactions on Power Systems*, *PWRS-2*(1), 8–16. doi:10.1109/TPWRS.1987.4335064

Ayan, K., & Kılıc, U. (2012). Artificial bee colony algorithm solution for optimal reactive power flow. *Applied Soft Computing*, *12*(5), 1477–1482. doi:10.1016/j.asoc.2012.01.006

Bellman, R. E., & Dreyfus, S. E. (1962). *Applied dynamic programming*. Princeton, NJ: Princeton University Press. doi:10.1515/9781400874651

Bozorg, H. O., Afshar, A., & Marino, M. A. (2006). Honey-bees mating optimization (HBMO) algorithm: A new heuristic approach for water resources optimization. *Water Resources Management*, *20*(5), 661–680. doi:10.100711269-005-9001-3

Burchett, B. T. (2013). Aerodynamic parameter identification for symmetric projectiles: An improved gradient based method. *Aerospace Science and Technology*, *30*(1), 119–127. doi:10.1016/j.ast.2013.07.010

Cai, H. R., Chung, C. Y., & Wong, K. P. (2008). Application of differential evolution algorithm for transient stability constrained optimal power flow. *IEEE Transactions on Power Systems*, *23*(2), 719–728. doi:10.1109/TPWRS.2008.919241

Cai, J. J., Li, Q., Li, L. X., Peng, H. P., & Yang, Y. X. (2012). A fuzzy adaptive chaotic ant swarm optimization for economic dispatch. *International Journal of Electrical Power & Energy Systems*, *34*(1), 154–160. doi:10.1016/j.ijepes.2011.09.020

Capitanescu, F., & Wehenkel, L. (2013). Experiments with the interior-point method for solving large scale optimal power flow problems. *Electric Power Systems Research*, *95*(2), 276–283. doi:10.1016/j.epsr.2012.10.001

Chaohua, D., Weirong, C., Yunfang, Z., & Xuexia, Z. (2009). Seeker optimization algorithm for optimal reactive power dispatch. *IEEE Transactions on Power Systems*, *24*(3), 1218–1231. doi:10.1109/TPWRS.2009.2021226

Das, D. B., & Patvardhan, C. (2003). Useful multi-objective hybrid evolutionary approach to optimal power flow. *IEE Proceedings. Generation, Transmission and Distribution*, *150*(3), 275–282. doi:10.1049/ip-gtd:20030188

Deeb, N. I., & Shahidehpour, S. M. (1988). An efficient technique for reactive power dispatch using a revised linear programming approach. *Electric Power Systems Research*, *15*(2), 121–134. doi:10.1016/0378-7796(88)90016-8

Dhillon, T. S. (1991). Artificial neural network applications to power systems and their relationships to symbolic methods. *International Journal of Electrical Power & Energy Systems*, *13*(2), 66–72. doi:10.1016/0142-0615(91)90028-T

Dutta, S., Roy, P. K., & Nandi, D. (2014). Artificial Bee Colony Optimization for Optimal Reactive Power Dispatch Incorporating FACTS Devices. *International Journal of Energy Optimization and Engineering*, *2*(2).

Dutta, S., Roy, P. K., & Nandi, D. (2015). Optimal location of UPFC controller in transmission network using hybrid chemical reaction optimization algorithm. *International Journal of Electrical Power & Energy Systems*, *64*, 194–211. doi:10.1016/j.ijepes.2014.07.038

Engles, L., Larson, R. E., Peschon, J., & Stanton, K. N. (1976). *Dynamic programming applied to hydro and thermal generation scheduling*. New York: IEEE Tutorial Course Text.

Gaing, Z. L. (2005). Constrained optimal power flow by mixed-integer particle swarm optimization. *IEEE Power Engineering Society General Meeting, 1*, 243-250.

Gandomi, A. H., & Alavi, A. H. (2012). Krill herd: A new bio-inspired optimization algorithm. *Communications in Nonlinear Science and Numerical Simulation*, *17*(12), 4831–4845. doi:10.1016/j.cnsns.2012.05.010

Ghoshal, S. P. (2004). Application of GA/GA-SA Based fuzzy automatic generation control of a multi-area thermal generating system. *Electric Power Systems Research*, *70*(2), 115–127. doi:10.1016/j.epsr.2003.11.013

Ghoshal, S. P., Chatterjee, A., & Mukherjee, V. (2009). Bio-inspired fuzzy logic based tuning of Power system stabilizer. *Expert Systems with Applications*, *36*(5), 9281–9292. doi:10.1016/j.eswa.2008.12.004

Granada, M., Rider, M. J., Mantovani, J. R. S., & Shahidehpour, M. A. (2012). Decentralized approach for optimal reactive power dispatch using a Lagrangian decomposition method. *Electric Power Systems Research*, *89*, 148–156. doi:10.1016/j.epsr.2012.02.015

Granville, S. (1995). Optimal reactive power dispatch through interior point methods. *IEEE Transactions on Power Systems*, *9*(1), 136–146. doi:10.1109/59.317548

Grudinin, N. (1998). Reactive power optimization using successive quadratic programming method. *IEEE Transactions on Power Systems*, *13*(4), 1219–1225. doi:10.1109/59.736232

Guo, C. X., Zhan, J. P., & Wu, Q. H. (2012). Dynamic economic emission dispatch based on group search optimizer with multiple producers. *Electric Power Systems Research, 86*, 8–16. doi:10.1016/j.epsr.2011.11.015

Habiabollahzadeh, H., Luo, X., & Semlyen, A. (1989). Hydrothermal optimal power flow based on a combined linear and non-linear programming methodology. *IEEE Transactions on Power Systems*, *4*(2), 530–537. doi:10.1109/59.193826

Hadi, N., Mahdi, N., & Patrick, S. (2012). A multi objective gravitational search algorithm based on non-dominated sorting. *International Journal of Swarm Intelligence Research*, *3*(3), 32–49. doi:10.4018/jsir.2012070103

Hobbs, W., Hermon, J. G., Warner, S., & Sheble, G. B. (1988). An enhanced dynamic approach for unit commitment. *IEEE Transactions on Power Systems*, *3*(3), 1201–1205. doi:10.1109/59.14582

Huang, S. J. (2001). Enhancement of hydroelectric generation scheduling using ant colony system based optimization approaches. *IEEE Transactions on Energy Conversion*, *16*(3), 296–301. doi:10.1109/60.937211

Javidan, J., & Ghasemi, A. (2012). Environmental/economic power dispatch using multiobjective honey bee mating optimization. *International Review Electrical Engineering*, *7*(1), 3667–3675.

Jiejin, C., Xiaoqian, M., Qiong, L., Lixiang, L., & Haipeng, P. (2010). A multi-objective chaotic ant swarm optimization for environmental/economic dispatch. *International Journal of Electrical Power & Energy Systems*, *32*(5), 337–344. doi:10.1016/j.ijepes.2010.01.006

Karaboga, D., & Basturk, B. (2007). A powerful and efficient algorithm for numerical function optimization: Artificial bee colony (ABC) algorithm. *Journal of Global Optimization*, *39*(3), 459–471. doi:10.100710898-007-9149-x

Keyan, L., Wanxing, S., & Yunhua, L. (2006). Research on reactive power optimization based on adaptive genetic simulated annealing algorithm. *International Conference on Power System Technology, PowerCon*, 1–6.

Khazali, A. H., & Kalantar, M. (2011). Optimal reactive power dispatch based on harmony search algorithm. *International Journal of Electrical Power & Energy Systems*, *33*(3), 684–692. doi:10.1016/j.ijepes.2010.11.018

Kim, D. H., Lee, J. H., Hong, S. H., & Kim, S. R. (1998). A mixed-integer programming approach for the linearized reactive power and voltage control-comparison with gradient projection approach. *Proceedings of International Conference on Energy Management and Power Delivery, EMPD, 1*, 67–72.

Kursat, A., & Ulas, K. (2013). Solution of transient stability-constrained optimal power flow using artificial bee colony algorithm. *Turkish Journal of Electrical Engineering and Computer Sciences, 21*(2), 360–372.

Lai, L. L., & Ma, J. T. (1997). Application of evolutionary programming to reactive power planning-comparison with nonlinear programming approach. *IEEE Transactions on Power Systems, 12*(1), 198–206. doi:10.1109/59.574940

Li, L., Wang, Y., Trautmann, H., Jing, N., & Emmerich, M. (2018). Multiobjective evolutionary algorithms based on target region preferences. *Swarm and Evolutionary Computation*, 1–20.

Lo, K. L., & Zhu, S. P. (1991). A decoupled quadratic programming approach for optimal power dispatch. *Electric Power Systems Research, 22*(1), 4760. doi:10.1016/0378-7796(91)90079-3

Lu, F. C., & Hsu, Y. Y. (1995). Reactive power/voltage control in a distribution substation using dynamic programming. *IEEE Proceedings- Generation, Transmission and Distribution, 142*(6), 639-645.

MacArthur, R., & Wilson, E. (1997). *The theory of biogeography*. Princeton, NJ: Princeton University Press.

Mirjalili, S. (2015). Moth-flame optimization algorithm: A novel nature-inspired heuristic paradigm. *Knowledge-Based Systems, 89*, 228–249. doi:10.1016/j.knosys.2015.07.006

Mirjalili, S., Mirjalili, S. M., & Lewis, A. (2014). Grey wolf optimizer. *Advances in Engineering Software, 69*, 46–61. doi:10.1016/j.advengsoft.2013.12.007

Mohammadi-ivatloo, B., Rabiee, A., Soroudi, A., & Ehsan, M. (2012). Imperialist competitive algorithm for solving non-convex dynamic economic power dispatch. *Energy, 44*(1), 228–240. doi:10.1016/j.energy.2012.06.034

Momoh, J. A., El-Hawary, M. E., & Adapa, R. (1999). A review of selected optimal power flow literature to 1993. II. Newton, linear programming and interior point methods. *IEEE Transactions on Power Systems, 14*(1), 105–111. doi:10.1109/59.744495

Mondal, S., Bhattacharya, A., & nee Dey, S. H. (2013). Multi-objective economic emission load dispatch solution using gravitational search algorithm and considering wind power penetration. *International Journal of Electrical Power & Energy Systems*, *44*(1), 282–292. doi:10.1016/j.ijepes.2012.06.049

Mukherjee, A., & Mukherjee, V. (2016). Solution of optimal power flow with FACTS devices using a novel oppositional krill herd algorithm. *International Journal of Electrical Power & Energy Systems*, *78*, 700–714. doi:10.1016/j.ijepes.2015.12.001

Nanda, J., Mishra, S., & Saikia, L. C. (2009). Maiden Application of Bacterial Foraging-Based Optimization Technique in Multiarea Automatic Generation Control. *IEEE Transactions on Power Systems*, *24*(2), 602–609. doi:10.1109/TPWRS.2009.2016588

Niknam, T., Mojarrad, H. D., Meymand, H. Z., & Firouzi, B. B. (2011). A new honey bee mating optimization algorithm for non-smooth economic dispatch. *Energy*, *36*(2), 896–908. doi:10.1016/j.energy.2010.12.021

Ozyon, S., Temurtas, P. H., Durmus, P. B., & Kuvat, G. (2012). Charged system search algorithm for emission constrained economic power dispatch problem. *Energy*, *46*(1), 420–430. doi:10.1016/j.energy.2012.08.008

Park, J. H., Kim, Y. S., Eom, I. K., & Lee, K. Y. (1993). Economic load dispatch for piecewise quadratic cost function using Hopfield neural network. *IEEE Transactions on Power Systems*, *8*(3), 1030–1038. doi:10.1109/59.260897

Petcharaks, N., & Ongsakul, W. (2007). Hybrid enhance Lagrangian relaxation and quadratic programming for hydrothermal scheduling. *Electric Power Components and Systems*, *35*(1), 19–42. doi:10.1080/15325000600815449

Pradhan, M., Roy, P. K., & Pal, T. (2016). Grey wolf optimization applied to economic load dispatch problems. *International Journal of Electrical Power & Energy Systems*, *83*, 325–334. doi:10.1016/j.ijepes.2016.04.034

Rajabioum, R. (2011). Cuckoo optimization algorithm. *Applied Soft Computing*, *11*(8), 5508–5518. doi:10.1016/j.asoc.2011.05.008

Rajasomashekar, S., & Aravindhababu, P. (2012). Biogeography based optimization technique for best compromise solution of economic emission dispatch. *Swarm and Evolutionary Computation*, *7*, 47–57. doi:10.1016/j.swevo.2012.06.001

Ramirez, J. M., & Onate, P. E. (2003). Short term hydrothermal generation scheduling model using a genetic algorithm. *IEEE Transactions on Power Systems*, *18*(4), 1256–1264. doi:10.1109/TPWRS.2003.819877

Rao, R. V., Savsani, V. J., & Vakharia, D. P. (2012). Teaching-learning- based optimization: An optimization method for continuous non-linear large scale problems. *Inform Sciences*, *183*(1), 1–15. doi:10.1016/j.ins.2011.08.006

Rizk-Allah, R. M., El-Sehiemy, R. A., & Wang, G.-G. (2018). A novel parallel hurricane optimization algorithm for secure emission/economic load dispatch solution. *Applied Soft Computing*, *63*, 206–222. doi:10.1016/j.asoc.2017.12.002

Roa-Sepulveda, C. A., & Pavez-Lazo, B. J. (2003). A solution to the optimal power flow using simulated annealing. *International Journal of Electrical Power & Energy Systems*, *25*(1), 47–57. doi:10.1016/S0142-0615(02)00020-0

Roy, P. K., Ghoshal, S. P., & Thakur, S. S. (2011). Optimal power flow using biogeography based optimization. *International Journal of Electrical Power and Energy Conversion, 2*(3).

Roy, P. K., Mandal, B., & Bhattacharya, K. (2012). Gravitational search algorithm based optimal reactive power dispatch for voltage stability enhancement. *Electric Power Components and Systems*, *40*(9), 956–976. doi:10.1080/15325008.2012.675405

Roy, S. (2018). The *maximum likelihood* optima for an economic load dispatch in presence of demand and generation variability. *Energy*, *147*, 915–923. doi:10.1016/j.energy.2018.01.044

Santos, A., & Costa, G. R. D. (1995). Optimal power flow solution by Newton's method applied to an augmented Lagrangian function. *IET Generation, Transmission & Distribution*, *142*(1), 33–36. doi:10.1049/ip-gtd:19951586

Shaheen, H. I., Rashed, G. I., & Cheng, S. J. (2011). Optimal location and parameter setting of UPFC for enhancing power system security based on Differential Evolution algorithm. *International Journal of Electrical Power & Energy Systems*, *33*(1), 94–105. doi:10.1016/j.ijepes.2010.06.023

Shen, C. M., & Laughton, M. A. (1970). Power system load scheduling with secuirity constraints using dual linear programming. *Proceeding IEE*, *117*(1), 2117-2127.

Shouzheng, W., Lixin, M., & Dashuai, S. (2010). Hybrid differential evolution particle swarm optimization algorithm for reactive power optimization. *Power and Energy Engineering Conference (APPEEC) Asia-Pacific*, 1–4.

Simon, D. (2008). Biogeography-based optimization. *IEEE Transactions on Evolutionary Computation*, *12*(6), 702–713. doi:10.1109/TEVC.2008.919004

Sivasubramani, S., & Swarup, K. S. (2011). Multi-objective harmony search algorithm for optimal power flow problem. *International Journal of Electrical Power & Energy Systems*, *33*(3), 745–752. doi:10.1016/j.ijepes.2010.12.031

Sivasubramani, S., & Swarup, K. S. (2012). Multiagent based differential evolution approach to optimal power flow. *Applied Soft Computing*, *12*(2), 735–740. doi:10.1016/j.asoc.2011.09.016

Storn, R., & Price, K. (1995). *Differential evolution – a simple and efficient adaptive scheme for global optimization over continuous spaces* (A Technical Report TR-95-012). International Computer Science Institute, Berkeley, CA.

Sun, I., Ashley, B., Brewer, B., Hughes, A., & Tinney, W. F. (1984). Optimal power flow by Newton approach. *IEEE Transactions on Power Apparatus and Systems*, *103*(10), 2864–2880. doi:10.1109/TPAS.1984.318284

Taher, S. A., & Amooshahi, M. K. (2012). New approach for optimal UPFC placement using hybrid immune algorithm in electric power systems. *International Journal of Electrical Power & Energy Systems*, *43*(1), 899–909. doi:10.1016/j.ijepes.2012.05.064

Tang, W. J., Li, M. S., He, S., Wu, Q. H., & Saunders, J. R. (2006). Optimal power flow with dynamic loads using bacterial foraging algorithm. *International Conference on Power System Technology, Powercon*, 1–5. 10.1109/ICPST.2006.321538

Tizhoosh, H. R. (2005). Opposition-Based Learning: A New Scheme for Machine Intelligence. *International Conference on Computational Intelligence for Modelling, Control and Automation (CIMCA)*, *1*, 695-701. 10.1109/CIMCA.2005.1631345

Tripathy, M., & Mishra, S. (2007). Bacteria foraging-based solution to optimize both real power loss and voltage stability limit. *IEEE Transactions on Power Systems*, *22*(1), 240–248. doi:10.1109/TPWRS.2006.887968

Urdaneta, A. J., Gomez, J. F., Sorrentino, E., Flores, L., & Diaz, R. (1999). A hybrid genetic algorithm for optimal reactive power planning based upon successive linear programming. *IEEE Transactions on Power Systems*, *14*(4), 1292–1298. doi:10.1109/59.801887

Vo, D. N., & Ongsakul, W. (2012). Economic dispatch with multiple fuel types by enhanced augmented Lagrange Hopfield network. *Applied Energy*, *91*(1), 281–289. doi:10.1016/j.apenergy.2011.09.025

Wei, Y., Fang, L., Chung, C. Y., & Wong, K. P. (2006). A hybrid genetic algorithm-interior point method for optimal reactive power flow. *IEEE Transactions on Power Systems*, *21*(3), 1163–1169. doi:10.1109/TPWRS.2006.879262

Wells, D. W. (1968). Methods for economic secure loading of a power system. *Proceedings of the Institution of Electrical Engineers*, *115*(8), 1190–1194. doi:10.1049/piee.1968.0210

Wennan, L., Yihua, L., Xingtao, X., & Maojun, I. (2008). Reactive power optimization in area power grid based on improved Tabu search algorithm. *Third Intinternational conference electric utility deregulation restructuring power technologies*, 1472–1477. 10.1109/DRPT.2008.4523637

Wong, K. P., & Wong, Y. W. (1994). Short term hydrothermal scheduling, Part I: Simulated annealing approach. *IEE Proceedings. Generation, Transmission and Distribution*, *141*(5), 497–501. doi:10.1049/ip-gtd:19941350

Wood, W. G. (1982). Spinning reserve constrained economic dispatch. *IEEE Transactions on Power Apparatus and Systems*, *101*(2), 381–388. doi:10.1109/TPAS.1982.317118

Xu, J., Lam, A. Y. S., & Li, V. O. K. (2011). Chemical reaction optimization for task scheduling in grid computing. *IEEE Transactions on Parallel and Distributed Systems*, *22*(10), 1624–1631. doi:10.1109/TPDS.2011.35

Yan, X., & Quintana, V. H. (1999). Improving an interior point based OPF by dynamic adjustments of step sizes and tolerances. *IEEE Transactions on Power Systems*, *14*(2), 709–717. doi:10.1109/59.761902

Yu, D. C., Fagan, J. E., Foote, B., & Aly, A. A. (1986). An optimal load flow study by the generalized reduced gradient approach. *Electric Power Systems Research*, *10*(1), 47–53. doi:10.1016/0378-7796(86)90048-9

Zhu, J. Z., & Xiong, X. F. (2003). Optimal reactive power control using modified interior point method. *Electric Power Systems Research*, *66*(2), 187–192. doi:10.1016/S0378-7796(03)00078-6

Zou, Y. (2010). Optimal reactive power planning based on improved Tabu search Algorithm. *International Conference on Electrical and Control Engineering (ICECE)*, 3945–8.

Chapter 2
Economic Load Dispatch:
Optimal Power Flow and Optimal Reactive Power Dispatch Concept

ABSTRACT

The main objective of the power system is to deliver electric energy to its loads economically and efficiently in a safe and reliable manner. Due to the complicated structure of the present power system network and competitive environment introduced by deregulation, optimal power flow (OPF) and optimal reactive power flow (ORPD) provide efficient exploitation of existing power generations. This chapter describes the detail problem formulation of OPF and ORPD problems. In this study, three different single objectives, namely fuel cost minimization, voltage profile improvement, and transmission loss minimization, are considered. Moreover, in order to judge the effectiveness of the proposed methods for multi-objective scenario, two bi-objectives, namely simultaneous minimization of fuel cost and voltage deviation; simultaneous minimization of fuel cost and transmission loss; and one tri-objective function, namely simultaneous minimization of fuel cost with voltage deviation and loss, are considered.

DOI: 10.4018/978-1-5225-6971-8.ch002

INTRODUCTION

The main objective of any electrical power utility is to ensure reliable and stable power supply economically to the consumers. For smaller size power systems, it is comparatively a trivial problem. In earlier days, individual power systems were used and operated separately using regional concept. However, the state of the art is to have interconnection of different supply agencies to form very large complex power systems in order to achieve several advantages. In an interconnected power system, the real and the reactive powers of the generators are required to vary within operating limits in order to meet the particular load demand with minimum fuel cost. In the generation plant, there are two factors which are to be considered at every load change. These are the division of load and economic factor. But due to the deregulation of the industry, optimal power flow (OPF) is used to handle such problems. OPF is one of the fundamental issues of power system operation and planning. It was first introduced by Dommel and Tinney (Dommel & Tinney,1968). Thereafter, several methods of OPF have been proposed by the power system research community. At present, OPF has become reliable enough for practical use and has taken a place on standard power system analysis tools. Thus, the main purpose of OPF solution is to schedule the power generation in such a way that minimizes the fuel cost while satisfying all the equality and inequality constraints. In addition to the minimization of fuel cost, the OPF may also be used to achieve the other benefits such as reduction of system loss, improvement of voltage profile and improvement of system security. Thus, the objective of the OPF is to find steady state operating point which minimizes generation cost, system loss, voltage deviation etc while maintaining an acceptable system performance in terms of limits on generators' real and reactive powers, line flows, outputs of various compensating devices etc.

Due to the continuous growth in the demand for electricity with unmatched generation and transmission capacity expansion, voltage instability is emerging as a new challenge to power system planning and operation. Unavailability of sufficient reactive power sources to maintain normal voltage profiles at heavily loaded buses are the main reasons for the voltage collapse. This problem may be overcome by reallocating reactive power generations in the system. This can be achieved by adjusting transformer taps, var injections of shunt compensators and generator voltages. In addition, the system losses can be minimized via redistribution of reactive powers in the system. Optimal reactive power dispatch (ORPD), which is one of the application functions of modern energy management system (EMS) is used to minimize total system loss and improve voltage profile.

ECONOMIC LOAD DISPATCH

Economic load dispatch (ELD) (Lee & Park, 1998; Lokeshgupta & Sivasubramani., 2018) problem is one of the fundamental issues of power system operation and planning. The economic operation is very important for a power generation system to return the maximum profit on the huge capital invested. For achieving maximum profit, so, it is important to operate the power system at maximum possible efficiency. Maximum efficiency not only minimizes the cost of per unit energy produced by the company but also extends benefits to the consumers who are utilizing it. Economic operation determines the power produced by each unit in such a way that minimizes the fuel cost to serve the specific system load. Thus, the main objective of constrained ELD is to determine optimal operation strategy for allocation of generations of committed generating units' so as to meet the load demand at minimum operating cost subject to a variety of constraints. In essence, it is a complex non-linear optimization problem with a mixture of linear and non-linear constraints.

Objective Function

The objective of ELD (Qin et al., 2017) is to minimize the generation cost while satisfying all the equality and inequality constraints and is given by (1)

$$\text{Minimize } FC(P_g) = \sum_{i=1}^{n} FC_i(P_{gi}) \tag{1}$$

where, $FC(P_{gi})$ is the total fuel cost; $FC_i(P_{gi})$, P_{gi} are the fuel cost and generating power, respectively, of the ith generating unit; n is the number of generators.

Constraints

Equality Constraints

The equality constraints are given below.

- The sum of the generated powers of all units must be equal to sum of the power demanded by the load and the total transmission loss, which is given by (2)

$$\sum_{i=1}^{n} P_{gi} = P_D + P_L \tag{2}$$

where P_L is the transmission loss, which is given by (3)

$$P_L = \sum_{i=1}^{n}\sum_{j=1}^{n} P_{gi} B_{ij} P_{gj} + \sum_{i=1}^{n} B_{i0} P_{gi} + B_{00} \tag{3}$$

where B_{ij}, B_{i0}, B_{00} are transmission loss coefficients.

Inequality Constraints

- The power output of each unit must be greater than or equal to the minimum power permitted and also be less than or equal to maximum power permitted on that specified unit. Thus the inequality constraint is expressed as:

$$P_{gi}^{\min} \le P_{gi} \le Pg_i^{\max} \tag{4}$$

where $P_{gi}^{\min}$, $P_{gi}^{\max}$ are the minimum and maximum power outputs, respectively, of the ith unit.

- The power output of each unit is limited by ramp up/down rate as given below:

$$Max\left(P_{gi}^{\min}, P_{gi}^{0} - DR_i\right) \le P_{gi} \le Min\left(P_{gi}^{\max}, P_{gi}^{0} + UR_i\right) \quad (5)$$

where P_{gi}^{0} is the previous operating point of the ith unit; DR_i, UR_i are the down rate and up rate limits of the ith unit.

- Due to some faults in the generating units or their accessories such as pumps, boilers etc. practical generating units might have prohibited operating zones and each unit must avoid operation in prohibited zones. The operating zone of the ith unit may be described as follows:

$$\begin{aligned} P_{gi}^{\min} &\le P_{gi} \le P_{gi,1}^{l} \\ P_{gi,j-1}^{u} &\le P_{gi} \le P_{gi,j}^{l}, \quad j = 2,3,\ldots\ldots,n_i \\ P_{gi,n_i}^{u} &\le P_{gi} \le P_{gi}^{\max} \end{aligned} \quad (6)$$

where n_i is the number of prohibited zones of the ith unit; $P_{gi,j}^{l}$ is the lower generation limit of prohibited zone j of the ith unit; $P_{gi,j-1}^{u}$ is the upper generation limit of prohibited zone j-1 of the ith unit.

Limitation of ELD

In practical system, generating units and loads are not connected to a fixed bus. Therefore, the economic dispatch may results in unacceptable power flows or voltages in the power system network.

Example of Network Limitation

Let us considered two alternators A, B and two loads L_A, L_B those are connected to two different buses through the double line show in Figure 1. The maximum power flow limit on each line is considered as 100 MW

Acceptable of ELD Solution

Let us assume the total demand as 300 MW and generators generation limit is also considered. The power flow in the line and the optimal generation scheduling of this economic dispatch is shown in Figure 2. It may be observed from Figure 2 that the feasible solution is achieved.

Figure 1. Two bus power system network

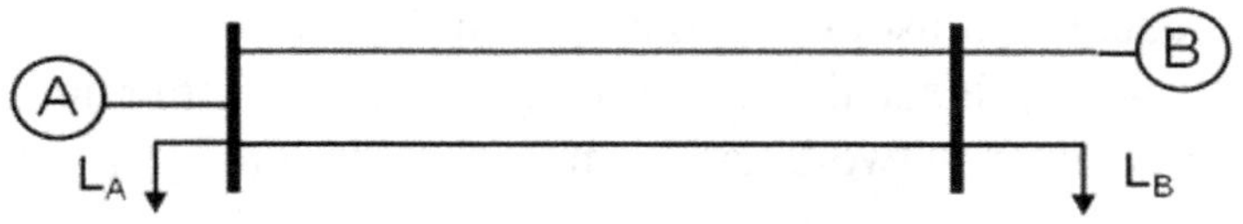

Figure 2. Power flow and generation scheduling of 2-bus power system for 300 MW load

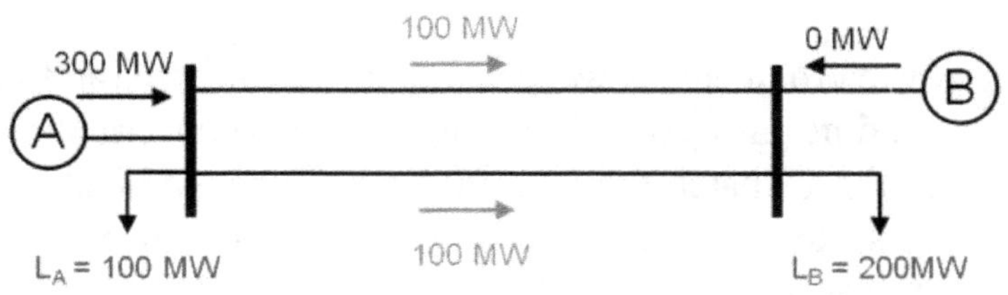

Unacceptable ELD Solution

Now, let us consider the total demand of 500 MW and generators generation limit is also taken into consideration. The optimal generation scheduling of the generators A and B for economic operation are found to be 500 MW and 0MW respectively. The optimal solution is illustrated in Figure 3. It is observed from Figure 3 that in this configuration line limit of the transmission line is violated. So, this economic dispatch solution is not acceptable.

Modified ELD Solution

In this case the solution of the ELD can be modified easily to produce acceptable flows.This could be done mathematically by adding the following

Figure 3. Unacceptable power flow of 2-bus power system for 500 MW load

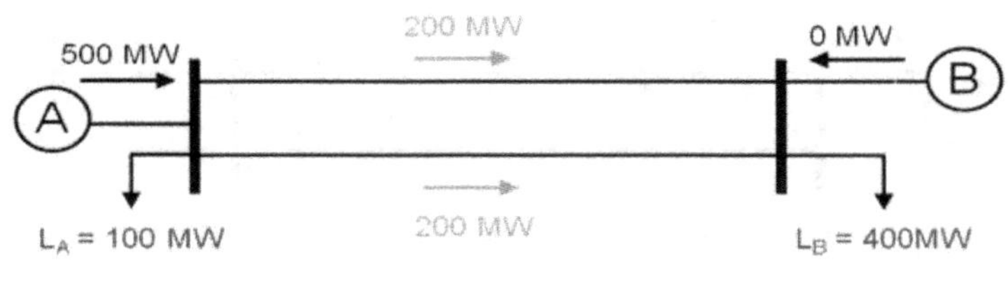

inequality constraint: $P_A - L_A \leq 200MW$. The optimal solution obtained using the inequality constraint is illustrated in Figure 4.

However, adding inequality constraints for each transmission line is not practical for complex network having more than hundred transmission line. Therefore, a general approach is very much essential. This can be achieved through solution of optimal power flow (OPF) problem.

OPTIMAL POWER FLOW

With increasing load demand, power generation, transmission and distribution network is expanding day by day. Therefore, the main concern of the complex power system is to dispatch the load in the most economical way as well as maintain the system efficiency. In consideration of several equality and inequality constraints, the main objective of optimal power flow (OPF) problem is to optimize various objectives like fuel cost reduction, voltage profile improvement, emission reduction or others. This can be achieved by tuning the various control variables such as active power generations, generator bus voltages, tap changing transformer ratios, shunt capacitor outputs. Moreover, the various equality and inequality constraints like power flow balance, generation and transmission limit, voltage profile maintenance etc are taken into consideration to improve the efficiency of the power system. The mathematical problem formulation of OPF is described below (Ghasemi et al., 2014; Mandal and Roy, 2014; Hörsch, et al., 2018):

Problem Formulation of OPF

The objective of OPF is to minimize the objective function while satisfying all the equality and inequality constraints. The OPF problem can be formulated as:

Figure 4. Acceptable power flow of 2-bus power system for 500 MW load

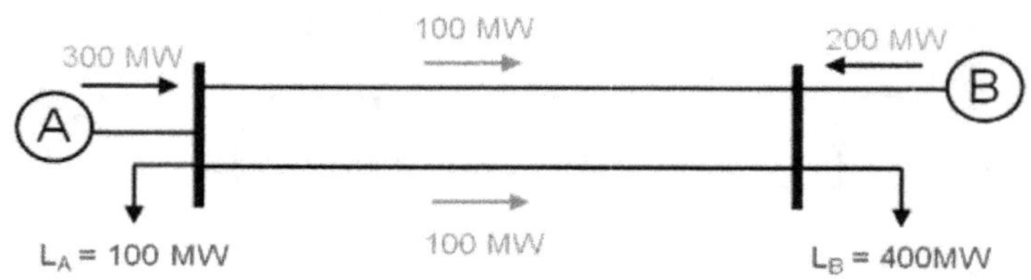

Minimize $OF\left(x,y\right)$ (7)

$$\text{subject to: } \begin{cases} e\left(x,y\right)=0 \\ ie_l \le ie\left(x,y\right) \le ie_u \end{cases} \tag{8}$$

where $OF\left(x,y\right)$ is the objective function; $e\left(x,y\right)$ are the equality constraints; $ie\left(x,y\right)$ are the inequality constraints; ie_l, ie_u are the lower and upper limits of the inequality constraints; x is the vector of dependent variables consisting of slack bus active power, load voltages, generators' reactive powers and transmission lines' loadings; y is the vector of independent variables consisting of generators' active powers except slack bus, generators' voltages, transformers' tap settings and reactive power injections. Hence, x and y can be expressed as:

$$x=[P_{g1}, V_{l_1} \cdots V_{l_{NL}}, Q_{g_1} \cdots Q_{g_{NG}}, S_{l_1} \cdots S_{l_{NTL}}] \tag{9}$$

$$y=[P_{g_2} \cdots P_{g_{NG}}, V_{g_1} \cdots V_{g_{NG}}, T_1 \cdots T_{NT}, Q_{c_1} \cdots Q_{c_{NC}}] \tag{10}$$

where NG and NL are the number of generator and load buses; NTL is number of transmission lines; NT and NC are the number of transformers and shunt compensators.

In ORPD problem also, the objective is to minimize the individual objective function of bus voltage deviation and transmission loss, satisfying all operational constraints. The dependent variable vector (x) consists of load voltages, generators' reactive powers and transmission lines' loadings and the independent variable vector (y) consists of generators' voltages, transformers' tap settings and reactive power injections.

Objective Function

Different single and multi-objective functions as listed below are used in this book to judge the superiority of various evolutionary optimization approaches.

Single Objective

In this book, three different single objective functions namely, fuel cost minimization, transmission loss minimization and voltage profile improvement are considered to determine the effectiveness of the proposed algorithms.

- **Cost reduction considering quadratic cost function**

This is the base case to minimize the fuel cost of generation which is defined by a quadratic function (Chaib et al., 2016). So the objective function for fuel cost minimization is:

$$FC = \sum_{i=1}^{NG} a_i P_{gi}^{\ 2} + b_i P_{gi} + c_i \tag{11}$$

$$F_1(\boldsymbol{x}, \boldsymbol{u}) = \left(\sum_{i=1}^{Ngen} \alpha_i + \beta_i P_{gi} + \gamma_i P_{gi}^2 \right) + Penalty$$

here, $a_i, b_i, and\, c_{i,}$ are cost coefficients of the ith generator.

- **Cost minimization giving consideration to valve-point effect**

In practical power system a ripple-like effect is observed in generating units having multi-valve steam turbines (Tan et al., 2015). To relfect this valve-point effect, the fuel cost function may be extended by adding a sinusoidal term to it (Niknam et al., 2012). Therefore, the objective function being non-convex can be formulated as below (Alsumait et al., 2010).

$$FC = \sum_{i=1}^{NG} a_i P_{gi}^{\ 2} + b_i P_{gi} + c_i + \left| d_i \times \sin\left(e_i \times \left(P_{gi\min} - P_{gi}\right)\right)\right| \tag{12}$$

where $d_i, and\, e_i$ represent the coefficients of the valve-point effect.

- **Voltage profile enhancement**

The voltage profile of any power system is related to the safety and service quality issues. The improvement of voltage profile of the power system means to minimize the voltage deviation from the unity at the load buses. This voltage deviation (VD) can be expressed by the following equation.

$$VD = \sum_{i=1}^{NL} \left|\left(V_i - V_{ref}\right)\right| \tag{13}$$

where V_{ref} is the reference voltage of the buses which is taken as 1 $p.u.$; V_i is the voltage of the ith bus; NL is the number of load buses.

- **Transmission loss minimization**

Power losses in the transmission lines sometime consider as the fitness function for OPF problem that can be mathematically formulated as follows:

$$P_L = \sum_{k=1}^{NTL} G_k \left[V_i^2 + V_j^2 - 2\left|V_i\right|\left|V_j\right|\cos\delta_{ij}\right] \tag{14}$$

where V_i, V_j are the voltage of the ith bus and j^{th} bus connected in the kth transmission line. NTL is the number of transmission line.

Multi-Objective OPF Problem

In multi-objective OPF work, three different objective functions are considered to determine the effectiveness of the proposed BBO technique. The objective functions are as follows:

- **Minimization of fuel cost and voltage deviation**

The objective of this type of problem is to minimize the voltage deviation of buses and fuel cost of the generating units that may be mathematically modeled as follows (Gaing, 2005):

$$FCVD(x,y) = \sum_{i=1}^{NG} a_i P_{gi}^{\,2} + b_i P_{gi} + c_i + \left|d_i \times \sin\left(e_i \times \left(P_{gi\min} - P_{gi}\right)\right)\right| + \sum_{i=1}^{NB} w_{1_i} \times \left|\left(V_i - V_{ref}\right)\right| \tag{15}$$

where $FCVD(x,y)$ is the combined fuel cost and voltage deviation objective function for minimization; V_{ref} is the reference voltage of the buses which is taken as 1 $p.u.$; V_i is the voltage of the ith bus; NB is the number of buses; NG is the number of generators and w_{1_i} is the weight factor of the ith unit. Since voltage deviations largely depend on system loads, w_{1_i} is taken in terms of system loads and used to transform the voltage deviation to a penalty cost, given by (16) (Gaing, 2005)

$$w_{1_i} = FC(x,y) \times \frac{P_{di}}{P_d} \tag{16}$$

where P_{di} is the active load of the ith bus and P_d is the total active load of the system.

- **Minimization of fuel cost and transmission loss**

To reduce the transmission losses in addition to fuel cost, mathematical formulation of the objective function is formulated by (17) (Gaing, 2005)

$$FCTL(x,y) = \sum_{i=1}^{NG} a_i P_{gi}^{\ 2} + b_i P_{gi} + c_i + \left| d_i \times \sin\left(e_i \times \left(P_{gi\min} - P_{gi} \right) \right) \right| + w_2 \times P_{loss} \tag{17}$$

where $FCTL(x,y)$ is the combined fuel cost and transmission loss minimization objective function. P_{loss} is the total power losses and w_2 is a weighting factor. Since transmission loss largely depends on system loads, w_2 is taken it terms of system loads and used to transform the transmission loss to a penalty cost. w_2 is given by (18).

$$w_2 = \frac{FC(x,y)}{P_d} \tag{18}$$

- **Minimization of fuel cost, transmission loss and voltage deviation**

To improve the voltage profile and to reduce the transmission loss, mathematical formulation of this type objective function is given by (19) (Gaing, 2005)

$$FCVDTL(x,y)=\sum_{i=1}^{NG} a_i P_{gi}^2 + b_i P_{gi} + c_i + \left| d_i \times \sin\left(e_i \times \left(P_{gi\min} - P_{gi}\right)\right)\right| + \sum_{i=1}^{NB} w_{1_i} \times \left|\left(V_i - V_{ref}\right)\right| + w_2 \times P_{loss} \quad (19)$$

where $FCVDTL(x,y)$ is the combined fuel cost, voltage deviation and transmission loss objective function for minimization.

Constraints

Equality Constraints

The equality constraints represent the load flow equations, which are as given below:

$$\begin{cases} P_{gi} - P_{Li} = \sum_{j=1}^{N} \left|V_i\right|\left|V_j\right|\left(G_{ij}\cos\delta_{ij} + B_{ij}\sin\delta_{ij}\right) \\ Q_{gi} - Q_{Li} = \sum_{j=1}^{N} \left|V_i\right|\left|V_j\right|\left(G_{ij}\sin\delta_{ij} - B_{ij}\cos\delta_{ij}\right) \end{cases} \quad (20)$$

where V_i, V_j are the voltages of the ith and the jth bus, respectively; P_{gi}, Q_{gi} are the active and reactive powers, respectively, of the ith generator; P_{Li}, Q_{Li} are active and reactive power, respectively, of the ith load bus; $G_{ij}, B_{ij}, \delta_{ij}$ are the conductance, admittance and phase difference of voltages between the ith and the jth bus; N is the number of buses.

Inequality Constraints

- **Generator constraints**

Generator voltage, active and reactive power of the ith bus lie between their upper and lower limits, respectively, as given by:

$$\begin{cases} V_{g_i}^{\min} \le V_{g_i} \le V_{g_i}^{\max} & i = 1, 2, \cdots, NG \\ P_{g_i}^{\min} \le P_{g_i} \le P_{g_i}^{\max} & i = 1, 2, \cdots, NG \\ Q_{g_i}^{\min} \le Q_{g_i} \le Q_{g_i}^{\max} & i = 1, 2, \cdots, NG \end{cases} \quad (21)$$

where $V_{g_i}^{\min}, V_{g_i}^{\max}$ are the minimum and maximum generator voltages, respectively, of the ith generating unit; $P_{g_i}^{\min}, P_{g_i}^{\max}$ are the minimum and maximum active powers, respectively, of the ith generating unit; $Q_{g_i}^{\min}, Q_{g_i}^{\max}$ are the minimum and maximum reactive powers, respectively, of the ith generating unit.

- **Load bus constraints**

$$V_{l_i}^{\min} \le V_{l_i} \le V_{l_i}^{\max} \qquad i = 1, 2, \cdots, NL \quad (22)$$

where $V_{l_i}^{\min}, V_{l_i}^{\max}$ are the minimum and maximum load voltages, respectively, of the ith generating unit.

- **Transmission line constraints**

$$S_{li} \le S_{l_i}^{\max} \qquad i = 1, 2, \cdots, NTL \quad (23)$$

where S_{l_i} is the apparent power flow of the ith branch; $S_{l_i}^{\max}$ is the maximum apparent power flow limit of the ith branch.

- **Transformer Tap Constraints**

Transformer tap settings are bounded between upper and lower limits as given below:

$$TC_i^{\min} \le TC_i \le TC_i^{\max} \qquad i = 1, 2, \cdots, NT \quad (24)$$

where $TC_i^{\min}, TC_i^{\max}$ are the minimum and maximum tap setting limits, respectively, of the ith transformer.

- **Shunt compensator constraints**

Shunt compensation is restricted by its limits as follows:

$$Q_{c_i}^{\min} \le Q_{c_i} \le Q_{c_i}^{\max} \qquad i = 1, 2, \cdots, NC \tag{25}$$

where $Q_{ci}^{\min}, Q_{ci}^{\max}$ are the minimum and maximum var injection limits, respectively, of the ith shunt capacitor.

OPTIMAL REACTIVE POWER DISPATCH

In the last two decades, the security of power system operation has received much attention among the power system researchers. This can be achieved through the study of optimal reactive power dispatch (ORPD) problem (Nuaekaew et al., 2017).The main objective of ORPD is to improve the voltage profile and minimizing system real power losses through proper distribution of reactive power in the system. Therefore, the problem of the ORPD can be optimized to enhance the voltage profile and minimize the system losses as well. The mathematical problem formulation of ORPD is discussed below.

Problem Formulation of ORPD

Minimization of Transmission Loss

Mathematical formulation of this type of objective function is given by (26)

$$Min \;\; P_L \tag{26}$$

where P_L is the total power losses.

Power losses can be mathematically formulated as follows:

$$P_L = \sum_{k=1}^{NTL} G_k \left[V_i^2 + V_j^2 - 2\left|V_i\right|\left|V_j\right| \cos \delta_{ij} \right] \tag{27}$$

where G_k is the conductance of the kth line connected between the ith bus and the jth bus

Minimization of Voltage Deviation

The objective of this type of problem is to minimize the voltage deviations of all load buses while satisfying all equality and inequality constraints and may be mathematically modeled as follows:

$$Min \;\; SVD = Min \sum_{i=1}^{NL} \left|\left(V_i - V_{ref}\right)\right| \tag{28}$$

Equality Constraints

The equality constraints represent the load flow equations, which are as given below:

$$\begin{cases} P_{gi} - P_{Li} = \sum_{j=1}^{N} |V_i||V_j|\left(G_{ij}\cos\delta_{ij} + B_{ij}\sin\delta_{ij}\right) \\ Q_{gi} - Q_{Li} = \sum_{j=1}^{N} |V_i||V_j|\left(G_{ij}\sin\delta_{ij} - B_{ij}\cos\delta_{ij}\right) \end{cases} \tag{29}$$

where V_i, V_j are the voltages of the ith and the jth bus, respectively; P_{gi}, Q_{gi} are the active and reactive powers, respectively, of the ith generator; P_{Li}, Q_{Li} are active and reactive power, respectively, of the ith load bus; $G_{ij}, B_{ij}, \delta_{ij}$ are the conductance, admittance and phase difference of voltages between the ith and the jth bus; N is the number of buses.

Inequality Constraints

- **Generator constraints**

Generator voltage of the various buses must lie between their upper and lower limits, respectively, as given by:

$$V_{g_i}^{\min} \le V_{g_i} \le V_{g_i}^{\max} \qquad i = 1, 2, \cdots, NG \tag{30}$$

where $V_{g_i}^{\min}, V_{g_i}^{\max}$ are the minimum and maximum generator voltages, respectively, of the ith generating unit;

- **Load bus constraints**

$$V_{l_i}^{\min} \le V_{l_i} \le V_{l_i}^{\max} \qquad i = 1, 2, \cdots, NL \tag{31}$$

where $V_{l_i}^{\min}, V_{l_i}^{\max}$ are the minimum and maximum load voltages, respectively, of the ith generating unit.

- **Transmission line constraints**

$$S_{li} \le S_{l_i}^{\max} \qquad i = 1, 2, \cdots, NTL \tag{32}$$

where S_{l_i} is the apparent power flow of the ith branch; $S_{l_i}^{\max}$ is the maximum apparent power flow limit of the ith branch.

- **Transformer tap constraints**

Transformer tap settings are bounded between upper and lower limits as given below:

$$TC_i^{\min} \le TC_i \le TC_i^{\max} \qquad i = 1, 2, \cdots, NT \tag{33}$$

where $TC_i^{\min}, TC_i^{\max}$ are the minimum and maximum tap setting limits, respectively, of the ith transformer.

- **Shunt compensator constraints**

Shunt compensation is restricted by its limits as follows:

$$Q_{c_i}^{\min} \le Q_{c_i} \le Q_{c_i}^{\max} \qquad i = 1, 2, \cdots, NC \tag{34}$$

where $Q_{ci}^{\min}, Q_{ci}^{\max}$ are the minimum and maximum var injection limits, respectively, of the ith shunt capacitor.

CONCLUSION

In this chapter, economic load dispatch (ELD), optimal power flow (OPF) and optimal reactive power flow (ORPD) problems are elaborately discussed. Moreover, the limitations of the ELD are discussed. Furthermore, both single and multi-objective OPF are presented in this chapter. The popularities of OPF and ORPD encourage the present authors to consider it for the present study.

In the recent past, various stochastic optimization techniques were proposed in the literature for fine tuning of OPF and ORPD problem. But recent researchers identified some shortcomings, such as poor convergence, large dimension, input parameter dependence etc. of the available algorithms for OPF and ORPD analysis. The previous chapter proposes some intelligent control algorithms for further analysis and betterment of the OPF and ORPD performances. A brief description of biogeography-based optimization (BBO), artificial bee colony (ABC) optimization, differential evolution (DE), teaching-learning-based optimization (TLBO), chemical reaction based optimization (CRO), krill herd algorithm (KHA), grey wolf optimization (GWO) are presented in Chapter 2 with their mathematical models. These algorithms are proposed as an effective optimization tool to solve constraint OPF and ORPD problems. The simulation results of OPF and ORPD in various test cases are discussed in the next chapter.

REFERENCES

Alsumait, J. S., Sykulski, J. K., & Al-Othman, A. K. (2010). 'A hybrid GA–PS–SQP method to solve power system valve-point economic dispatch problems'. *Applied Energy*, *87*(5), 1773–1781. doi:10.1016/j.apenergy.2009.10.007

Chaib, A. E., Bouchekara, H. R. E. H., Mehasni, R., & Abido, M. A. (2016). Optimal power flow with emission and non-smooth cost functions using backtracking search optimization algorithm. *International Journal of Electrical Power & Energy Systems*, *81*, 64–77. doi:10.1016/j.ijepes.2016.02.004

Dommel, H. W., & Tinney, W. F. (1968). Optimal power flow solutions. *IEEE Transactions on Power System Apparatus and Systems*, *PAS-87*(10), 1866–1876. doi:10.1109/TPAS.1968.292150

Gaing, Z. L. (2005). Constrained optimal power flow by mixed-integer particle swarm optimization. *IEEE Power Engineering Society General Meeting*, *1*, 243-250.

Ghasemi, M., Ghavidel, S., Rahmani, S., Roosta, A., & Falah, H. (2014). A novel hybrid algorithm of imperialist competitive algorithm and teaching learning algorithm for optimal power flow problem with non-smooth cost functions. *Engineering Applications of Artificial Intelligence*, *29*, 54–69. doi:10.1016/j.engappai.2013.11.003

Hörsch, J., Ronellenfitsch, H., Witthaut, D., & Brown, T. (2018). Linear optimal power flow using cycle flows. *Electric Power Systems Research*, *158*, 126–135. doi:10.1016/j.epsr.2017.12.034

Lee, K. Y., & Park, J. H. (1998). Adaptive Hopfield neural networks for economic load dispatch. *IEEE Transactions on Power Systems*, *13*(2), 519–526. doi:10.1109/59.667377

Lokeshgupta, B., & Sivasubramani, S. (2018). Multi-objective dynamic economic and emission dispatch with demand side management. *International Journal of Electrical Power & Energy Systems*, *97*, 334–343. doi:10.1016/j.ijepes.2017.11.020

Mandal, B., & Roy, P. K. (2014). Multi-objective optimal power flow using quasi-oppositional teaching learning based optimization. *Applied Soft Computing*, *21*, 590–606. doi:10.1016/j.asoc.2014.04.010

Niknam, T., Narimani, M. R., & Azizipanah-Abarghooee, R. (2012). A new hybrid algorithm for optimal power flow considering prohibited zones and valve point effect. *Energy Conversion and Management*, *58*, 197–206. doi:10.1016/j.enconman.2012.01.017

Nuaekaew, K., Artrit, P., Pholdee, N., & Bureerat, S. (2017). Optimal reactive power dispatch problem using a two-archive multi-objective grey wolf optimizer. *Expert Systems with Applications*, *87*, 79–89. doi:10.1016/j.eswa.2017.06.009

Qin, Q., Cheng, S., Chu, X., Lei, X., & Shi, Y. (2017). Solving non-convex/non-smooth economic load dispatch problems via an enhanced particle swarm optimization. *Applied Soft Computing*, *59*, 229–242. doi:10.1016/j.asoc.2017.05.034

Tan, Y., Li, C., Cao, Y., Lee, K. Y., Li, L., Tang, S., & Zhou, L. (2015). Improved group search optimization method for optimal power flow problem considering valve-point loading effects. *Neurocomputing*, *148*, 229–239. doi:10.1016/j.neucom.2013.09.065

Chapter 3
Evolutionary Algorithm Applied to Economic Load Dispatch

ABSTRACT

This chapter introduces various evolutionary algorithms, namely grey wolf optimization (GWO), teaching-learning-based optimization (TLBO), biogeography-based optimization (BBO), krill herd algorithm (KHA), chemical reaction optimization (CRO) algorithms, for solving the economic load dispatch (ELD) problem of various power systems. To demonstrate the superiority of the proposed approaches in solving non-convex, non-linear and constrained ELD problem, the aforesaid approaches are implemented on 10-unit, 15-unit, 40-unit, 80-unit, and 140-unit test systems. It is observed from the simulation results that HCRO exhibits significantly better performance in terms of solution quality and convergence speed for all the cases compared to other discussed algorithms. Furthermore, the statistical results confirm the robustness of the proposed HCRO algorithm.

INTRODUCTION

Economic load dispatch (ELD) (Lokeshgupta & Sivasubramani 2018) is one of the most important tasks to be performed in the operation and planning of a power system that decides the generation schedule of generating units with an objective of minimizing the total fuel cost. During the last decades, the

DOI: 10.4018/978-1-5225-6971-8.ch003

electrical power market became more and more liberal and highly competitive. The main objective of ELD is to optimize the operation cost while fulfill the load demands of customers. The economic operation of the generating systems has always occupied an important position in the electric power industry. ELD is defined as the process of allocating generation levels to the generating units in such a manner so that the system load is supplied entirely and most economically. A good load dispatch reduces the production cost, increases the system reliability, and maximizes the energy capability of thermal units. But, for practical purposes, it is necessary to go for alternate sources of energy. The fuel cost characteristics of modern generating units are highly nonlinear with demand for solution techniques having no restrictions on to the shape of the fuel cost curves. The purpose of this work is to present a solution strategy to solve ELD problem in an efficient way while considering several aspects of ELD. From the viewpoint of these perspectives, numerous investigations have been undertaken and, time-to-time, reported in the literature. In essence, it is a complex non-linear optimization problem with a mixture of linear and non-linear constraints.

LITERATURE SURVEY

The ELD problem is one of the most practical subjects in electrical power engineering and is always under high attention of several researchers. A lot of optimization methods including classical and stochastic search approaches have been applied to solve ELD problem. A bibliographical survey of existing literature on ELD methods reveals that various numerical optimization techniques are employed to solve the ELD problem. Previously, a number of derivative-based approaches such as gradient method (GM) (Dodu et al., 1972), lambda iteration method (LIM) (Chen & Wang, 1993; Aravindhababu & Nayar, 2002), linear programming (LP) (Parikh & Chattopadhyay, 1996), Newton's method (Ramanathan, 1985), quadratic programming (QP) (Fan & Zhang, 1998), Lagrangian multiplier method (Nanda et al., 1994), interior point algorithm (IPA) (Yan & Quintana, 1997), classical technique based on co-ordination equations (El-Keib et al., 1994) were applied to solve ELD problems. Because of highly nonlinear characteristics of the problem with many local optimum solutions and a large number of constraints, the classical calculus-based methods cannot perform satisfactorily for solving ELD problems. These methods ignore the portions of incremental cost curve that are not continuous or monotonically increasing. Because of highly nonlinear characteristics of the problem with many local optimum solutions and a large number of constraints, the classical calculus-based methods cannot perform satisfactorily for solving ELD problems. Numerous mathematical programming optimization methods have been applied to solve the ELD problems in the past decades. LP methods are fast and reliable, but have a disadvantage associated with the piecewise linear cost approximation. So in the classical method, the fuel cost curve is approximated according to their requirement but use of such approximation may lead to a huge loss of revenue over the time. Recently, Yang *et al.* (Yang det al., 2013) presented a mathematical programming based method named quadratically constrained programming (QCP) to solve the non-smooth and non-convex ELD problem. Some traditional algorithms such as dynamic programming (DP) (Liang & Glover, 1992) impose no restrictions on the nature of the cost curves and therefore it can solve ELD problems with inherently nonlinear and discontinuous cost curves. Dhillon proposed artificial neural network (ANN) (Dhilon, 1991) to solve power system optimization problem. Hopfield neural networks (HNN) (Park et al., 1993) based method is another alternative to solve non-linear ELD problems. A large number of iterations and oscillation are the major concern for solving the ELD problem using the Hopfield neural network. An augmented Lagrange Hopfield network (ALHN) (Dieu & Schegnerb, 2013) solving ELD problem was proposed by Dieu and Schegnerb (2013) Furthermore, Dieu and Ongsakul (2012) (Dieu & Ongsakul, 2012) introduced a hybrid method based on QP and ALHN for solving ELD problem with piecewise quadratic cost functions and prohibited zones. However, these methods suffer from excessive numerical iterations, resulting in huge computations. Thus, developing a reliable, fast and efficient algorithm is still an active area of research in power systems. Various investigations on ELD have been explored till date, as better solutions would result in significant economical benefits.

Evolutionary optimization approaches such as particle swarm optimization (PSO) (Wu & Guo, 2017; Chen, 2008; Chaturvedi et al., 2009; Gaing, 2003; Kumar et al., 2011; Kuo, 2008; Meng et al., 2010; Niknam, 2010; Park et al., 2010; Panigrahi et al., 2008; Shayeghi & Ghasemi, 2012; Wang et al., 2010), genetic algorithm (GA) (Shang et al., 2017; Basu, 2008; Deb et al., 2002; Ciornei & Kyriakides, 2012; Subbaraj et al., 2009; Subbaraj et al., 2011; Yalcinoz & Altun, 2005), DE (Amjady & Sharifzadeh, 2010; Coelho & Mariani, 2006; Nima & Hossein, 2010; Storn & Price, 1997; Wang et al., 2012; Wu et al., 2010; Yuan et al., 2008), ABC (Hemamalini & Simon, 2010; Karaboga & Basturk, 2007; Shayeghi & Ghasemi, 2011; Shayeghi et al., 2012), EP (Basu, 2007; Jayabarati & Sadasivam, 2000; Lee & Yao, 2004;

Sinha et al., 2003; Venkatesh et al., 2003; Wong & Yuryevich, 1998), BFO (Hooshmand et al., 2012; Hota et al., 2010; Pandit et al., 2012), ant colony optimization (ACO) (Pothiya et al., 2010), charged system search algorithm (CSSA) (Ozyon et al., 2012), honey bee mating optimization (HBMO) (Bozorg et al., 2006; Javidan & Ghasemi, 2012; Niknam et al., 2011), ongroup search optimizer (OSO) (Guo et al., 2012), harmony search algorithm (HSA) (Pandi et al., 2011), immune algorithm (IA) (Hemamalini & Simon, 2011), imperialist competitive algorithm (ICA) (Mohammadi-ivatloo et al., 2012), ant swarm optimization (ASO) (Cai et al., 2012; Jiejin et al., 2010), biogeography based optimization (BBO) (Bhattacharya & Chattopadhyay, 2010; Li et al., 2011; Lohokare et al., 2012), evolutionary algorithms (EA) (Qu et al., 2018) etc. are proposed for solving ELD problems having non-differentiable cost functions, due to their simplicity, the absence of convexity assumptions, and excellent random parallel search capability. The major challenge faced by evolutionary techniques is to maintain a proper balance between exploration (global search) and exploitation (local search). Recent research endeavors have been directed toward the application of efficient and near-optimal evolutionary algorithms (EAs). These algorithms are search algorithms based on the simulated evolutionary process of natural selection, variation, and genetics. But as the evolution progresses, later solutions may be nearer to the global point and the step size should be reduced gradually to enable quick convergence. Quantum-inspired PSO (QPSO) was proposed by Meng *et al.* (Meng et al., 2010) which had stronger search ability and quicker convergence speed, not only because of the introduction of quantum computing theory, but also due to two special implementations like self-adaptive probability selection and chaotic sequences mutation. The proposed approach was tested on three power system cases consisting of 3, 13 and 40 thermal units. Basu developed ABC (Basu, 2013) optimization technique to solve multi-area ELD (MAELD) problem considering tie line constraints and nonlinearities like multiple fuels, valve-point loading and prohibited operating zones. The PSO algorithm is quite impressive for solving ELD problem because of its first convergence feature. Kumar and Mohan (2011) proposed a multi-objective directed bee colony optimization algorithm (MODBC) (Kumar & Mohan, 2011) and successfully applied for solving a multi-objective problem of optimizing the conflicting economic and emission dispatch while satisfying both equality and inequality constraints. Cai *et al.* developed a chaotic PSO (CPSO) (Cai et al., 2007) algorithm for solving the ELD problems of thermal generators in power systems. Recently, modifications of conventional population-based methods are made for getting better quality solutions of ELD problem. Hota et al. implemented an improved PSO (IPSO) (Hota et al., 2009) to solve the non-smooth HTS problem. Hosseinnezhad, and Babaei (2013) implemented θ -PSO (Hosseinnezhad & Babaei, 2013) method on 6, 13, 15 and 40 generating units test systems to solve ELD problems considering various nonlinear constraints like ramp rate limits, valve point loadings, prohibited operating zone and proposed method was compared with some modified versions of PSO and other optimization techniques. Shaw *et al.* applied OBL concept to accelerate the performance of the GSA (Shaw et al., 2012). Chatterjee *et al.* proposed opposition-based harmony search algorithm (OHSA) (Chatterjee et al., 2012) to employ OBL for harmony memory initialization and also for the generation jumping. Liao *et al.* presented isolation Niche immune-based GA algorithm (Liao, 2012). for solving dynamic ELD (DELD) problem. Iteration based PSO (IBPSO) (Safari & Shayeghi, 2011) invented by Safari and Shayeghi (2011) was applied to determine the feasible optimal solution of the ELD problem considering various generator constraints. Two test systems, including 6 and 15 unit generating, were applied to compare the performance of the proposed algorithm with PSO, chaotic PSO, hybrid GAPSO, self-organizing hierarchical PSO (SOH_PSO) methods. Panigrahi *et al.* presented a novel optimization approach named clonal algorithm (CA) (Panigrahi et al., 2007). to solve constrained ELD problem using artificial immune system (AIS). Simulations were performed on various systems with different number of generating units and comparisons were performed with other prevalent approaches. Afzalan and Joorabian (2013) addressed a novel method namely, an epsilon-multi-objective genetic algorithm (Afzalan & Joorabian, 2013) to solve multi-objective economic load dispatch problem with non-smooth and non-convex cost. The proposed method was tested on 6, 10, 13 and 40 generating units, and the simulation results of four power systems demonstrated the advantages of the proposed method for reducing the cost function. Chandram *et al.* (Chandram et al., 2011) presented equal embedded algorithm (EEA) to solve the ELD problem with quadratic and cubic fuel cost functions and transmission losses. The proposed method was implemented and tested by considering 3, 15 and 26 generators to solve the ELD problem. Nandaand and Narayanan (2002) presented the solution of ELD with line flow constraints through the application of GA (Nandaand & Narayanan 2002). Two representative systems, i.e. IEEE 14 bus and IEEE 30 bus systems were considered for the investigations. Various researchers were looking for better alternatives, and therefore, hybrid methods combining different conventional techniques were identified. Recently, Roy *et al.* proposed modified shuffled frog leaping algorithm (MSFLA) (Roy et al., 2013) with

GA for the ELD problem. Saber *(2012)* presented a novel modified PSO (MPSO) (Saber, 2012) which includes advantages of bacterial foraging (BF) and PSO for constrained DELD problem. Bhattacharya *et al.* proposed a hybrid technique combining differential evolution with biogeography-based optimization (DE/BBO) algorithm (Bhattacharya & Chattopadhyay, 2010) to solve both convex and non-convex ELD problems of thermal power units considering transmission losses and constraints such as ramp rate limits, valve-point loading and prohibited operating zones. Vaisakh *et al.* implemented bacterial foraging based PSO-DE (BPSO-DE) algorithm (Vaisakh et al., 2012) by integrating BFO, PSO and DE for solving static and dynamic ELD problems of various test systems.

Furthermore, in order to enhance the search efficiency, two or more basic evolutionary algorithms have successfully been hybridized by various researchers in the last two decades. Some of the hybrid evolutionary algorithms are hybrid genetic algorithm and pattern search based sequential quadratic programming (GA–PS–SQP) (Alsumait et al., 2010), hybrid interior point based differential evolution (HIPDE) (Duvvuru & Swarup, 2011), hybrid CPSO–SQP (Jiejin et al., 2012), (Khamsawang & Jiriwibhakorn, 2010). hybrid dynamic search PSO with tabu search algorithm (DSPSO–TSA). Recently, Hooshmand et al. have solved ELD problem by hybridizing bacterial foraging technique with Nelder mead algorithm technique (Hooshmand et al., 2012). A hybrid evolutionary algorithm based on the hybridization of bee colony optimization with sequential quadratic programming was employed by Basu (Basu, 2013) to solve dynamic ELD problem. Bhattacharya, and Chattopadhyay developed hybrid DE/BBO approach to solve ELD problem of small, medium and large-scale power systems.

MOTIVATION OF THE PRESENT WORK

Conventional optimization methods that make use of derivatives and gradients, in general, are not able to locate or identify the global optimum. Classical optimization methods are highly sensitive to starting points and frequently converge to the local optimum solution or diverge together. The reason for this is that classical evolutionary approaches were developed for general purposes and their performance may not be good for specific applications. In the classical method based ELD problem, the fuel cost curve is approximated according to their requirement but the use of such approximation may lead to a huge loss of revenue over the time. Because of highly nonlinear characteristics of the practical system with many local optimum solutions and a large number of constraints, the classical calculus-based methods cannot perform satisfactorily. In large practical thermal generation plants with large size turbines, the input–output characteristics of generators are highly nonlinear and discontinuous due to the presence of valve-point loading effect. Therefore, for solving practical ELD problems classical methods are infeasible because of non-linear characteristics like ramp rate limits, prohibited operating zones and non-smooth cost functions. For handling nonlinear problems, various evolutionary algorithms mentioned in the previous section have emerged as a useful optimization tool. However, each of these methods has its own strengths and weakness, but long computational time is a common drawback of most of them.

In view of the above, the present chapter of this dissertation will survey the potential impact on the solution quality of the constrained ELD problem for different cost functions/multiple fuels/dynamic operational constraints (ramp rate limits) with valve point loading discontinuities of generating units. In the present work, KHA, TLBO and QOTLBO techniques which are the most recent development in the category of heuristic optimization technique are successfully applied for getting improved optimal solution of the constraint ELD problem.

The organization of this chapter is as follows: Mathematical modeling of ELD is given in the section titled "Mathematical problem formulation." The various algorithms applied to ELD are mentioned in the section titled "Various algorithms applied to ELD." Then the simulation results and discussions are presented in the next section. Finally, a conclusion is drawn at the end of this chapter.

MATHEMATICAL PROBLEM FORMULATION

Objective Function

The ELD problem is addressed as to simultaneously minimize the overall cost rate and meet the load demand of a power system while satisfying an equality and inequality constraints. Main production cost of thermal generator is fuel cost. Therefore, the objective function of ELD problem may be expressed as under:

$$F_t = \sum_{i=1}^{m} F_{gi}(P_{gi}) \tag{1}$$

where F_t is the total fuel cost; $F_i(P_{gi})$ is the cost of the ith thermal generator and *m* is the number of committed generators to the operating system.

Usually, fuel cost of a thermal unit is expressed as a quadratic polynomial of its output as below.

$$F_i(P_{gi}) = a_i + b_i P_{gi} + c_i P_{gi}^2 \tag{2}$$

where a_i, b_i, $\mathrm{c_i}$ are the cost coefficients and P_{gi} is the power output of the ith generator.

However, for an accurate modeling valve-point effect is added in the fuel input-power output cost function. The valve-point effects introduce ripples in the heat-rate curves and make the objective function discontinuous, non-convex and with multiple minima. Considering the valve-point effects, the fuel cost function of the ith thermal generating unit is expressed as the sum of a quadratic and a sinusoidal function in the following form:

$$F_i(P_{gi}) = a_i + b_i P_{gi} + c_i P_{gi}^2 + \left| d_{\mathrm{i}} \sin(e_i (P_{gi}^{\min} - P_{gi})) \right| \tag{3}$$

where d_i, e_i are the fuel cost coefficients of the ith unit reflecting the effect of valve point loading; $p_{gi}^{\min}$ is the minimum active power of the ith generating unit.

Due to escalating fuel prices, displacement of scarce and expensive fuels by more abundant fuels, thermal units use multiple fuels like coal, natural gas and oil. ELD solution the valve-point effects multi-fuel options should be included in the cost function which may be express as follows:

$$F_i(P_{gi}) = \begin{cases} a_{i1} + b_{i1} P_{gi} + c_{i1} P_{gi}^2 + \left| d_{\mathrm{i1}} \sin(e_{i1} (P_{gi}^{\min} - P_{gi})) \right| & \text{if } \mathrm{P_{gi}^{min}} \leq P_{gi} \leq P_{gi1} \\ a_{i2} + b_{i2} P_{gi} + c_{i2} P_{gi}^2 + \left| d_{\mathrm{i2}} \sin(e_{i2} (P_{gi1} - P_{gi})) \right| & \text{if } \mathrm{P_{gi1}} \leq P_{gi} \leq P_{gi2} \\ \quad\quad . & \\ \quad\quad . & \\ a_{ik} + b_{ik} P_{gi} + c_{ik} P_{gi}^2 + \left| d_{\mathrm{ik}} \sin(e_{ik} (P_{gik} - P_{gi})) \right| & \text{if } \mathrm{P_{gik}} \leq P_{gi} \leq P_{gi}^{\max} \end{cases} \tag{4}$$

Constraints

The ELD problem is subjected to the following equality and inequality constraints.

Real Power Balance Constraints

The sum of the generated powers of all units must be equal to the sum of the power demanded by the load and the total transmission loss. Total electric power generation must be distributed between power demand and the power loss in the transmission lines. In the power balance criterion, the equality constraint should be satisfied as:

$$\sum_{i=1}^{m} P_{gi} - P_d - P_{loss} = 0 \tag{5}$$

where, m represents the total number of generators. So, the total generated real power should be the same with the total load demand P_d plus transmission losses P_{loss} of the system. The P_{loss} is calculated using power flows coefficients B_{00}, B_{i0}, B_{ij} by the following formula:

$$P_{loss} = \sum_{i=1}^{m}\sum_{j=1}^{m} P_{gi} B_{ij} P_{gj} + \sum_{i=1}^{m} P_{gi} B_{0i} + B_{00} = 0 \tag{6}$$

where B_{00}, B_{i0}, B_{ij} are loss coefficients which may be taken as constant under normal condition.

Inequality Constraints or Generator Capacity Constraints

The power output of each unit must be greater than or equal to the minimum power permitted and also be less than or equal to maximum power permitted on that specified unit. The power generated by each generator must be within its lower and upper operating limits. It may mathematically be formulated by

$$P_{gi}^{\min} \leq P_{gi} \leq P_{gi}^{\max} \qquad i = 1, 2, \ldots, m \tag{7}$$

where $P_{gi}^{\min}$, $P_{gi}^{\max}$ are the minimum and maximum active power, respectively, of the ith unit.

Prohibited Operating Zone

Due to vibration in its shaft bearing, practical generating units might have prohibited operating zone in the input-output curve of the generator and each unit must avoid operation in prohibited zones. The operating zone of the ith unit may be described as follows:

$$\begin{aligned} &P_{gi}^{\min} \le P_{gi} \le P_{g_{i,1}} \\ &P_{g_{i,j-1}} \le P_{gi} \le P_{g_{i,j}} \qquad j = 2,3,\ldots,n_i \\ &P_{g_{i,n_i}} \le P_{gi} \le P_{gi}^{\max} \end{aligned} \tag{8}$$

where n_i represents the number of prohibited operating zones of the ith unit; $P_{gi,j-1}, P_{gi,j}$ are the generation limits of the jth prohibited zone of the ith unit.

VARIOUS ALGORITHMS APPLIED TO ELD

Biogeography Based Optimization Technique Applied to ELD

In this chapter, BBO technique has been employed to solve constrained ELD problem and to find optimal or near optimal generation satisfying both equality and inequality constraints. The algorithm of the proposed method is as enumerated below.

Step 1: Initialize the input parameters of BBO as follows: After several runs, with different combinations of BBO input parameters, the following parameters are found to be best for the optimal performance of the technique. Habitat modification probability, $P^{\text{mod}} = 1$; mutation probability $= 0.005$; maximum immigration rate, $I_{\max} = 1$; maximum emigration rate $E_{\max} = 1$; step size fornumerical integration, $dt = 1$; number of iterations = 50 (for systems I, II, III, V and VI); 100 (for system IV); elitism parameter=2; number of *SIV*s = number of generating units; number of habitats =50.

Step 2: The initial values of *SIV* of each habitat should be randomly selected while satisfying different equality and inequality constraints of ELD problems. Several numbers of habitats depending upon the population size are being generated. Each habitat represents a potential solution to the given problem.

Step 3: Calculate the *HSI* i.e. value of the objective function for each habitat of the population set for given emigration rate, E_r , immigration rate, I_r and species, N_s . Here, in ELD problem, HSI^i indicates the fuel cost due to the ith set of power output (i.e. i^{th} habitat) in \$/hr. SIV^{iq} represents the value of power output of the qth generator of the ith habitat set, H^i. In this work, each habitat is a vector with m generating units. Each individual habitat within the total of *H* habitats represents a candidate solution for solving the ELD problem. For example, if there are *m* units that must be operated to provide powers to loads, then, the ith individual H^i can be defined as follows:

$$H^i = SIV^{iq} = [SIV^{i1}, SIV^{i2}, ..., SIV^{im}];\ i = 1,2,...,S; q = 1,2,...m \tag{11}$$

where SIV^{iq} is the power output of the qth unit of the ith individual. The matrix represents the total population of the habitat set.

Step 4: Based on the *HSI* values elite habitats are identified.

Step 5: Each non-elite habitat is modified by performing probabilistically migration operation and *HSI* of each modified set is recalculated. Feasibility of a solution is verified i.e. each *SIV* should satisfy equality and inequality constraints as mentioned in the specific problem.

Step 6: Species count probability of each habitat is updated using equation (17) of chapter 2. Mutation operation is performed on the non-elite habitat and *HSI* value of each new habitat is computed.

Step 7: Feasibility of a problem solution is verified i.e. each *SIV* should satisfy equality and inequality constraints of generator as mentioned in the specific problem.

Step 8: Go to step (3) for the next iteration.

Step 9: Stop iterations after a predefined number of iterations.

The flowchart for the proposed technique is given in Figure 1.

Figure 1. Flowchart of ELD using BBO technique

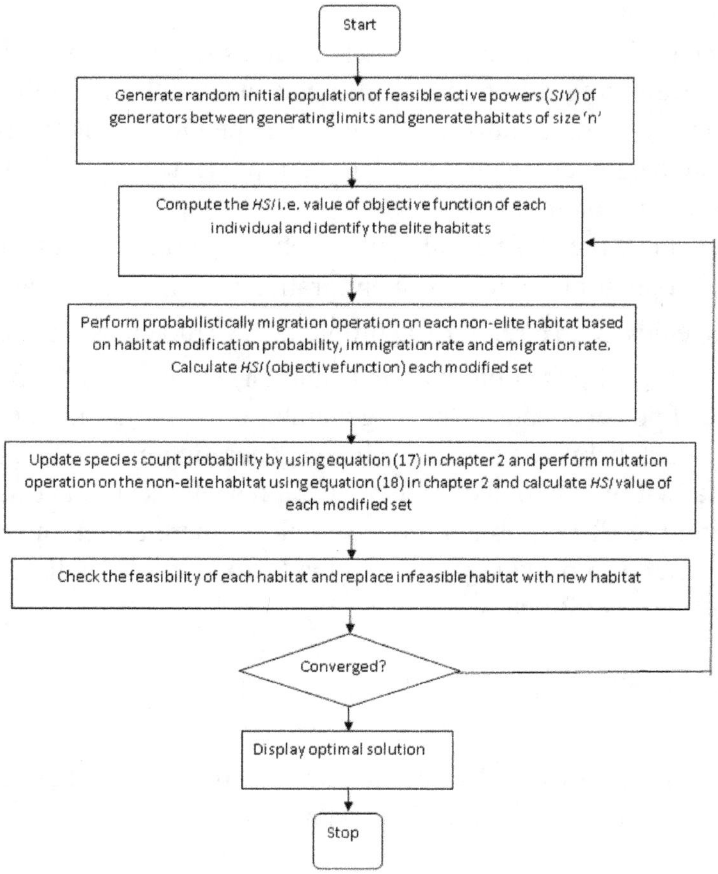

KHA Applied to ELD

In this research work, the proposed KHA algorithm is developed and successfully implemented to solve different types of ELD problems. The different steps of KHA algorithm for solving ELD problems are described below

Step 1: Active generation of all generating units except the last unit is initialized randomly within their effective real power operating limits i.e. each element must satisfy generator capacity constraint of (7). The active power generation of the last unit is evaluated using (5) and check whether it satisfy the inequality constraint (7) or not. If it does not satisfy the inequality constraint, the corresponding solution set is reinitialized.

Several initial solutions set depending upon the population size are generated. Each feasible solution set (control variables) represents the position of different krill individual. Depending upon the population size (n_p), initial krill position matrix P is created which is given by

$$P = \begin{bmatrix} P_{g1}^{1}, P_{g2}^{1}, \ldots, P_{gi}^{1} \ldots, P_{gn}^{1} \\ P_{g1}^{2}, P_{g2}^{2}, \ldots, P_{gi}^{2} \ldots, P_{gn}^{2} \\ P_{g1}^{3}, P_{g2}^{3}, \ldots, P_{gi}^{3} \ldots, P_{gn}^{3} \\ \ldots \quad \ldots \quad \ldots \\ P_{g1}^{n}, P_{g2}^{n}, \ldots, P_{gi}^{n} \ldots, P_{gn}^{n} \end{bmatrix} \tag{9}$$

Step 2: Evaluate the fitness of each individual of the current population using (1).

Step 3: Evaluate induced motion, forging motion and diffusion motion).

Step 4: Modify the position each krill individual.

Step 5: Applying crossover and mutation of DE to modify the position of each individual krill. The position of each krill individual represents a potential solution comprises of active power generation of ELD problem.

Step 6: Check whether the active power of all generating units except the last unit violates the operating limits or not. If any power generation is less than the minimum level it is made equal to the minimum value and if it is greater than the maximum level it is made equal to the maximum level. Afterward, the power generation of the last unit is evaluated using (5) and check whether it satisfies the inequality constraints or not. The infeasible solutions are replaced by the best feasible solutions.

Step 7: Go to Step 2 until stopping criteria is met. The KHA is stopped running when there is no significant improvement in the solution or the maximum number of iteration (generation) is reached. In this study, the stopping criterion is the maximum number of iteration for which most of the krill individuals are idle.

TLBO Algorithm Applied to ELD

Step 1: Active power generation of all the thermal units except the last unit of the ELD problem are randomly selected between their operating limits. Afterward, the thermal generation of the last unit is calculated using the equality constraint (5) and check whether it satisfies the inequality constraint (7) or not. The infeasible solutions are reinitialized.

Step 2: Compute the objective function using (1).

Step 3: Calculate the mean grade of all the subjects offered to the students of the class.

Step 4: Identify the best solution (Teacher).

Step 5: Modify all the independent variables (i.e., thermal power generation of the units except the slack unit in all time interval) based on the teacher knowledge using (2.1).

Step 6: Update the independent variable (thermal generation of the units except the slack unit in all time interval) based on the learners' knowledge by utilizing the knowledge of some other learner of the same group.

Step 7: The thermal generation of the slack unit is evaluated using (5) and inequality constraint (7-8) is verified. If power generation is less than the minimum level it is made equal to the minimum value and if it is greater than the maximum level it is made equal to the maximum level.

Step 8: Go to step 3 until the current iteration reaches the pre-specified iteration number.

HCRO Algorithm Applied to ELD

The HCRO algorithm is applied to seek out the optimal or near-optimal solutions for the ELD problem. Its computational procedure can be illustrated using the following steps:

Step 1: Active power generation of all the thermal units except the last unit of the ELD problem are randomly selected between their operating limits. Afterward, the thermal generation of the last unit is calculated using the equality constraint (5) and check whether it satisfies the inequality constraint (7) or not. The infeasible solutions are reinitialized.

Step 2: Evaluate the fitness value using (14). Based on the objective value, sort the molecules and few are kept as elite molecules.

Step 3: Assign potential energy same as fitness value for each molecule of the population set.

Step 4: Apply on-wall ineffective collision operations, decomposition operation, inter-molecular ineffective collision and synthesis collision operation on non elite molecules as described in Steps 4, 5, and 6 to modify the independent variables i.e generators' active power, voltage of generator buses, reactive power injection, tap settings of transformers.

Step 5: To check the feasibility, run load flow and find the dependent variables. If any dependent variable violates the operating limit, replace that solution set by new feasible solution.

Step 6: Compute fitness value for each newly generated feasible solution set.

Step 7: If the stopping criteria is satisfied, go to Step 10. Otherwise, go to Step 4.

Step 8: Output optimal settings of the control variables and the corresponding cost.

GWO Algorithm Applied to ELD

Step 1: The input parameters such as active power generation of all the thermal units except the last unit of the ELD are initialized.

Step 2: Evaluate the fitness function of the initial solution set and elite solutions are identified.

Step 4: Find the updated best position of alpha, beta and delta wolves using (45) - (47) of chapter 2 and independent variables of ELD are modified. The feasibility of the controlled variables is checked. If any one of them does not satisfy the inequality constraints, the particular set is replaced by the best feasible solution set.

Step 5: Evaluate the fitness value of the feasible population set.

Step 6: Check if the number of iteration is less than the maximum specified iteration or not.

Step 7: When the maximum iteration is reached display the fitness value and the updated final position of the wolves denoted by X(t+1) from the prey which is generally a circle around the prey.

Step 8: If the maximum iteration is not reached, iteration cycle is incremented.

Step 9: Go to Step 4 until a stopping criterion is satisfied.

SIMULATION RESULTS AND DISCUSSIONS

In the simulation study, BBO, TLBO, KHA, GWO, CRO, and HCRO techniques are carried out on five various cases, namely, 10-unit with multiple fuel, 15-unit with quadratic cost and prohibited operating zone, 40-unit with transmission loss, 80-unit without transmission loss and 140-unit with ramp rate and prohibited operating zone of ELD problems. To verify the superiority, the performances of the proposed methods are compared to other optimization techniques like SDE (Wang et al., 2007), GA (Noman & Iba 2008; Khamsawang & Jiriwibhakorn 2010; Pattanaik et al., 2018), NPSO-LRS (Selvakumar & Thanushkodi, 2007), SOH_PSO (Chaturvedi et al., 2008), GAAPI (Ciornei & Kyriakides 2012), NPSO (Niknam et al., 2011), FAPSO (Niknam et al., 2011), PSO (Niknam et al., 2011; Khamsawang & Jiriwibhakorn. 2010), DE (Noman & Iba 2008), SPSO (Chaturvedi et al., 2008.), PC_PSO (Chaturvedi et al., 2008), MTS (Khamsawang & Jiriwibhakorn. 2010), SA (Pothiya et al., 2008) etc. which were tested earlier by the various researchers and reported in the literature. The programming of proposed ELD based KHA, TLBO, GWO, BBO, CRO and HCRO methods are developed in MATLAB software, using MATLAB 7.1 and run on PC with 2.5 GHz core duo processor of 2 GB RAM.

Test System 1

Initially, the proposed BBO, TLBO, KHA, GWO, CRO, and HCRO techniques are applied on a small test system consisting of 10 generating units for a load demand of 2700 MW. In this case, valve-point loading, multiple fuels cost functions without transmission loss is used to demonstrate how good the proposed approaches work for this type of nonlinear and non-convex ELD problem. The system input data including fuel cost for this test system are taken from (Chiang, 2005). The test data are also illustrated in Tables 11-12 in the Appendix section. The best fuel cost, fuel type and generation value of each unit obtained in 50 trials for KHA, TLBO, GWO, BBO, CRO and HCRO techniques are tabulated in Table 1. It can be inferred from the results that the solution obtained by the proposed methods doesn't violate any system constraints. From the results, it is also clear that the proposed KHA, TLBO, BBO, GWO, CRO and HCRO generate cost output of 605.7493 \$/hr, 606.7213 \$/hr, 608.3492 \$/hr, 605.6764 \$/hr, 605.6353 \$/hr, and 605.6286 \$/hr, respectively. It clearly shows that the solution obtained from the proposed CRO and HCRO algorithms are better than those obtained by other techniques. Moreover, for this 10-unit system, the best, worst and mean cost obtained by the proposed CRO and HCRO methods in 50 different independent trials are compared with those obtained by other optimization methods available in the literature and are summarized in Table 2. The results furnished in Table 2 demonstrate that the proposed KHA, GWO, TLBO, BBO, CRO and HCRO methods are better than new adaptive PSO (NAPSO) (Niknam et al., 2011), FAPSO (Niknam et al., 2011), improved GA with multiplier updating (IGA-

MU) (Amjady & Nasiri-Rad, 2010), PSO with local random search (PSO-LRS) (Amjady & Nasiri-Rad, 2010), NPSO (Niknam et al., 2011), NPSO-LRS (Selvakumar & Thanushkodi, 2007), TSA (Khamsawang & Jiriwibhakorn, 2010), distributed PSO and tabu search algorithm (DSPSO-TSA) (Khamsawang & Jiriwibhakorn, 2010), PSO (Khamsawang & Jiriwibhakorn, 2010), GA (Khamsawang & Jiriwibhakorn, 2010) and adaptive real-coded GA (ARCGA) (Amjady & Nasiri-Rad, 2010).

Test System 2

In this case, the same algorithms are applied to a medium size test system consisting of the 15 generating units. The transmission losses and prohibited operating zone are taken into consideration for this case study. The total power demand of the system is 2630 MW. The generating units 2, 5, 6, and 12 have prohibited operating zones whose data are taken from (Niknam et al., 2011). The rest of the system data are adopted from. The test data and transmission loss data are also listed in Tables 13 and 14, respectively, in the Appendix section. The optimal cost, power loss and optimal generation scheduling obtained by the proposed KHA, GWO, TLBO, BBO, CRO and HCRO methods are presented in Table 3 which clearly demonstrates that the simulation results satisfy all the operational constraints. The minimum

Table 1. Optimal generation and cost obtained BBO, TLBO, KHA, GWO, CRO, and HCRO for test system 1 (10-unit without loss and with multiple fuels)

Unit No.	Fuel type	KHA	GWO	TLBO	BBO	CRO	HCRO
P_{g1} (MW)	2	212.5101	213.4779	212.1750	211.8961	213.4586	214.4801
P_{g2} (MW)	3	209.2365	209.2353	208.7692	211.4077	208.9901	209.9787
P_{g3} (MW)	2	332.6694	332.0054	336.2187	344.5229	332.0272	332.0204
P_{g4} (MW)	3	238.9635	238.9665	232.1530	243.3970	237.7577	238.2956
P_{g5} (MW)	1	269.2000	269.2514	275.2444	289.9861	269.1449	269.1417
P_{g6} (MW)	3	237.4970	238.1586	239.2152	241.7133	238.2954	238.2956
P_{g7} (MW)	1	280.4050	280.7272	286.4042	310.4244	280.5957	280.6060
P_{g8} (MW)	3	238.5634	237.6265	236.4921	238.9379	237.8923	237.8926
P_{g9} (MW)	3	416.1820	414.2881	412.7265	328.0414	415.4582	412.9062
P_{g10} (MW)	1	264.7731	266.2631	260.6017	279.6732	266.3799	266.3831
Cost($/hr)		605.7493	605.6764	606.7213	608.3492	605.6353	605.6286

Table 2. Statistical results of various algorithms for test system 1 (10-uint without loss and with multiple fuels)

Algorithms	Best fuel cost ($/day)	Average fuel Cos($/day)	Worst fuel cost ($/ day)
NAPSO	623.62170	623.6335	623.67543
FAPSO	624.2189	624.2782	624.2951
IGA-MU	624.5178	627.6087	630.8705
PSO-LRS	624.2297	625.7887	628.3214
NPSO	624.1624	625.2180	627.4237
NPSO-LRS	624.1273	624.9985	626.9981
DSPSO-TSA	623.8375	623.8625	623.9001
PSO	624.3045	624.5054	625.9252
GA	624.5050	624.7419	624.8169
TSA	624.3078	635.0623	624.8285
ARCGA	623.8281	623.8431	623.8550
KHA	605. 7493	605.8043	605.9426
GWO	605.6764	605.7136	605.8235
TLBO	606.7213	606.8227	606.8701
BBO	608.3492	609.4124	610.5349
CRO	605.6353	606.1036	606.2713
HCRO	605.6286	605.7003	605.7963

generation cost and transmission loss obtained by HCRO are 32547.5943 $/hr and 26.7750 MW, respectively. But, generation cost obtained by KHA, GWO, TLBO, BBO and CRO techniques are 32549.0060 $/hr, 32548.8607 $/hr, 32549.0739 $/hr, 32549.4280 $/hr and 32548.0500 $/hr, respectively and the transmission loss obtained using these techniques are 26.8566 MW, 26.8891 MW, 26.8653 MW, 26.8564 MW and 26.7765 MW, respectively. Furthermore, to judge the superiority and robustness, the statistical results of the proposed methods for 50 independent trials are compared with those obtained by other optimization techniques such as FAPSO (Niknam et al., 2011), PSO (Niknam et al., 2011), GA (Noman & Iba, 2008), ESO (Noman & Iba, 2008), DE (Noman & Iba, 2008), SPSO (Chaturvedi et al., 2008), PC_PSO (Chaturvedi et al., 2008), SOH_PSO (Chaturvedi et al, 2008), MTS (Khamsawang & Jiriwibhakorn 2010), SA (Selvakumar & Thanushkodi, 2007), society-civilization algorithm (SCA) (Selvakumar & Thanushkodi, 2007), APSO (Panigrahi et al., 2008), civilized swarm optimization (CSO)

(Selvakumar & Thanushkodi, 2007) and GAAPI (Ciornei & Kyriakides, 2012) available in the literature. The statistical results listed in Table 4 clearly show that the proposed KHA, GWO, TLBO, BBO, CRO and HCRO methods dominate almost all other referred methods. However, the statistical results achieved by the HCRO method are found to be the best. This implies that, on average, the quality of solutions obtained by HCRO method is better than other methods.

Test System 3

Furthermore, to demonstrate the efficiency of the proposed methods, simulation studies are carried out on a comparatively large test system consisting of 40 generating unit. The transmission losses are taken into account to convert it a more realistic practical system because the power stations are situated far from each other and therefore, the economic scheduling must take transmission losses into account. During the transmission of electricity, some energy is 'lost' from the transmission system. The B-coefficient data of this system are adopted from (Ciornei & Kyriakides, 2012). The test data are also furnished in Table 15 in the Appendix section. The optimal generation scheduling of all 40 units obtained by the different methods is listed in Table 5. In order to judge the superiority of the obtained solutions, the result obtained by the proposed CRO and HCRO algorithms are compared with TLBO, BBO, KHA, GWO, QOTLBO (Roy & Bhui, 2013), GAAPI (Ciornei & Kyriakides, 2012) and shuffled DE (SDE) (Wang et al., 2007) and Table 6 represents the comparative statistical results of the aforementioned methods. From the presented results, it can be observed that fuel cost obtained by HCRO is better than those obtained by other methods.

Test System 4

In order to further investigate the effectiveness and robustness, the proposed methods are further applied to a large system consisting of 80 generating units. This case has a larger and more complex solution space than all the previous cases, and so the difference between different ELD solution methods can be better revealed in this case. This large-scale system is created by expanding the 40-generating unit system. The system load demand is considered as 21,000 MW. The transmission loss is not considered in this simulation study. The cost coefficients, generation limits, and valve point coefficients are derived from

Table 3. Optimal generation and cost obtained by various algorithms for test system 2 (15-unit with loss and with POZ)

Unit	KHA	GWO	TLBO	BBO	CRO	HCRO
P_{g1} (MW)	454.9127	454.9728	454.9873	454.6546	455.0000	454.9938
P_{g2} (MW)	454.8624	454.8991	454.9812	454.9960	454.9644	454.9992
P_{g3} (MW)	129.7936	129.6998	129.9998	129.4999	129.9887	130.0000
P_{g4} (MW)	129.8136	129.9299	130.0000	130.0000	130.0000	129.9836
P_{g5} (MW)	237.3107	238.4476	237.2537	237.2658	233.8004	233.7883
P_{g6} (MW)	459.9922	459.9759	459.9983	459.9976	459.8325	459.9996
P_{g7} (MW)	464.8143	464.9077	464.9983	464.9977	464.9742	464.9913
P_{g8} (MW)	60.0028	60.0158	60.0004	60.0169	60.0903	60.1003
P_{g9} (MW)	25.0092	25.0023	25.0000	25.0212	25.1318	25.0026
P_{g10} (MW)	30.6803	29.0361	29.7528	30.6744	31.2693	31.2713
P_{g11} (MW)	74.3042	74.6322	74.0096	74.3329	76.4892	76.5922
P_{g12} (MW)	79.9329	79.9691	79.9999	79.8072	80.0000	79.9903
P_{g13} (MW)	25.0000	25.0000	25.0001	25.0000	25.0000	25.0000
P_{g14} (MW)	15.0041	15.0002	15.0000	15.0041	15.0000	15.0000
P_{g15} (MW)	15.4236	15.4006	15.8839	15.5881	15.2357	15.0625
Cost ($/hr)	32549.0060	32548.8607	32549.0739	32549.4280	32548.0500	32547.5943
Transmission loss(MW)	26.8566	26.8891	26.8653	26.8564	26.7765	26.7750

(Niknam et al., 2011) with modifications to consider prohibited operating zones of units 10–14 and 50–54. The system data of prohibited operating zone are adopted from (Niknam et al., 2011). Table 7 shows the optimal generation of 80 generating units and fuel cost obtained by the proposed KHA, GWO, TLBO, BBO, CRO and HCRO approaches. The simulation results suggest that all the discussed methods generate feasible solutions without violating any operational constraint. The simulation results demonstrate the effectiveness of the proposed methods in handling ELD with all practical constraints. To validate the superiority, a comparison is among the best solution obtained from the proposed methods and the solutions obtained by the PSO (Niknam et al., 2011), NPSO (Niknam et al., 2011) and FAPSO (Niknam et al., 2011) is made. The results of this comparison are shown in Table 8. The comparison shows that HCRO provides the best generation cost among all the algorithms, which demonstrate that the proposed HCRO method outperforms the other techniques in terms of optimal solutions.

Table 4. Statistical results of various algorithms for test system 2 (15- unit with loss and with POZ)

Algorithms	Best fuel cost ($/hr)	Average fuel Cos($/hr)	Worst fuel cost ($/hr)
FAPSO	32659.794	32663.19	32676.07
PSO	32,858	32,989	33,031
GA	33063.54	33,228	33,337
ESO	32640.86	32,620	32,710
DE	32588.865	32609.851	32641.419
SPSO	32798.69	NA	NA
PC_PSO	32775.36	NA	NA
SOH_PSO	32751.39	32,878	32,945
MTS	32716.87	32767.4	32796.13
SA	32786.4	32869.51	33028.95
SCA	32867.025	33138.3020	33381.0607
APSO	32742.77	2976.6812	NA
CSO	32588.9189	32679.8775	32796.7792
GAPPI	32732.95	NA	NA
KHA	32549.0060	32551.0602	32552.4818
GWO	32548.8607	32550.1663	32551.2920
TLBO	32549.0739	32550.8275	32552.9311
BBO	32549.4280	32551.6759	32554.00243
CRO	32548.0500	32549.1036	32549.8546
HCRO	32547.5943	32547.7137	32548.1005

Test System 5

Finally, the proposed BBO, TLBO, KHA, GWO, CRO, and HCRO techniques are applied on a large test system consisting of 140 generating units for a load demand of 49342 MW. The ramp rate limits and the prohibited operating zone is considered to demonstrate how good the proposed approaches work for this type of system. The transmission loss is not considered for this simulation study. The total system data of this test case are available in (Park et al., 2010;

Table 5. Optimal generation and cost obtained by various algorithms for test system 3 (40-unit with loss)

Unit	KHA	GWO	TLBO	BBO	CRO	HCRO
P_{g1} (MW)	114.0000	114.0000	114.0000	114.0000	114.0000	113.5367
P_{g2} (MW)	114.0000	114.0000	114.0000	114.0000	114.0000	113.6954
P_{g3} (MW)	120.0000	120.0000	120.0000	120.0000	120.0000	120.0000
P_{g4} (MW)	190.0000	180.7326	190.0000	190.0000	188.3893	180.7326
P_{g5} (MW)	87.8139	88.5789	93.6032	87.8242	97.0000	88.4359
P_{g6} (MW)	140.0000	140.0000	140.0000	140.0000	140.0000	140.0000
P_{g7} (MW)	300.0000	300.0000	300.0000	300.0000	300.0000	300.0000
P_{g8} (MW)	300.0000	299.5611	300.0000	300.0000	300.0000	299.5611
P_{g9} (MW)	300.0000	299.9798	300.0000	300.0000	300.0000	300.0000
P_{g10} (MW)	279.7163	279.9786	279.7210	279.6852	280.5082	279.9786
P_{g11} (MW)	168.7140	244.0524	240.6129	243.8910	247.2746	243.9604
P_{g12} (MW)	169.0198	100.0889	94.6350	101.3210	94.0396	100.0889
P_{g13} (MW)	483.7522	482.5948	480.7207	482.1035	483.5643	483.5532
P_{g14} (MW)	484.0391	482.9409	483.2913	484.0252	483.6735	482.9409
P_{g15} (MW)	484.4595	483.8711	483.1263	484.6940	484.7031	483.8711
P_{g16} (MW)	484.5491	483.7963	483.6973	483.9803	484.6966	483.7963
P_{g17} (MW)	489.7833	489.7968	488.4392	489.2688	490.1511	489.7968
P_{g18} (MW)	489.7888	489.3228	489.3488	490.4314	490.5288	489.3228
P_{g19} (MW)	511.4795	511.8249	511.3076	513.4334	512.6490	511.1120
P_{g20} (MW)	511.9583	511.8301	512.9300	513.9697	511.1470	511.8301
P_{g21} (MW)	524.4630	523.4767	523.6251	523.6353	523.3253	523.4767
P_{g22} (MW)	544.1778	528.3479	531.6414	526.9131	549.9789	528.3479
P_{g23} (MW)	523.3576	524.2751	524.6266	524.3000	523.3543	524.2743
P_{g24} (MW)	523.3092	522.8295	523.1267	521.6905	523.2492	523.1350
P_{g25} (MW)	523.4250	523.7182	523.3862	523.5878	525.4197	523.7182
P_{g26} (MW)	524.2864	524.3740	523.3009	524.0070	523.9691	524.3740
P_{g27} (MW)	10.2949	10.2899	10.2757	10.3370	11.2685	10.1471
P_{g28} (MW)	10.4727	10.6337	10.1153	10.1679	11.3901	10.6337
P_{29} (MW)	10.2033	10.6496	10.1194	11.3318	11.1179	10.6496
P_{g30} (MW)	88.1073	87.8188	87.8070	87.8836	87.9252	87.8463
P_{g31} (MW)	190.0000	190.0000	190.0000	190.0000	190.0000	190.0000
P_{g32} (MW)	190.0000	190.0000	190.0000	190.0000	190.0000	190.0000
P_{g33} (MW)	190.0000	190.0000	190.0000	190.0000	190.0000	190.0000
P_{g34} (MW)	200.0000	200.0000	200.0000	200.0000	200.0000	200.0000

continued on following page

Table 5. Continued

Unit	KHA	GWO	TLBO	BBO	CRO	HCRO
P_{g35} (MW)	200.0000	171.2759	200.0000	200.0000	165.4323	171.2759
P_{g36} (MW)	164.7653	164.8722	164.7780	165.6940	164.7948	164.8722
P_{g37} (MW)	110.0000	110.0000	110.0000	110.0000	110.0000	110.0000
P_{g38} (MW)	110.0000	110.0000	110.0000	110.0000	110.0000	110.0000
P_{g39} (MW)	110.0000	110.0000	110.0000	110.0000	110.0000	110.0000
P_{g40} (MW)	512.5734	549.4370	521.4887	514.3487	511.6064	549.9970
Cost ($/hr)	136594.7185	136581.8805	136600.2814	136623.8729	136567.7850	136563.4847
Loss (MW)	982.5097	968.9485	973.7243	976.5244	969.1568	968.9607

Table 6. Statistical results of various algorithms for test system 3 (40-unit with loss)

Algorithms	Best fuel cost ($/hr)	Average fuel cost($/hr)	Worst fuel cost ($/hr)
GAPPI	139864.96	NA	NA
SDE	138157.46	NA	NA
QOTLBO	137329	137334.1526	137335.5489
KHA	136594.7185	136597.1491	136601.2826
GWO	136581.8805	136583.5428	136587.1345
TLBO	136600.2814	136604.2296	136609.2938
BBO	136623.8729	136628.4520	136634.6213
CRO	136567.7850	136569.3243	136573.0291
HCRO	136563.4847	136563.9917	136564.7308

Reddy & Vaisakh, 2013). The test data are also illustrated in Tables 16 and 17 in the Appendix section. The optimal generation scheduling and cost obtained by the HCRO, CRO, GWO, BBO, TLBO and KHA methods are illustrated in Table 9. It is found from the results listed in Tables 9 that the minimum cost achieved by the proposed HCRO method is 1659184.6226 $/hr which is significantly better than that obtained by other methods. Moreover, by studying the results of Table 9, it can be found that the proposed algorithm succeeded in finding optimal results without violating any operating constraint. In order

Table 7. Optimal generation and cost obtained by various algorithms for test system 4 (80-unit without loss and with POZ)

Unit	KHA	GWO	TLBO	BBO	CRO	HCRO	Unit	KHA	GWO	TLBO	BBO	CRO	HCRO
P_{g1} (MW)	110.9696	111.2015	114.0000	114.0000	114.0000	114.0000	P_{g41} (MW)	114.0000	114.0000	114.0000	114.0000	114.0000	114.0000
P_{g2} (MW)	114.0000	114.0000	114.0000	114.0000	114.0000	114.0000	P_{g42} (MW)	114.0000	111.1643	114.0000	114.0000	111.1050	114.0000
P_{g3} (MW)	97.4689	97.5470	97.4733	97.7809	97.4308	97.9476	P_{g43} (MW)	97.4687	120.0000	100.3313	97.8726	97.5328	97.8024
P_{g4} (MW)	179.7640	179.7555	180.1350	179.8276	179.7548	180.0064	P_{g44} (MW)	179.7470	179.7275	179.9940	179.7975	179.7359	179.8032
P_{g5} (MW)	97.0000	94.4563	97.0000	97.0000	97.0000	97.0000	P_{g45} (MW)	97.0000	97.0000	97.0000	97.0000	97.0000	97.0000
P_{g6} (MW)	140.0000	140.0000	140.0000	140.0000	140.0000	140.0000	P_{g46} (MW)	140.0000	140.0000	140.0000	140.0000	140.0000	140.0000
P_{g7} (MW)	259.6383	259.6470	260.2244	300.0000	259.5859	261.8433	P_{g47} (MW)	259.7098	259.8925	300.0000	259.9638	259.6657	259.6125
P_{g8} (MW)	284.6458	284.8765	284.9304	285.4137	284.6078	285.6066	P_{g48} (MW)	284.7588	284.6409	287.3374	287.0950	284.7715	287.0351
P_{g9} (MW)	284.6091	284.7802	288.4796	285.6344	284.8051	286.2867	P_{g49} (MW)	284.6734	284.6159	287.6162	285.3049	284.8030	285.3348
P_{g10} (MW)	204.8035	204.7920	130.0419	204.7849	130.0046	130.0229	P_{g50} (MW)	130.0011	130.0051	130.1735	130.1190	204.8007	130.0293
P_{g11} (MW)	94.0006	168.8400	168.3177	94.0034	168.7999	94.0035	P_{g51} (MW)	94.0177	168.8084	94.5811	168.8082	94.0020	94.1057
P_{g12} (MW)	243.6116	94.0128	94.3845	168.7359	168.8233	168.8134	P_{g52} (MW)	168.8019	168.8413	168.9543	168.8260	168.8032	94.0173
P_{g13} (MW)	304.5267	214.9018	213.0524	214.7708	214.7576	214.7543	P_{g53} (MW)	125.0085	304.5856	214.8041	214.8210	304.5168	304.5172
P_{g14} (MW)	304.5218	304.5355	304.7431	394.2427	304.5391	304.5177	P_{g54} (MW)	304.5278	304.4893	304.4011	304.5264	394.2877	394.2536
P_{g15} (MW)	394.2870	304.5359	394.2713	304.5805	304.5184	304.5325	P_{g55} (MW)	304.5277	394.2993	394.3798	394.3593	304.5318	304.5247
P_{g16} (MW)	394.2815	304.5149	304.4897	214.7923	304.5213	394.2757	P_{g56} (MW)	304.5352	394.2897	394.2989	304.5513	304.5253	394.3032
P_{g17} (MW)	489.2647	489.3900	489.7117	489.5809	489.2853	489.3100	P_{g57} (MW)	489.2966	489.2878	489.3668	489.3311	489.2800	489.2981
P_{g18} (MW)	489.2728	489.3132	489.2595	489.2962	489.2990	489.3352	P_{g58} (MW)	489.2817	489.2933	489.0292	489.4309	489.2938	489.3390
P_{g19} (MW)	511.2918	511.2980	511.6016	511.2847	511.2823	511.5213	P_{g59} (MW)	511.2966	511.3779	510.8102	511.3266	511.2832	511.3118
P_{g20} (MW)	511.2951	511.2799	511.4145	511.3243	511.2893	511.4196	P_{g60} (MW)	511.2859	511.3243	511.4770	511.3276	511.2824	511.3146
P_{g21} (MW)	523.3086	523.3220	525.0962	523.8414	523.3375	523.5249	P_{g61} (MW)	523.3622	523.4706	525.2464	523.4345	523.3562	523.7831
P_{g22} (MW)	523.2698	523.3171	523.3474	523.4647	523.3024	523.4737	P_{g62} (MW)	523.3202	523.3906	523.6330	523.5307	523.2979	523.3759
P_{g23} (MW)	523.3451	523.2965	524.3620	523.5668	523.2887	523.3121	P_{g63} (MW)	523.3750	523.2880	523.7570	523.3110	523.3848	523.7813

continued on following page

Table 7. Continued

Unit	KHA	GWO	TLBO	BBO	CRO	HCRO	Unit	KHA	GWO	TLBO	BBO	CRO	HCRO
P_{g24} (MW)	523.3542	523.5381	523.9286	523.8298	523.2971	523.4489	P_{g64} (MW)	523.2965	524.0779	524.2466	523.4515	523.2824	523.3864
P_{g25} (MW)	523.3171	523.3233	523.3332	523.3025	523.3372	523.6005	P_{g65} (MW)	523.3165	523.4726	523.5247	523.5786	523.2915	523.3974
P_{g26} (MW)	523.3744	523.5109	524.7817	523.5425	523.2863	523.4877	P_{g66} (MW)	523.2946	523.4199	523.4313	523.6454	523.3131	523.5270
P_{g27} (MW)	10.0016	10.0113	10.0628	10.0007	10.0051	10.0075	P_{g67} (MW)	10.0012	10.0092	10.1287	10.0289	10.0033	10.0462
P_{g28} (MW)	10.0002	10.2038	10.0320	10.0012	10.0002	10.0039	P_{g68} (MW)	10.0041	10.0248	10.1449	10.0097	10.0009	10.0340
P_{29} (MW)	10.0008	10.0019	10.2084	10.0478	10.0036	10.0033	P_{69} (MW)	10.0030	10.0481	10.1278	10.0094	10.0008	10.0050
P_{g30} (MW)	97.0000	97.0000	97.0000	97.0000	97.0000	97.0000	P_{g70} (MW)	97.0000	89.9054	97.0000	97.0000	97.0000	92.6028
P_{g31} (MW)	190.0000	190.0000	190.0000	190.0000	190.0000	190.0000	P_{g71} (MW)	190.0000	190.0000	190.0000	190.0000	190.0000	190.0000
P_{g32} (MW)	190.0000	190.0000	190.0000	190.0000	190.0000	190.0000	P_{g72} (MW)	190.0000	190.0000	190.0000	190.0000	190.0000	190.0000
P_{g33} (MW)	190.0000	190.0000	190.0000	190.0000	190.0000	190.0000	P_{g73} (MW)	190.0000	190.0000	190.0000	190.0000	190.0000	190.0000
P_{g34} (MW)	165.0162	165.5015	169.8626	200.0000	165.1798	200.0000	P_{g74} (MW)	165.2551	165.7860	184.4702	167.2983	164.9346	164.7847
P_{g35} (MW)	200.0000	165.4174	200.0000	200.0000	200.0000	200.0000	P_{g75} (MW)	200.0000	164.8854	200.0000	200.0000	200.0000	200.0000
P_{g36} (MW)	200.0000	200.0000	200.0000	200.0000	200.0000	200.0000	P_{g76} (MW)	200.0000	165.9195	169.5364	200.0000	200.0000	200.0000
P_{g37} (MW)	110.0000	110.0000	110.0000	110.0000	110.0000	110.0000	P_{g77} (MW)	110.0000	110.0000	110.0000	110.0000	110.0000	110.0000
P_{g38} (MW)	110.0000	110.0000	110.0000	110.0000	110.0000	110.0000	P_{g78} (MW)	110.0000	110.0000	110.0000	110.0000	110.0000	110.0000
P_{g39} (MW)	110.0000	110.0000	110.0000	110.0000	110.0000	110.0000	P_{g79} (MW)	110.0000	110.0000	110.0000	110.0000	110.0000	110.0000
P_{g40} (MW)	511.3133	511.2561	511.3695	511.3105	511.2835	511.3246	P_{g80} (MW)	511.2791	511.2810	511.2831	511.2797	511.2878	511.2900
		KHA		GWO		TLBO		BBBO		CRO		HCRO	
Cost ($/hr)		243038.7352		242972.8268		243112.5486		243142.2195		242956.2721		242945.0039	

Table 8. Statistical results of various algorithms for test system 4 (80-unit without loss and with POZ)

Algorithms	Best fuel cost ($/hr)	Average fuel cost($/hr)	Worst fuel cost ($/hr)
FAPSO	244273.5429	NA	NA
PSO	249248.3751	NA	NA
KHA	243038.7352	243046.6503	243058.1237
GWO	242972.8268	242977.3771	242981.4729
TLBO	243112.5486	243123.4282	243137.0073
BBO	243142.2195	243154.1801	243173.8827
CRO	242956.2721	242960.8327	242966.2827
HCRO	242945.0039	242946.1325	242947.9920

to judge the robustness of the proposed HCRO method, its statistical results are compared with those obtained by CRO, GWO, TLBO, BBO and KHA. These statistical results are illustrated in Table 10. It can be observed that the minimum, maximum and the average cost obtained from the proposed HCRO method is better than CRO, GWO, TLBO, BBO and KHA. This clearly suggests that the HCRO approach is more consistent than the other optimization techniques discussed in the simulation study. From the results, it is clear that the proposed method is robust and it is practically applicable to the real system.

CONCLUSION

In this chapter, several stochastic search techniques named BBO, TLBO, KHA, GWO, CRO, and HCRO are proposed to solve the different types of non-linear ELD problems. Five different types of non-smooth and non-convex ELD problems with the nonlinearities of valve-point loading, prohibited operating zones, ramp rate constraints and multiple fuels options are used to evaluate the performance of BBO, TLBO, KHA, GWO, CRO, and HCRO algorithms. To test the effectiveness, the proposed six solution methods are

Table 9. Optimal generation and cost obtained by various algorithms for test system 5 (140-unit with ramp rate and with POZ)

Unit	GWO	CRO	HCRO	Unit	GWO	CRO	HCRO	Unit	GWO	CRO	HCRO
P_{g1} (MW)	117.8677	116.4339	116.6934	P_{g48} (MW)	249.4106	248.5935	248.2601	P_{g95} (MW)	837.5000	837.5000	837.4999
P_{g2} (MW)	164.0000	164.0000	164.0000	P_{g49} (MW)	245.1759	247.4113	247.2097	P_{g96} (MW)	681.9969	681.9986	681.9997
P_{g3} (MW)	189.4000	189.7186	189.7762	P_{g50} (MW)	246.5507	245.6897	248.1214	P_{g97} (MW)	719.9994	719.9993	719.9996
P_{g4} (MW)	189.4027	189.5286	189.9567	P_{g51} (MW)	168.1448	165.8956	258.1829	P_{g98} (MW)	717.9993	717.9997	717.9977
P_{g5} (MW)	168.5306	168.4640	168.4301	P_{g52} (MW)	310.9545	457.2635	311.1048	P_{g99} (MW)	719.9995	719.9991	719.9990
P_{g6} (MW)	188.2446	188.9113	188.5407	P_{g53} (MW)	168.1903	214.1220	186.0433	P_{g100} (MW)	963.9933	963.9999	963.9996
P_{g7} (MW)	489.8800	489.6404	489.9375	P_{g54} (MW)	195.8751	173.6883	250.1176	P_{g101} (MW)	957.9964	957.9994	957.9942
P_{g8} (MW)	489.9213	489.9604	489.7800	P_{g55} (MW)	180.3298	183.0632	181.0761	P_{g102} (MW)	947.9000	947.9000	947.9000
P_{g9} (MW)	495.6212	495.9519	495.7974	P_{g56} (MW)	180.0306	180.9749	180.4969	P_{g103} (MW)	934.0000	934.0000	934.0000
P_{g10} (MW)	495.6035	495.8458	494.8765	P_{g57} (MW)	113.6087	119.1665	113.3240	P_{g104} (MW)	935.0000	934.9997	934.9974
P_{g11} (MW)	495.9300	495.9929	495.8086	P_{g58} (MW)	199.5870	200.0383	198.6119	P_{g105} (MW)	876.5000	876.5000	876.5000
P_{g12} (MW)	495.8904	495.8347	495.7735	P_{g59} (MW)	311.5861	309.9753	304.8855	P_{g106} (MW)	880.9000	880.9000	880.9000
P_{g13} (MW)	505.9579	505.7655	505.9954	P_{g60} (MW)	304.0442	292.3180	303.3416	P_{g107} (MW)	873.7000	873.7000	873.7000
P_{g14} (MW)	508.8750	508.7418	508.9688	P_{g61} (MW)	165.3727	163.0017	163.8317	P_{g108} (MW)	877.3998	877.4000	877.4000
P_{g15} (MW)	505.9992	505.9960	505.9960	P_{g62} (MW)	95.2757	95.0006	95.4500	P_{109} (MW)	871.7000	871.7000	871.7000
P_{g16} (MW)	504.8599	504.8970	504.9702	P_{g63} (MW)	379.8162	453.5344	443.8125	P_{g110} (MW)	864.8000	864.8000	864.8000
P_{g17} (MW)	505.9836	505.8564	505.7512	P_{g64} (MW)	202.7958	418.1921	247.9443	P_{g111} (MW)	882.0000	882.0000	882.0000
P_{g18} (MW)	505.8488	505.9823	505.9557	P_{g65} (MW)	471.0863	483.2669	480.2309	P_{g112} (MW)	94.0010	94.0104	94.0488
P_{g19} (MW)	504.3259	504.8405	504.7602	P_{g66} (MW)	224.5037	208.3491	213.3740	P_{g113} (MW)	94.0103	94.0634	94.0532
P_{g20} (MW)	504.9250	504.7966	504.8732	P_{g67} (MW)	477.1727	486.5100	481.4809	P_{g114} (MW)	94.0138	94.0003	94.0043

continued on following page

Table 9. Continued

Unit	GWO	CRO	HCRO	Unit	GWO	CRO	HCRO	Unit	GWO	CRO	HCRO
P_{g21} (MW)	504.9985	504.9467	504.9347	P_{g68} (MW)	483.1519	466.0591	476.6101	P_{g115} (MW)	244.2497	244.0123	244.6732
P_{g22} (MW)	504.9981	505.0000	504.9998	P_{69} (MW)	130.5892	130.7960	131.7900	P_{g116} (MW)	244.2299	244.2127	244.0591
P_{g23} (MW)	504.9747	504.8879	504.8448	P_{g70} (MW)	234.7048	339.2042	339.4228	P_{g117} (MW)	244.0337	244.0470	245.9710
P_{g24} (MW)	504.9204	504.9277	504.8505	P_{g71} (MW)	138.8982	139.9576	146.7406	P_{g118} (MW)	95.0691	95.0193	95.1916
P_{g25} (MW)	536.9662	536.8421	536.8198	P_{g72} (MW)	388.2203	325.3144	325.5733	P_{g119} (MW)	95.0180	95.0037	95.0094
P_{g26} (MW)	536.6740	536.8832	536.9998	P_{g73} (MW)	297.2904	196.2629	360.3990	P_{g120} (MW)	116.0004	116.0038	116.0005
P_{g27} (MW)	548.9705	548.8165	548.8599	P_{g74} (MW)	406.0561	218.4115	329.3658	P_{g121} (MW)	175.0738	175.0233	175.0041
P_{g28} (MW)	548.7894	548.9327	548.8905	P_{g75} (MW)	418.5286	180.7786	191.4169	P_{g122} (MW)	2.0104	2.0001	2.0031
P_{29} (MW)	500.8579	500.8934	500.9661	P_{g76} (MW)	191.6034	329.7123	206.7757	P_{g123} (MW)	4.0003	4.0001	4.0000
P_{g30} (MW)	499.0000	499.0000	499.0000	P_{g77} (MW)	409.4276	232.8954	309.2609	P_{g124} (MW)	15.0519	15.0011	15.0050
P_{g31} (MW)	505.8633	505.8974	505.9077	P_{g78} (MW)	330.2591	403.4313	332.0047	P_{g125} (MW)	9.0001	9.0004	9.0002
P_{g32} (MW)	505.9848	505.8974	505.8889	P_{g79} (MW)	530.2952	530.7779	530.7405	P_{g126} (MW)	12.0000	12.0000	12.0005
P_{g33} (MW)	505.9268	505.7938	505.9288	P_{g80} (MW)	530.7365	530.2666	530.1680	P_{g127} (MW)	10.0003	10.0012	10.0000
P_{g34} (MW)	505.8167	505.5887	505.6576	P_{g81} (MW)	508.7616	522.4682	532.4046	P_{g128} (MW)	113.5854	112.2214	112.1362
P_{g35} (MW)	499.8491	499.8918	499.9898	P_{g82} (MW)	56.0318	56.0206	56.0814	P_{g129} (MW)	4.0004	4.0002	4.0001
P_{g36} (MW)	499.6199	499.9934	499.9228	P_{g83} (MW)	115.0549	115.0221	115.0007	P_{g130} (MW)	5.0002	5.0006	5.0016
P_{g37} (MW)	240.9746	240.9325	240.9309	P_{g84} (MW)	115.2224	115.1084	115.0010	P_{g131} (MW)	5.0001	5.0001	5.0000
P_{g38} (MW)	240.9278	240.9317	240.9606	P_{g85} (MW)	115.3043	115.0369	115.0076	P_{g132} (MW)	50.0006	50.0106	50.0039
P_{g39} (MW)	773.5591	773.8585	773.9163	P_{g86} (MW)	207.0013	207.0011	207.0052	P_{g133} (MW)	5.0000	5.0000	5.0000
P_{g40} (MW)	768.9996	768.9928	768.9960	P_{g87} (MW)	207.0007	207.0013	207.0062	P_{g134} (MW)	42.0002	42.0000	42.0003
P_{g41} (MW)	3.0001	3.0012	3.0005	P_{g88} (MW)	177.8032	175.6159	175.2371	P_{g135} (MW)	42.0009	42.0033	42.0006

continued on following page

Table 9. Continued

Unit	GWO	CRO	HCRO	Unit	GWO	CRO	HCRO	Unit	GWO	CRO	HCRO
P_{g42} (MW)	3.0004	3.0003	3.0001	P_{g89} (MW)	177.7475	178.6029	189.2157	P_{g136} (MW)	41.0009	41.0005	41.0025
P_{g43} (MW)	246.6074	239.9946	239.5426	P_{g90} (MW)	175.6151	176.4821	175.2340	P_{g137} (MW)	17.0004	17.0003	17.0000
P_{g44} (MW)	246.8863	249.1404	240.3384	P_{g91} (MW)	177.6667	180.6357	176.3383	P_{g138} (MW)	7.0004	7.0040	7.0001
P_{g45} (MW)	243.3516	249.4069	249.4322	P_{g92} (MW)	575.4000	575.4000	575.4000	P_{g139} (MW)	7.0005	7.0005	7.0289
P_{g46} (MW)	249.6391	249.3445	247.4615	P_{g93} (MW)	547.5000	547.5000	547.5000	P_{g140} (MW)	32.7774	37.2046	27.0642
P_{g47} (MW)	249.0796	238.1861	245.2692	P_{g94} (MW)	836.8000	836.8000	836.8000				
			GWO			CRO			HCRO		
Cost ($/hr)			1659384.1739			1659250.8530			1659184.6226		

Table 10. Statistical results of various algorithms for test system 5 (140-unit with ramp rate and with POZ)

Algorithms	Best fuel cost ($/hr)	Average fuel cost($/hr)	Worst fuel cost ($/hr)
KHA	1659403.1773	1659411.2826	1659428.3113
GWO	1659384.1739	165393.4226	165402.0075
TLBO	1659458.1762	1659468.1326	1659481.3729
BBO	1659528.3662	1659588.1040	1659613.2773
CRO	1659250.8530	1659278.3293	1659319.3629
HCRO	1659184.6226	1659191.3880	1659199.3801

compared with some of the most recently published techniques available in the literature. Moreover, to better illustrate the robustness of the proposed methods, the statistical analysis is carried out on all the test systems. The simulation results show that the proposed BBO, TLBO, KHA, GWO, CRO, and HCRO are promising methods for solving various non-convex ELD problems. This clearly suggests that the proposed methods are promising and encouraging for further research in this direction. However, among these proposed techniques, HCRO method outperforms the other methods in terms of solution quality, computational efficiency and robustness.

REFERENCES

Afzalan, E., & Joorabian, M. (2013). Emission, reserve and economic load dispatch problem with non-smooth and non-convex cost functions using epsilon-multi-objective genetic algorithm variable. *Electrical Power and Energy Systems*, *52*, 55–67. doi:10.1016/j.ijepes.2013.03.017

Alsumait, J. S., Sykulski, J. K., & Al-Othman, A. K. (2010). A hybrid GA–PS–SQP method to solve power system valve-point economic dispatch problems. *Applied Energy*, *87*(5), 1773–1781. doi:10.1016/j.apenergy.2009.10.007

Amjady, N., & Nasiri-Rad, H. (2010). Solution of nonconvex and nonsmooth economic dispatch by a new adaptive real coded genetic algorithm. *Expert Systems with Applications*, *37*(7), 5239–5245. doi:10.1016/j.eswa.2009.12.084

Amjady, N., & Sharifzadeh, H. (2010). Solution of non-convex economic dispatch problem considering valve loading effect by a new modified differential evolution algorithm. *International Journal of Electrical Power & Energy Systems*, *32*(8), 893–903. doi:10.1016/j.ijepes.2010.01.023

Aravindhababu, P., & Nayar, K. R. (2002). Economic dispatch based on optimal lambda using radial basis function network. *International Journal of Electrical Power & Energy Systems*, *24*(7), 551–556. doi:10.1016/S0142-0615(01)00063-1

Basu, M. (2007). Dynamic economic emission dispatch using evolutionary programming and fuzzy satisfying method. *International Journal of Electrical Power & Energy Systems*, *8*(4), 1–15.

Basu, M. (2008). Dynamic economic emission dispatch using non-dominated sorting genetic algorithm-II. *International Journal of Electrical Power & Energy Systems*, *30*(2), 140–149. doi:10.1016/j.ijepes.2007.06.009

Basu, M. (2013). Artificial bee colony optimization for multi-area economic dispatch. *Electrical Power and Energy Systems*, *49*, 181–187. doi:10.1016/j.ijepes.2013.01.004

Basu, M. (2013). Hybridization of bee colony optimization and sequential quadratic programming for dynamic economic dispatch. *International Journal of Electrical Power & Energy Systems*, *44*(1), 591–596. doi:10.1016/j.ijepes.2012.08.026

Bhattacharya, A., & Chattopadhyay, P. K. (2010). Biogeography-based optimization for different economic load dispatch problems. *IEEE Transactions on Power Systems*, *25*(2), 1064–1077. doi:10.1109/TPWRS.2009.2034525

Bhattacharya, A., & Chattopadhyay, P. K. (2010). Hybrid differential evolution with biogeography-based optimization for solution of economic load dispatch. *IEEE Transactions on Power Systems*, *25*(4), 1955–1964. doi:10.1109/TPWRS.2010.2043270

Bozorg, H. O., Afshar, A., & Marino, M. A. (2006). Honey-bees mating optimization (HBMO) algorithm: A new heuristic approach for water resources optimization. *Water Resources Management*, *20*(5), 661–680. doi:10.100711269-005-9001-3

Cai, J., Ma, X., Li, L., & Peng, H. (2007). Chaotic particle swarm optimization for economic dispatch considering the generator constraints. *Energy Conversion and Management*, *48*(2), 645–653. doi:10.1016/j.enconman.2006.05.020

Cai, J. J., Li, Q., Li, L. X., Peng, H. P., & Yang, Y. X. (2012). A fuzzy adaptive chaotic ant swarm optimization for economic dispatch. *International Journal of Electrical Power & Energy Systems*, *34*(1), 154–160. doi:10.1016/j.ijepes.2011.09.020

Chandram, K., Subrahmanyam, N., & Sydulu, M. (2011). Equal embedded algorithm for economic load dispatch problem with transmission losses. *International Journal of Electrical Power & Energy Systems*, *33*(3), 500–507. doi:10.1016/j.ijepes.2010.12.002

Chatterjee, A., Ghoshal, S. P., & Mukherjee, V. (2012). Solution of combined economic and emission dispatch problems of power systems by an opposition-based harmony search algorithm. *International Journal of Electrical Power & Energy Systems*, *39*(1), 9–20. doi:10.1016/j.ijepes.2011.12.004

Chaturvedi, K. T., Pandit, M., & Srivastava, L. (2008). Self-organizing hierarchical particle swarm optimization for nonconvex economic dispatch. *IEEE Transactions on Power Systems*, *23*(3), 1079–1087. doi:10.1109/TPWRS.2008.926455

Chaturvedi, K. T., Pandit, M., & Srivastava, L. (2009). Particle swarm optimization with time varying acceleration coefficients for non-convex economic power dispatch. *International Journal of Electrical Power & Energy Systems*, *31*(6), 249–257. doi:10.1016/j.ijepes.2009.01.010

Chen, C. L., & Wang, S. C. (1993). Branch-and bound scheduling for thermal generating units. *IEEE Transactions on Energy Conversion*, *8*(2), 184–189. doi:10.1109/60.222703

Chen, G. (2008). Power station short-term generation optimal dispatch based on local random search PSO algorithm. *Electric Power Automation Equipment*, *5*(28), 52–55.

Chiang, C. L. (2005). Improved genetic algorithm for power economic dispatch of units with valve-point effects and multiple fuels. *IEEE Transactions on Power Systems*, *20*(4), 1690–1699. doi:10.1109/TPWRS.2005.857924

Ciornei, I., & Kyriakides, E. (2012). A GA-API solution for the economic dispatch of generation in power system operation. *IEEE Transactions on Power Systems*, *27*(1), 233–242. doi:10.1109/TPWRS.2011.2168833

Coelho, L. S., & Mariani, V. C. (2006). Combining of chaotic differential evolution and quadratic programming for economic dispatch optimization with valve-point effect. *IEEE Transactions on Power Systems*, *21*(2), 989–996. doi:10.1109/TPWRS.2006.873410

Deb, K., Pratap, A., Agarwal, S., & Meyarivan, T. (2002). A fast and elitist multi-objective genetic algorithm: NSGA-II. *IEEE Transactions on Evolutionary Computation*, *6*(2), 182–197. doi:10.1109/4235.996017

Dhillon, T. S. (1991). Artificial neural network applications to power systems and their relationships to symbolic methods. *International Journal of Electrical Power & Energy Systems*, *13*(2), 66–72. doi:10.1016/0142-0615(91)90028-T

Dieu, V. N., & Ongsakul, W. (2012). Economic dispatch with multiple fuel types by enhanced augmented Lagrange Hopfield network. *Applied Energy*, *91*(1), 281–289. doi:10.1016/j.apenergy.2011.09.025

Dieu, V. N., & Schegnerb, P. (2013). Augmented Lagrange Hopfield network initialized by quadratic programming for economic dispatch with piecewise quadratic cost functions and prohibited zones. *Applied Soft Computing*, *13*(1), 292–301. doi:10.1016/j.asoc.2012.08.026

Dodu, J. C., Martin, P., Merlin, A., & Pouget, J. (1972). An optimal formulation and solution of short-range operating problems for a power system with flow constraints. *Proceedings of the IEEE*, *60*(1), 54–63. doi:10.1109/PROC.1972.8557

Duvvuru, N., & Swarup, K. S. (2011). A hybrid interior point assisted differential evolution algorithm for economic dispatch. *IEEE Transactions on Power Systems*, *26*(2), 541–549. doi:10.1109/TPWRS.2010.2053224

El-Keib, A. A., Ma, H., & Hart, J. L. (1994). Environmentally constrained economic dispatch using the LaGrangian relaxation method. *IEEE Transactions on Power Systems*, *9*(4), 1723–1729. doi:10.1109/59.331423

Fan, J. Y., & Zhang, L. (1998). Real-time economic dispatch with line flow and emission constraints using quadratic programming. *IEEE Transactions on Power Systems*, *13*(2), 320–325. doi:10.1109/59.667345

Gaing, Z. L. (2003). Particle swarm optimization to solving the economic dispatch considering the generator constraints. *IEEE Transactions on Power Systems*, *18*(3), 1187–1195. doi:10.1109/TPWRS.2003.814889

Guo, C. X., Zhan, J. P., & Wu, Q. H. (2012). Dynamic economic emission dispatch based ongroup search optimizer with multiple producers. *Electric Power Systems Research*, *86*, 8–16. doi:10.1016/j.epsr.2011.11.015

Hemamalini, S., & Simon, S. P. (2010). Artificial bee colony algorithm for economic load dispatch problem with non-smooth cost functions. *Electric Power Components and Systems*, *38*(7), 786–803. doi:10.1080/15325000903489710

Hemamalini, S., & Simon, S. P. (2011). Dynamic economic dispatch using artificial immune system for units with valve-point effect. *International Journal of Electrical Power & Energy Systems*, *33*(4), 868–874. doi:10.1016/j.ijepes.2010.12.017

Hooshmand, R. A., Parastegari, M., & Morshed, M. J. (2012). Emission, reserve and economic load dispatch problem with non-smooth and non-convex cost functions using the hybrid bacterial foraging-Nelder-Mead algorithm. *Applied Energy*, *89*(1), 443–453. doi:10.1016/j.apenergy.2011.08.010

Hosseinnezhad, V., & Babaei, E. (2013). Economic load dispatch using θ-PSO. *International Journal of Electrical Power & Energy Systems*, *49*(1), 160–169. doi:10.1016/j.ijepes.2013.01.002

Hota, P. K., Barisal, A. K., & Chakrabarti, R. (2009). An improved PSO technique for short-term optimal hydrothermal scheduling. *Electric Power Systems Research*, *79*(7), 1047–1053. doi:10.1016/j.epsr.2009.01.001

Hota, P. K., Barisal, A. K., & Chakrabarti, R. (2010). Economic emission load dispatch through fuzzy based bacterial foraging algorithm. *International Journal of Electrical Power & Energy Systems*, *32*(7), 794–803. doi:10.1016/j.ijepes.2010.01.016

Javidan, J., & Ghasemi, A. (2012). Environmental/economic power dispatch using multiobjective honey bee mating optimization. *International Review of Electrical Engineering*, *7*(1), 3667–3675.

Jayabarati, T., & Sadasivam, G. (2000). Evolutionary programming based economic dispatch for units with multiple fuel cost functions. *IEEE Transactions on Power Systems*, *10*(3), 167–170.

Jiejin, J. (2012). A hybrid CPSO–SQP method for economic dispatch considering the valve-point effects. *Energy Conversion and Management*, *53*(1), 175–181. doi:10.1016/j.enconman.2011.08.023

Jiejin, C., Xiaoqian, M., Qiong, L., Lixiang, L., & Haipeng, P. (2010). A multi-objective chaotic ant swarm optimization for environmental/economic dispatch. *International Journal of Electrical Power & Energy Systems*, *32*, 337–44.

Karaboga, D., & Basturk, B. (2007). A powerful and efficient algorithm for numerical function optimization: Artificial bee colony (ABC) algorithm. *Journal of Global Optimization*, *39*(3), 459–471. doi:10.100710898-007-9149-x

Khamsawang, S., & Jiriwibhakorn, S. (2010). DSPSO–TSA for economic dispatch problem with non-smooth and non-continuous cost functions. *Energy Conversion and Management*, *51*(2), 365–375. doi:10.1016/j.enconman.2009.09.034

Kumar, R., Sharma, D., & Sadu, A. (2011). A hybrid multi-agent based particle swarm optimization algorithm for economic power dispatch. *International Journal of Electrical Power & Energy Systems*, *33*(1), 115–123. doi:10.1016/j.ijepes.2010.06.021

Kumar, V. S., & Mohan, M. R. (2011). A genetic algorithm solution to the optimal short-term hydrothermal scheduling. *International Journal of Electrical Power & Energy Systems*, *33*(4), 827–835. doi:10.1016/j.ijepes.2010.11.008

Kuo, C. C. (2008). A novel coding scheme for practical economic dispatch by modified particle swarm approach. *IEEE Transactions on Power Systems*, *23*(4), 1825–1835. doi:10.1109/TPWRS.2008.2002297

Lee, C. Y., & Yao, X. (2004). Evolutionary programming using mutations based on the Lévy probability distribution. *IEEE Transactions on Evolutionary Computation*, *8*(1), 1–13. doi:10.1109/TEVC.2003.816583

Li, X., Wang, J., Zhou, J., & Yin, M. (2011). A perturb biogeography based optimization with mutation for global numerical optimization. *Applied Mathematics and Computation*, *218*(2), 598–609. doi:10.1016/j.amc.2011.05.110

Liang, Z. X., & Glover, J. D. (1992). A zoom feature for a dynamic programming solution to economic dispatch including transmission losses. *IEEE Transactions on Power Systems*, *7*(2), 544–550. doi:10.1109/59.141757

Liao, G. C. (2012). Integrated Isolation Niche and Immune Genetic Algorithm for solving Bid-Based Dynamic Economic Dispatch. *International Journal of Electrical Power & Energy Systems*, *42*(1), 264–275. doi:10.1016/j.ijepes.2012.03.005

Lohokare, M. R., Panigrahi, B. K., Pattnaik, S. S., Devi, S., & Mohapatra, A. (2012). Neighborhood search-driven accelerated biogeography-based optimization for optimal load dispatch. *IEEE Transactions on Systems, Man and Cybernetics. Part C, Applications and Reviews*, *42*(5), 641–652. doi:10.1109/TSMCC.2012.2190401

Lokeshgupta, B., & Sivasubramani, S. (2018). Multi-objective dynamic economic and emission dispatch with demand side management. *International Journal of Electrical Power & Energy Systems*, *97*, 334–343. doi:10.1016/j.ijepes.2017.11.020

Ma, H. (2010). An analysis of the equilibrium of migration models for biogeography based optimization. *Inform Sciences*, *180*(18), 3444–3464. doi:10.1016/j.ins.2010.05.035

Ma, H., & Simon, D. (2011). Blended biogeography-based optimization for constrained optimization. *Engineering Applications of Artificial Intelligence*, *24*(3), 517–525. doi:10.1016/j.engappai.2010.08.005

Meng, K., Wang, H. G., Dong, Z. Y., & Wong, K. P. (2010). Quantum-inspired particle swarm optimization for valve-point economic load dispatch. *IEEE Transactions on Power Systems*, *25*(1), 215–222. doi:10.1109/TPWRS.2009.2030359

Mohammadi-ivatloo, B., Rabiee, A., Soroudi, A., & Ehsan, M. (2012). Imperialist competitive algorithm for solving non-convex dynamic economic power dispatch. *Energy*, *44*(1), 228–240. doi:10.1016/j.energy.2012.06.034

Mohammadi-Ivatloo, B., Rabiee, A., Soroudi, A., & Ehsan, M. (2012). Iteration PSO with time varying acceleration coefficients for solving non-convex economic dispatch problems. *International Journal of Electrical Power & Energy Systems*, *42*(1), 508–516. doi:10.1016/j.ijepes.2012.04.060

Nanda, J., Hari, L., & Kothari, M. L. (1994). Economic emission dispatch with line flow constraints using a classical technique. *IEE Proceedings. Generation, Transmission and Distribution*, *141*(1), 1–10. doi:10.1049/ip-gtd:19949770

Nanda, J., & Narayanan, R. B. (2002). Application of genetic algorithm to economic load dispatch with linefow constraints. *International Journal of Electrical Power & Energy Systems*, *24*(9), 723–729. doi:10.1016/S0142-0615(02)00002-9

Niknam, T. (2010). A new fuzzy adaptive hybrid particle swarm optimization algorithm for non-linear, non-smooth and non-convex economic dispatch problem. *Applied Energy*, *87*(1), 327–339. doi:10.1016/j.apenergy.2009.05.016

Niknam, T., Mojarrad, H. D., & Meymand, H. Z. (2011). Non-smooth economic dispatch computation by fuzzy and self-adaptive particle swarm optimization. *Applied Soft Computing*, *11*(2), 2805–2817. doi:10.1016/j.asoc.2010.11.010

Niknam, T., Mojarrad, H. D., Meymand, H. Z., & Firouzi, B. B. (2011). A new honey bee mating optimization algorithm for non-smooth economic dispatch. *Energy*, *36*(2), 896–908. doi:10.1016/j.energy.2010.12.021

Nima, A., & Hossein, S. (2010). Solution of non-convex economic dispatch problem considering valve loading effect by a new modified differential evolution algorithm. *International Journal of Electrical Power & Energy Systems*, *32*(8), 893–903. doi:10.1016/j.ijepes.2010.01.023

Noman, N., & Iba, H. (2008). Differential evolution for economic load dispatch problems. *Electric Power Systems Research*, *78*(8), 1322–1331. doi:10.1016/j.epsr.2007.11.007

Ozyon, S., Temurtas, P. H., Durmus, P. B., & Kuvat, G. (2012). Charged system search algorithm for emission constrained economic power dispatch problem. *Energy*, *46*(1), 420–430. doi:10.1016/j.energy.2012.08.008

Pandi, V. R., Panigrahi, B. K., Mohapatra, A., & Mallick, M. K. (2011). Economic load dispatch solution by improved harmony search with wavelet mutation. *International Journal on Computer Science and Engineering*, *6*(1), 122–131. doi:10.1504/IJCSE.2011.041220

Pandit, N., Tripathi, A., Tapaswi, S., & Pandit, M. (2012). An improved bacterial foraging algorithm for combined static/dynamic environmental economic dispatch. *Applied Soft Computing*, *12*(11), 3500–3513. doi:10.1016/j.asoc.2012.06.011

Panigrahi, B. K., Ravikumar, P. V., & Das, S. (2008). Adaptive particle swarm optimization approach for static and dynamic economic load dispatch. *Energy Conversion and Management*, *49*(6), 1407–1415. doi:10.1016/j.enconman.2007.12.023

Panigrahi, B. K., Yadav, S. R., Agrawal, S., & Tiwari, M. K. (2007). A clonal algorithm to solve economic load dispatch. *Electric Power Systems Research*, *77*(10), 1381–1389. doi:10.1016/j.epsr.2006.10.007

Parikh, J., & Chattopadhyay, D. (1996). A multi-area linear programming approach for analysis of economic operation of the Indian power system. *IEEE Transactions on Power Systems*, *11*(1), 52–58. doi:10.1109/59.485985

Park, J. B., Jeong, Y. W., Shin, J. R., & Lee, K. Y. (2010). An improved particle swarm optimization for nonconvex economic dispatch problems. *IEEE Transactions on Power Systems*, *25*(1), 156–166. doi:10.1109/TPWRS.2009.2030293

Park, J. B., Jeong, Y. W., Shin, J. R., & Lee, K. Y. (2010). An improved particle swarm optimization for nonconvex economic dispatch problems. *IEEE Transactions on Power Systems*, *25*(1), 156–166. doi:10.1109/TPWRS.2009.2030293

Park, J. H., Kim, Y. S., Eom, I. K., & Lee, K. Y. (1993). Economic load dispatch for piecewise quadratic cost function using Hopfield neural network. *IEEE Transactions on Power Systems*, *8*(3), 1030–1038. doi:10.1109/59.260897

Pattanaik, J. K., Basu, M., & Dash, D. P. (2018). (Article in Press). Improved real coded genetic algorithm for dynamic economic dispatch. *Journal of Electrical Systems and Information Technology*. doi:10.1016/j.jesit.2018.03.002

Pothiya, S., Ngamroo, I., & Kongprawechnon, W. (2008). Application of multiple tabu search algorithm to solve dynamic economic dispatch considering generator constraints. *Energy Conversion and Management*, *49*(4), 506–516. doi:10.1016/j.enconman.2007.08.012

Pothiya, S., Ngamroo, I., & Kongprawechnon, W. (2010). Ant colony optimisation for economicdispatch problem with non-smoothcost functions. *International Journal of Electrical Power & Energy Systems*, *32*(5), 478–487. doi:10.1016/j.ijepes.2009.09.016

Qu, B. Y., Zhu, Y. S., Jiao, Y. C., Wu, M. Y., Suganthan, P. N., & Liang, J. J. (2018). A survey on multi-objective evolutionary algorithms for the solution of the environmental/economic dispatch problems. *Swarm and Evolutionary Computation*, *38*, 1–11. doi:10.1016/j.swevo.2017.06.002

Ramanathan, R. (1985). Fast economic dispatch based on the penalty factors from Newton's method. *IEEE Transactions on Power Apparatus and Systems*, *104*(7), 1624–1629. doi:10.1109/TPAS.1985.319191

Reddy, A. S., & Vaisakh, K. (2013). Shuffled differential evolution for large scale economic dispatch. *Electric Power Systems Research*, *96*, 237–245. doi:10.1016/j.epsr.2012.11.010

Roy, P., Roy, A., & Chakraborty, P. (2013). Modified shuffled frog leaping algorithm with genetic algorithm crossover for solving economic load dispatch problem with valve-point effect. *Applied Soft Computing*, *13*(11), 4244–4252. doi:10.1016/j.asoc.2013.07.006

Roy, P. K., & Bhui, S. (2013). Multi-objective quasi-oppositional teaching learning based optimization for economic emission load dispatch problem. *International Journal of Electrical Power & Energy Systems*, *53*, 937–948. doi:10.1016/j.ijepes.2013.06.015

Saber, A. Y. (2012). Economic dispatch using particle swarm optimization with bacterial foraging effect. *International Journal of Electrical Power & Energy Systems*, *34*(1), 38–46. doi:10.1016/j.ijepes.2011.09.003

Safari, A., & Shayeghi, H. (2011). Iteration particle swarm optimization procedure for economic load dispatch with generator constraints. *Expert Systems with Applications*, *38*(5), 6043–6048. doi:10.1016/j.eswa.2010.11.015

Selvakumar, A., & Thanushkodi, K. (2009). Optimization using civilized swarm: Solution to economic dispatch with multiple minima. *Electric Power Systems Research*, *79*(1), 8–16. doi:10.1016/j.epsr.2008.05.001

Selvakumar, A. I., & Thanushkodi, K. (2007). A new particle swarm optimization solution to nonconvex economic dispatch problems. *IEEE Transactions on Power Systems*, *22*(1), 42–51. doi:10.1109/TPWRS.2006.889132

Shang, Y., Lu, S., Gong, J., Liu, R., Li, X., & Fan, Q. (2017). Improved genetic algorithm for economic load dispatch in hydropower plants and comprehensive performance comparison with dynamic programming method. *Journal of Hydrology (Amsterdam)*, *554*, 306–316. doi:10.1016/j.jhydrol.2017.09.029

Shaw, B., Mukherjee, V., & Ghoshal, S. P. (2012). A novel opposition-based gravitational search algorithm for combined economic and emission dispatch problems of power systems. *International Journal of Electrical Power & Energy Systems*, *35*(1), 21–33. doi:10.1016/j.ijepes.2011.08.012

Shayeghi, H., & Ghasemi, A. (2011). Solving economic load dispatch problems with valve point effects using artificial bee colony algorithm. *Iranian Journal of Electrical and Computer Engineering*, *3*(3), 2569–2577.

Shayeghi, H., & Ghasemi, A. (2012). Application of MOPSO for economic load dispatch solution with transmission losses. *International Journal on Technical and Physical Problems of Engineering*, *4*(10), 27–34.

Shayeghi, H., Shayanfar, H. A., & Ghasemi, A. (2012). Application of ABC algorithm for action based dispatch in the restructured power systems. *International Journal on Technical and Physical Problems of Engineering*, *4*(1), 114–119.

Simon, D. (2008). Biogeography-based optimization. *IEEE Transactions on Evolutionary Computation*, *12*(6), 702–713. doi:10.1109/TEVC.2008.919004

Sinha, N., Chakrabarti, R., & Chattopadhyay, P. K. (2003). Evolutionary programming techniques for economic load dispatch. *IEEE Transactions on Evolutionary Computation*, *7*(1), 83–94. doi:10.1109/TEVC.2002.806788

Storn, R., & Price, K. (1997). Differential evolution – a simple and efficient heuristic for global optimization over continuous spaces. *Journal of Global Optimization*, *11*(4), 341–359. doi:10.1023/A:1008202821328

Subbaraj, P., Rengaraj, R., & Salivahanan, S. (2009). Enhancement of combined heat and power economic dispatch using self adaptive real-coded genetic algorithm. *Applied Energy*, *86*(6), 915–921. doi:10.1016/j.apenergy.2008.10.002

Subbaraj, P., Rengaraj, R., & Salivahanan, S. (2011). Enhancement of self-adaptive real coded genetic algorithm using Taguchi method for economic dispatch problem. *Applied Soft Computing*, *11*(1), 83–92. doi:10.1016/j.asoc.2009.10.019

Vaisakh, K., Praveena, P., Rao, S. R. M., & Meah, K. (2012). Solving dynamic economic dispatch problem with security constraints using bacterial foraging PSO-DE algorithm. *International Journal of Electrical Power & Energy Systems*, *39*(1), 56–67. doi:10.1016/j.ijepes.2012.01.005

Venkatesh, P., Gnanadass, R., & Padhy, N. P. (2003). Comparison and application of evolutionary programming techniques to combined economic emission dispatch with line flow constraints. *IEEE Transactions on Power Systems*, *18*(2), 688–696. doi:10.1109/TPWRS.2003.811008

Wang, S. K., Chiou, J. P., & Liu, C. W. (2007). Non-smooth/non-convex economic dispatch by a novel hybrid differential evolution algorithm. *IET Generation, Transmission & Distribution*, *1*(5), 793–803. doi:10.1049/iet-gtd:20070183

Wang, Y., Cai, Z., & Zhang, Q. (2012). Enhancing the search ability of differential evolution through orthogonal crossover. *Information Sciences*, *85*(1), 153–177. doi:10.1016/j.ins.2011.09.001

Wang, Y., Li, B., & Yuan, B. (2010). Hybrid of comprehensive learning particle swarm optimization and SQP algorithm for large scale economic load dispatch optimization of power system. *Science China. Information Sciences*, *53*(8), 1566–1573. doi:10.100711432-010-4034-5

Wong, K. P., & Yuryevich, J. (1998). Evolutionary-programming based algorithm for environmentally constrained economic dispatch. *IEEE Transactions on Power Systems*, *13*(2), 301–306. doi:10.1109/59.667339

Wu, L. H., Wang, Y. N., Yuan, X. F., & Zhou, S. W. (2010). Environmental/economic power dispatch problem using multi-objective differential evolution algorithm. *Electric Power Systems Research*, *80*(9), 1171–1181. doi:10.1016/j.epsr.2010.03.010

Wu, Y., & Guo, J. (2017). Particle Swarm Optimization Using Levy Distribution for Economic Load Dispatch Problem. *Proceedings of the 2017 International Conference on Industrial Informatics- Computing Technology, Intelligent Technology, Industrial Information Integration (ICIICII)*.

Yalcinoz, T., & Altun, H. (2005). A new genetic algorithm with arithmetic crossover to economic and environmental economic dispatch. *International Journal of Engineering Intelligent Systems for Electrical Engineering and Communications*, *13*(1), 45–52.

Yan, X., & Quintana, V. H. (1997). An efficient predictor–corrector interior point algorithm for security-constrained economic dispatch. *IEEE Transactions on Power Systems*, *12*(2), 803–810. doi:10.1109/59.589693

Yang, L., Fraga, E. S., & Papageorgiou, L. G. (2013). Mathematical programming formulations for non-smooth and non-convex electricity dispatch problems. *Electric Power Systems Research*, *95*, 302–308. doi:10.1016/j.epsr.2012.09.015

Yuan, X., Wang, L., Yuan, T., Zhang, Y., Cao, B., & Yang, B. A. (2008). A modified differential evolution approach for dynamic economic dispatch with valve-point effects. *Energy Conversion and Management*, *49*(12), 3447–3453. doi:10.1016/j.enconman.2008.08.016

APPENDIX

Table 11. Generating units' capacities and quadratic cost coefficient data of Test system 1

Unit No.	P_{min} (MW)	P_{max} (MW)	a ($/MW²h)	b ($/MWh)	c ($/ hr.)	Unit No.	P_{min} (MW)	P_{max} (MW)	a ($/MW²h)	b ($/MWh)	c ($/hr.)
1	100	196	0.002176	-0.3975	26.97		85	138	0.001049	-0.03114	1.983
	196	250	0.001861	-0.3059	21.13	6	138	200	0.002758	-0.6348	52.85
	50	114	0.001138	-0.03988	1.865		200	265	0.005935	-2.338	266.8
2	114	157	0.001620	-0.1980	13.65		200	331	0.001107	-0.1325	18.93
	157	230	0.004194	-1.269	118.4	7	331	391	0.001165	-0.2267	43.77
	200	332	0.001457	-0.3116	39.79		391	500	0.0002454	0.3559	-43.35
3	332	388	0.0008035	0.03389	2.876		99	138	0.001049	-0.03114	1.983
	388	500	0.00001176	0.4864	-59.14	8	138	200	0.002758	-0.6348	52.85
	99	138	0.001049	-0.03114	1.983		200	265	0.005935	-2.3380	266.8
4	138	200	0.002758	-0.6348	52.85		130	213	0.0006121	-0.01817	14.23
	200	265	0.005935	2.338	266.8	9	213	370	0.001554	-0.5675	88.53
	190	338	0.001066	-0.08733	13.92		370	440	0.007033	-0.04514	15.30
5	338	407	0.001597	-0.5206	99.76		200	362	0.001102	-0.09938	13.97
	407	490	0.0001498	0.4462	53.99	10	362	407	0.001137	-0.2024	46.71
							407	490	0.00004164	0.5084	46.71

Table 12. Valve point coefficient data of Test system1

Unit No.	P_{min} (MW)	P_{max} (MW)	d ($/hr.)	e (rad/ MW)	Unit No.	P_{min} (MW)	P_{max} (MW)	d ($/hr.)	e (rad/ MW)
1	100	196	0.02697	-3.975	6	85	138	0.001983	-0.3114
	196	250	0.02113	-3.059		138	200	0.05285	-6.348
2	50	114	0.001865	-0.3988		200	265	0.2668	-23.38
	114	157	0.01365	-1.980	7	200	331	0.01893	-1.325
	157	230	0.1184	-12.69		331	391	0.04377	-2.267
3	200	332	0.03979	-3.116		391	500	-0.04335	3.559
	332	388	-0.002876	-23.38	8	99	138	0.001983	-0.3114
	388	500	-0.05914	4.864		138	200	0.05285	-6.348
4	99	138	0.001983	-0.3114		200	265	0.2668	-23.38
	138	200	0.05285	-6.348	9	130	213	0.01423	-0.1817
	200	265	0.2668	-23.38		213	370	0.08853	-0.05675
5	190	338	0.01392	-0.8733		370	440	0.01423	-0.1817
	338	407	0.09976	-5.206	10	200	362	0.01397	-0.9938
	407	490	-0.05399	4.462		362	407	0.04671	-2.024
						407	490	-0.06113	5.084

Table 13. Generation limits, fuel cost coefficients, ramping rate and prohibited operating zones data (Test system 2)

Unit No.	P_{min} (MW)	P_{max} (MW)	a ($/MW²h)	b ($/MWh)	c ($/h)	Prohibited zones (MW)
1	150	455	0.000299	10.1000	671	-
2	150	455	0.000183	10.2000	574	[185 225] [305 335] [420 450]
3	20	130	0.001126	8.8000	374	-
4	20	130	0.001126	8.8000	374	-
5	150	470	0.000205	10.4000	461	[180 200] [305 335] [390 420]
6	135	460	0.000301	10.1000	630	[230 255] [365 395] [430 455]
7	135	465	0.000364	9.8000	548	-
8	60	300	0.000338	11.2000	227	-
9	25	162	0.000807	11.2000	173	-
10	25	160	0.001203	10.7000	175	-
11	20	80	0.003586	10.2000	186	-
12	20	80	0.005513	9.9000	230	[30 40] [55 65]
13	25	85	0.000 371	13.1000	225	-
14	15	55	0.001929	12.1000	309	-
15	15	55	0.004447	12.4000	323	-

Table 14. B-loss coefficients of the Test system 2

		14	12	7	-1	-3	-1	-1	-1	-3	-5	-3	-2	4	3	-1
		12	15	13	0	-5	-2	0	1	-2	-4	-4	0	4	10	-2
		7	13	76	-1	-13	-9	-1	0	-8	-12	-17	0	-26	111	-28
		-1	0	-1	34	-7	-4	11	50	29	32	-11	0	1	1	-26
		-3	-5	-13	-7	90	14	-3	-12	-10	-13	7	-2	-2	-24	-3
		-1	-2	-9	-4	14	16	0	-6	-5	-8	11	-1	-2	-17	3
		-1	0	-1	11	-3	0	15	17	15	9	-5	7	0	-2	-8
$B =$	10^{-6} *	-1	1	0	50	-12	-6	17	168	82	79	-23	-36	1	5	-78
		-3	-2	-8	29	-10	-5	15	82	129	116	-21	-25	7	-12	-72
		-5	-4	-12	32	-13	-8	9	79	116	200	-27	-34	9	-11	-88
		-3	-4	-17	-11	7	11	-5	-23	-21	-27	140	1	4	-38	168
		-2	0	0	0	-2	-1	7	-36	-25	-34	1	54	-1	-4	28
		4	4	-26	1	-2	-2	0	1	7	9	4	-1	103	-101	28
		3	10	111	1	-24	-17	-2	5	-12	-11	-38	-4	-101	578	-94
		-1	-2	-28	-26	-3	3	-8	-78	-72	-88	168	28	28	-94	1283
$B_0 =$	10^{-3} *	-0.1	-0.2	2.8	-0.1	0.1	-0.3	-0.2	-0.2	0.6	3.9	-1.7	0	-3.2	6.7	-6.4
$B_{00} =$								0.0055								

Table 15. Generating units' capacities and cost coefficient data of Test system 3

Unit No.	P_{min} (MW)	P_{max} (MW)	a ($/MW²h)	b ($/MWh)	c ($/h.)	d ($/h)	e (rad/MW)	Unit No.	P_{min} (MW)	P_{max} (MW)	a ($/MW²h)	b ($/MWh)	c ($/h)	d ($/h)	e (rad/MW)
1	36	114	0.00690	6.73	94.705	100	0.084	21	254	550	0.00298	6.63	785.96	300	0.035
2	36	114	0.00690	6.73	94.705	100	0.084	22	254	550	0.00298	6.63	785.96	300	0.035
3	60	120	0.02028	7.07	309.54	100	0.084	23	254	550	0.00284	6.66	794.53	300	0.035
4	80	190	0.00942	8.18	369.03	150	0.063	24	254	550	0.00284	6.66	794.53	300	0.035
5	47	97	0.01140	5.35	148.89	120	0.077	25	254	550	0.00277	7.1	801.32	300	0.035
6	68	140	0.01142	8.05	222.33	100	0.084	26	254	550	0.00277	7.1	801.32	300	0.035
7	110	300	0.00357	8.03	287.71	200	0.042	27	10	150	0.52124	3.33	1055.1	120	0.077
8	135	300	0.00492	6.99	391.98	200	0.042	28	10	150	0.52124	3.33	1055.1	120	0.077
9	135	300	0.00573	6.60	455.76	200	0.042	29	10	150	0.52124	3.33	1055.1	120	0.077
10	130	300	0.00605	12.9	722.82	200	0.042	30	47	97	0.01140	5.35	148.89	120	0.077
11	94	375	0.00515	12.9	635.20	200	0.042	31	60	190	0.00160	6.43	222.92	150	0.063
12	94	375	0.00569	12.8	654.69	200	0.042	32	60	190	0.00160	6.43	222.92	150	0.063
13	125	500	0.00421	12.5	913.40	300	0.035	33	60	190	0.00160	6.43	222.92	150	0.063
14	125	500	0.00752	8.84	1760.4	300	0.035	34	90	200	0.00010	8.95	107.87	200	0.042
15	125	500	0.00708	9.15	1728.3	300	0.035	35	90	200	0.00010	8.62	116.58	200	0.042
16	125	500	0.00708	9.15	1728.3	300	0.035	36	90	200	0.00010	8.62	116.58	200	0.042
17	220	500	0.00313	7.97	647.85	300	0.035	37	25	110	0.01610	5.88	307.45	80	0.098
18	220	500	0.00313	7.95	649.69	300	0.035	38	25	110	0.01610	5.88	307.45	80	0.098
19	242	550	0.00313	7.97	647.83	300	0.035	39	25	110	0.01610	5.88	307.45	80	0.098
20	242	550	0.00313	7.97	647.81	300	0.035	40	242	550	0.00313	7.97	647.83	300	0.035

Table 16. Cost coefficient data of Test system 5

Unit No.	a ($/MW²h)	b ($/MWh)	c ($/h.)	d ($/h)	e (rad/MW)	Unit No.	a ($/MW²h)	b ($/MWh)	c ($/h)	d ($/h)	e (rad/MW)	Unit No.	a ($/MW²h)	b ($/MWh)	c ($/h)	d ($/h)	e (rad/MW)
1	0.032888	1.242	1220.6	0	0	48	0.00754	61.52	3138.8	0	0	95	0.000022	3.04	154.48	0	0
2	0.00828	41.095	1315.1	0	0	49	0.01643	58.635	3453.1	0	0	96	0.000203	1.709	332.83	0	0
3	0.003849	46.31	874.29	0	0	50	0.045934	44.647	5119.3	0	0	97	0.000198	1.668	326.6	0	0
4	0.003849	46.31	874.29	0	0	51	0.000044	71.584	1898.4	0	0	98	0.000215	1.789	345.31	0	0
5	0.042468	54.242	1976.5	700	0.08	52	0.000044	71.584	1898.4	1100	0.043	99	0.000218	1.815	350.37	0	0
6	0.014992	61.215	1338.1	0	0	53	0.000044	71.584	1898.4	0	0	100	0.000193	2.726	370.38	0	0
7	0.007039	11.791	1818.3	0	0	54	0.000044	71.584	1898.4	0	0	101	0.000197	2.732	367.07	0	0
8	0.003079	15.055	1134	0	0	55	0.002528	85.12	2473.4	0	0	102	0.000324	2.651	124.88	0	0
9	0.005063	13.226	1320.6	0	0	56	0.000131	87.682	2781.7	0	0	103	0.000344	2.798	130.79	0	0
10	0.005063	13.226	1320.6	600	0.055	57	0.010372	69.532	5515.5	0	0	104	0.00069	1.595	878.75	0	0
11	0.005063	13.226	1320.6	0	0	58	0.007627	78.339	3478.3	0	0	105	0.00065	1.503	827.96	0	0
12	0.003552	14.498	1106.5	0	0	59	0.012464	58.172	6240.9	0	0	106	0.000233	2.425	432.01	0	0
13	0.003901	14.651	1176.5	0	0	60	0.039441	46.636	9960.1	0	0	107	0.000239	2.499	445.61	0	0
14	0.003901	14.651	1176.5	0	0	61	0.007278	76.947	3672	0	0	108	0.000261	2.674	467.22	0	0
15	0.003901	14.651	1176.5	800	0.06	62	0.000044	80.761	1837.4	0	0	109	0.000259	2.692	475.94	0	0
16	0.003901	14.651	1176.5	0	0	63	0.000044	70.136	3108.4	0	0	110	0.000707	1.633	899.46	0	0
17	0.002393	15.669	1017.4	0	0	64	0.000044	70.136	3108.4	0	0	111	0.000786	1.816	1000.4	0	0
18	0.002393	15.669	1017.4	0	0	65	0.018827	49.84	7095.5	0	0	112	0.014355	89.83	1269.1	0	0
19	0.003684	14.656	1229.1	0	0	66	0.010852	65.404	3392.7	0	0	113	0.014355	89.83	1269.1	0	0
20	0.003684	14.656	1229.1	0	0	67	0.018827	49.84	7095.5	0	0	114	0.014355	89.83	1269.1	0	0

continued on following page

Table 16. Continued

Unit No.	a ($/MW²h)	b ($/MWh)	c ($/h.)	d ($/h)	e (rad/MW)	Unit No.	a ($/MW²h)	b ($/MWh)	c ($/h)	d ($/h)	e (rad/MW)	Unit No.	a ($/MW²h)	b ($/MWh)	c ($/h)	d ($/h)	e (rad/MW)
21	0.003684	14.656	1229.1	0	0	68	0.018827	49.84	7095.5	0	0	115	0.030266	64.125	4965.1	0	0
22	0.003684	14.656	1229.1	600	0.05	69	0.03456	66.465	4288.3	0	0	116	0.030266	64.125	4965.1	0	0
23	0.004004	14.378	1267.9	0	0	70	0.08154	22.941	13813	1200	0.03	117	0.030266	64.125	4965.1	0	0
24	0.003684	14.656	1229.1	0	0	71	0.023534	64.314	4435.5	0	0	118	0.024027	76.129	2243.2	0	0
25	0.001619	16.261	975.93	0	0	72	0.035475	45.017	9750.8	1000	0.05	119	0.00158	81.805	2290.4	600	0.07
26	0.005093	13.362	1532.1	0	0	73	0.000915	70.644	1042.4	0	0	120	0.022095	81.14	1681.5	0	0
27	0.000993	17.203	641.99	0	0	74	0.000044	70.959	1159.9	0	0	121	0.07681	46.665	6743.3	1200	0.043
28	0.000993	17.203	641.99	0	0	75	0.000044	70.959	1159.9	0	0	122	0.95344	78.412	394.4	0	0
29	0.002473	15.274	911.53	0	0	76	0.001307	70.302	1304	0	0	123	4.4e-05	112.09	1243.2	0	0
30	0.002547	15.212	910.53	0	0	77	0.000392	70.662	1156.2	0	0	124	0.072468	90.871	1454.7	0	0
31	0.003542	15.033	1074.8	0	0	78	8.7e-05	71.101	2119	0	0	125	0.000448	97.116	1011.1	0	0
32	0.003542	15.033	1074.8	0	0	79	0.000521	37.854	779.52	0	0	126	0.59911	83.244	909.27	0	0
33	0.003542	15.033	1074.8	600	0.043	80	0.000498	37.768	829.89	0	0	127	0.24471	95.665	689.38	0	0
34	0.003542	15.033	1074.8	0	0	81	0.001046	67.983	2333.7	0	0	128	0.000042	91.202	1443.8	0	0
35	0.003132	13.992	1278.5	0	0	82	0.13205	77.838	2029	0	0	129	0.085145	104.5	535.55	0	0
36	0.001323	15.679	861.74	0	0	83	0.096968	63.671	4412	0	0	130	0.52472	83.015	617.73	0	0
37	0.00295	16.542	408.83	0	0	84	0.054868	79.458	2982.2	1000	0.05	131	0.17652	127.8	90.966	0	0
38	0.00295	16.542	408.83	0	0	85	0.054868	79.458	2982.2	0	0	132	0.063414	77.929	974.45	0	0
39	0.000991	16.518	1288.8	0	0	86	0.014382	93.966	3174.9	0	0	133	2.7405	92.779	263.81	0	0
40	0.001581	15.815	1436.3	600	0.043	87	0.013161	94.723	3218.4	0	0	134	0.11244	80.95	1335.6	0	0

continued on following page

Table 16. Continued

Unit No.	a ($/MW²h)	b ($/MWh)	c ($/h.)	d ($/h)	e (rad/MW)	Unit No.	a ($/MW²h)	b ($/MWh)	c ($/h)	d ($/h)	e (rad/MW)	Unit No.	a ($/MW²h)	b ($/MWh)	c ($/h)	d ($/h)	e (rad/MW)
41	0.90236	75.464	669.99	0	0	88	0.016033	66.919	3723.8	0	0	135	0.041529	89.073	1033.9	0	0
42	0.1103	129.54	134.54	0	0	89	0.013653	68.185	3551.4	0	0	136	0.000911	161.29	1391.3	0	0
43	0.024493	56.613	3427.9	0	0	90	0.028148	60.821	4322.6	0	0	137	0.005245	161.83	4477.1	0	0
44	0.029156	54.451	3751.8	0	0	91	0.01347	68.551	3493.7	0	0	138	0.23479	84.972	57.794	0	0
45	0.024667	54.736	3918.8	0	0	92	0.000064	2.842	226.8	0	0	139	0.23479	84.972	57.794	0	0
46	0.016517	58.034	3379.6	0	0	93	0.000252	2.946	382.93	0	0	140	1.1119	16.087	1258.4	0	0
47	0.026584	55.981	3345.3	0	0	94	0.000022	3.096	156.99	0	0				154.48	0	0

Table 17. Generating units' capacitieramp rate data of Test system 5

Unit No.	P_{min} (MW)	P_{max} (MW)	P0 (MW/hr)	UR (MW/hr)	DR (MW/hr)	Unit No.	P_{min} (MW)	P_{max} (MW)	P0 (MW/hr)	UR (MW/hr)	DR (MW/hr)	Unit No.	P_{min} (MW)	P_{max} (MW)	P0 (MW/hr)	UR (MW/hr)	DR (MW/hr)
1	71	119	98.4	30	120	**48**	160	250	232.9	702	702	**95**	795	978	801.5	36	36
2	120	189	134	30	120	**49**	160	250	168	702	702	**96**	578	682	582.7	138	204
3	125	190	141.5	60	60	**50**	160	250	208.4	702	702	**97**	615	720	680.7	144	216
4	125	190	183.3	60	60	**51**	165	504	443.9	1350	1350	**98**	612	718	670.7	144	216
5	90	190	125	150	150	**52**	165	504	426	1350	1350	**99**	612	720	651.7	144	216
6	90	190	91.3	150	150	**53**	165	504	434.1	1350	1350	**100**	758	964	921	48	48
7	280	490	401.1	180	300	**54**	165	504	402.5	1350	1350	**101**	755	958	916.8	48	48
8	280	490	329.5	180	300	**55**	180	471	357.4	1350	1350	**102**	750	1007	911.9	36	54
9	260	496	386.1	300	510	**56**	180	561	423	720	720	**103**	750	1006	898	36	54
10	260	496	427.3	300	510	**57**	103	341	220	720	720	**104**	713	1013	905	30	30
11	260	496	412.2	300	510	**58**	198	617	369.4	2700	2700	**105**	718	1020	846.5	30	30
12	260	496	370.1	300	510	**59**	100	312	273.5	1500	1500	**106**	791	954	850.9	30	30
13	260	506	301.8	600	600	**60**	153	471	336	1656	1656	**107**	786	952	843.7	30	30
14	260	509	368	600	600	**61**	163	500	432	2160	2160	**108**	795	1006	841.4	36	36
15	260	506	301.9	600	600	**62**	95	302	220	900	900	**109**	795	1013	835.7	36	36
16	260	505	476.4	600	600	**63**	160	511	410.6	1200	1200	**110**	795	1021	828.8	36	36
17	260	506	283.1	600	600	**64**	160	511	422.7	1200	1200	**111**	795	1015	846	36	36
18	260	506	414.1	600	600	**65**	196	490	351	1014	1014	**112**	94	203	179	120	120
19	260	505	328	600	600	**66**	196	490	296	1014	1014	**113**	94	203	120.8	120	120
20	260	505	389.4	600	600	**67**	196	490	411.1	1014	1014	**114**	94	203	121	120	120

continued on following page

Table 17. Continued

Unit No.	P_{min} (MW)	P_{max} (MW)	P0 (MW/hr)	UR (MW/hr)	DR (MW/hr)	Unit No.	P_{min} (MW)	P_{max} (MW)	P0 (MW/hr)	UR (MW/hr)	DR (MW/hr)	Unit No.	P_{min} (MW)	P_{max} (MW)	P0 (MW/hr)	UR (MW/hr)	DR (MW/hr)
21	260	505	354.7	600	600	**68**	196	490	263.2	1014	1014	**115**	244	379	317.4	480	480
22	260	505	262	600	600	**69**	130	432	370.3	1350	1350	**116**	244	379	318.4	480	480
23	260	505	461.5	600	600	**70**	130	432	418.7	1350	1350	**117**	244	379	335.8	480	480
24	260	505	371.6	600	600	**71**	137	455	409.6	1350	1350	**118**	95	190	151	240	240
25	280	537	462.6	300	300	**72**	137	455	412	1350	1350	**119**	95	189	129.5	240	240
26	280	537	379.2	300	300	**73**	195	541	423.2	780	780	**120**	116	194	130	120	120
27	280	549	530.8	360	360	**74**	175	536	428	1650	1650	**121**	175	321	218.9	180	180
28	280	549	391.9	360	360	**75**	175	540	436	1650	1650	**122**	2	19	5.4	90	90
29	260	501	480.1	180	180	**76**	175	538	428	1650	1650	**123**	4	59	45	90	90
30	260	501	319	180	180	**77**	175	540	425	1650	1650	**124**	15	83	20	300	300
31	260	506	329.5	600	600	**78**	330	574	497.2	1620	1620	**125**	9	53	16.3	162	162
32	260	506	333.8	600	600	**79**	160	531	510	1482	1482	**126**	12	37	20	114	114
33	260	506	390	600	600	**80**	160	531	470	1482	1482	**127**	10	34	22.1	120	120
34	260	506	432	600	600	**81**	200	542	464.1	1668	1668	**128**	112	373	125	1080	1080
35	260	500	402	600	660	**82**	56	132	118.1	120	120	**129**	4	20	10	60	60
36	260	500	428	900	900	**83**	115	245	141.3	180	180	**130**	5	38	13	66	66
37	120	241	178.4	180	180	**84**	115	245	132	120	180	**131**	5	19	7.5	12	6
38	120	241	194.1	180	180	**85**	115	245	135	120	180	**132**	50	98	53.2	300	300
39	423	774	474	600	600	**86**	207	307	252	120	180	**133**	5	10	6.4	6	6

continued on following page

Table 17. Continued

Unit No.	P_{min} (MW)	P_{max} (MW)	P0 (MW/hr)	UR (MW/hr)	DR (MW/hr)	Unit No.	P_{min} (MW)	P_{max} (MW)	P0 (MW/hr)	UR (MW/hr)	DR (MW/hr)	Unit No.	P_{min} (MW)	P_{max} (MW)	P0 (MW/hr)	UR (MW/hr)	DR (MW/hr)
40	423	769	609.8	600	600	**87**	207	307	221	120	180	**134**	42	74	69.1	60	60
41	3	19	17.8	210	210	**88**	175	345	245.9	318	318	**135**	42	74	49.9	60	60
42	3	28	6.9	366	366	**89**	175	345	247.9	318	318	**136**	41	105	91	528	528
43	160	250	224.3	702	702	**90**	175	345	183.6	318	318	**137**	17	51	41	300	300
44	160	250	210	702	702	**91**	175	345	288	318	318	**138**	7	19	13.7	18	30
45	160	250	212	702	702	**92**	360	580	557.4	18	18	**139**	7	19	7.4	18	30
46	160	250	200.8	702	702	**93**	415	645	529.5	18	18	**140**	26	40	28.6	72	120
47	160	250	220	702	702	**94**	795	984	800.8	36	36						

Chapter 4
Optimal Power Flow and Optimal Reactive Power Dispatch Using Different Evolutionary Optimization Techniques

ABSTRACT

This chapter describes grey wolf optimization (GWO), teaching-learning-based optimization (TLBO), biogeography-based optimization (BBO), krill herd algorithm (KHA), chemical reaction optimization (CRO), and hybrid CRO (HCRO) algorithms to solve both single and multi-objective optimal power flow (MOOPF) and optimal reactive power dispatch (ORPD) problems while satisfying various operational constraints. The proposed HCRO approach along with GWO, TLBO, BBO, KHA, and CRO algorithms are implemented on IEEE 30-bus system to solve four different single objectives: fuel cost minimization, system power loss minimization, voltage stability index minimization, and voltage deviation minimization; two bi-objectives optimization, namely minimization of fuel cost and transmission loss; minimization of fuel cost and voltage profile; and one tri-objective optimization, namely minimization of fuel cost, minimization of transmission losses, and improvement of voltage profile simultaneously. The simulation results clearly suggest that the proposed is able to provide a better solution than other approaches.

DOI: 10.4018/978-1-5225-6971-8.ch004

INTRODUCTION

The main objective of any electrical power utility is to ensure reliable and stable power supply economically to the consumers. For smaller size power systems, it is comparatively a trivial problem and also individual power systems were used and operated separately in regional concept earlier. However, the state of the art is to have interconnection of different supply agencies to form very large complex power systems in order to achieve several advantages. In an interconnected power system, the real and the reactive powers of the generators are required to vary within operating limits in order to meet the particular load demand with minimum fuel cost. In the generation plant, there are two factors which are to be considered at every load change. These are the division of load and economic factor. But due to the deregulation in the industry, optimal power flow (OPF) (Dommel & Tinney, 1968; El-Fergany & Hasanien, 2018) is used to handle such problems. At present, OPF has become reliable enough for practical use and has taken a place on standard power system analysis tools. Thus, the main purpose of OPF solution is to schedule the power generation in such a way that minimizes the fuel cost while satisfying all the equality and inequality constraints. In addition to the minimization of fuel cost, the OPF may also be used to achieve the other benefits such as reduction of system loss, improvement of voltage profile and improvement of system security. Thus, the objective of the OPF is to find steady state operating point which minimizes generation cost, system loss, voltage deviation etc while maintaining an acceptable system performance in terms of limits on generators' real and reactive powers, line flows, outputs of various compensating devices etc.

Due to the continuous growth in the demand of electricity with unmatched generation and transmission capacity expansion, voltage instability is emerging as a new challenge to power system planning and operation. Unavailability of sufficient reactive power sources to maintain normal voltage profiles at heavily loaded buses are the main reasons for the voltage collapse. This problem may be overcome by reallocating reactive power generations in the system. This can be achieved by adjusting transformer taps, var injections of shunt compensators and generator voltages. In addition, the system losses can be minimized via redistribution of reactive powers in the system.

Optimal reactive power dispatch (ORPD) (Mei et al., 2017; Mouassa et al., 2017; Nuaekaew et al., 2017) which is one of the application functions of modern energy management system (EMS) is used to minimize total system loss and improve voltage profile. Traditionally OPF and ORPD are solved with the help of various mathematical programming based optimization techniques such as linear programming (LP) (Aoki et al., 1988; Deeb & Shahidehpour, 1988; Momoh et al., 1999), dynamic programming (DP) (Bechert & Chen, 1997; Lu & Hsu, 1995), quadratic programming (QP), (Grudinin, 1998; Lo & Zhu, 1991; Quintana & Santos-Nieto, 1989), Newton method (Santos & Costa, 1995; Sun et al., 1984), Lagrangian decomposition method (Granada et al., 2012), interior point Method (IPM) (Yan & Quintana, 1999; Granville, 1995; Capitanescu & Wehenkel, 2013; Zhu & Xiong, 2003), non-linear programming (NLP) (Habiabollahzadeh et al., 1989), Tabu search (TS) (Abido, 2002; Wennan et al., 2008) gradient method (GM) (Burchett, 2013; Yu et al., 1986) and so on. Such classical optimization techniques have some drawbacks because its take long to for showing the optimal result. In recent years, many meta-heuristic algorithms such as simulated annealing (SA) (Roa-Sepulveda & Pavez-Lazo, 2003), evolutionary programming (EP) (Yan et al. 2004), differential evolution (DE) (Cai et al., 2008; Liang et al., 2007; Sivasubramani & Swarup, 2012), artificial bee colony optimization (ABC) (Ayan & Kılıc, 2012; Karaboga & Basturk, 2007; Kursat & Ulas, 2013; Rajabioum, 2011), genetic algorithm (GA) (Chan et al., 2007), krill herd algorithm (KHA) (Gandomi & Alavi, 2012), biogeography based optimization (BBO) (Roy et al., 2011), harmony search algorithm (HSA) (Khazali & Kalantar, 2011), gravitational search algorithm (GSA) (Roy et al., 2012; Hadi et al., 2012), Bacteria foraging optimization (BFO) (Tripathy & Mishra, 2007; Nandaet al., 2009), teaching learing based optimization (TLBO) (Rao et al., 2011), Seeker optimization algorithm (SOA) (Chaohua et al., 2009), multi-verse Optimizer (MVO) (Mirjalili, 2015; Sivasubramani & Swarup, 2011) have been implemented successfully to solve the non-linear OPF and ORPD problems. In recent years, many hybrid techniques like genetic algorithm hybridization with successive linear programming (GA/LP) (Urdaneta et al., 1999), evolutionary programming hybridization with nonlinear programming (EP/NLP) (Lai & Ma 1997), hybrid genetic algorithm based interior point method (GA/IPM) (Wei et al., 2006), hybrid differential evolution based particle swarm optimization (DE/PSO) (Shouzheng et al., 2010), simulated annealing based particle swarm algorithm (SA/PSO) (Mao et al., 2008) have been used to improve searching ability of the basic algorithms.

LITERATURE SURVEY

Lobato et al. (2001) proposed linear programming (LP) based OPF for minimization of transmission losses and generator reactive margins of the Spanish power system. In this method, the discrete nature of shunt reactors and capacitors were modeled by integer variables. Both the objective function and the constraints were linearized in each iteration and the method was found to be better than one which linearized only objective function once. (Momoh, 1989) presented an extension of basic Kuhn-Tucker conditions and employed a generalized quadratic-based model for the OPF. The conditions of feasibility, convergence and optimality were included in the proposed OPF technique. It was also capable of using hierarchical structures to include multiple objective functions and selectable constraints. The generalized technique using the sensitivity of objective functions with optimal adjustments in the constraints yielded a global optimal solution. Computational memory and execution time required were reduced. (Pudjianto et al., 2002) used LP and non-linear programming (NLP) based reactive OPF for allocating reactive powers among competing generators in a deregulated environment. It was concluded that the overall cost associated with the system reactive requirement as calculated by LP method was reasonably accurate, but the individual generator commitment might vary considerably. Whereas, NLP offered a faster computation speed and accuracy for the solution but the convergence could not be guaranteed for every condition. (Paranjothi & Anburaja, 2002) proposed a refined genetic algorithm to solve OPF. A real coded mixed integer genetic algorithm (MIGA) was reported by (Gaing & Huang, 2004) to solve OPF problems of 26-bus system with prohibited operating zone constraints. The results were compared to those obtained by a simple genetic algorithm (SGA) and the better results were reported. (Gaing & Chang, 2006) applied MIGA with arithmetic crossover to solve security constrained OPF problems. The algorithm was validated on 26-bus and IEEE 57-bus systems with contingency and without contingency conditions. The computational results were compared to the results of EP and encouraging results were obtained from the proposed method. (Lai et al., 1997) adopted an enhanced binary GA to solve OPF problem with the objective function being the quadratic fuel cost function for a given power system under normal and contingent operation conditions. In their formulation, the authors accounted for both discrete and continuous optimization variables. (Roa-Sepulveda & Pavez-Lazo, 2001) proposed SA technique to solve OPF problem. The main purpose of this work was to verify the viability of using SA to solve OPF problems simultaneously composed by load flow and economic dispatch problems. The technique was tested on 6-bus system and IEEE 30-bus system. The results showed superiority of the proposed method over classical economic dispatch, standard load flow analysis and Newton based optimal power flow. A PSO based OPF problem with piecewise linear cost function was reported by (Abido, 2002). Fuel cost minimization, voltage profile improvement and voltage stability enhancement were considered as competing objective functions in this paper. The proposed technique was implemented on IEEE 30-bus system. The environmental-economic transaction planning problem in the electricity market was formulated as a multi-objective OPF in (Ma et al., 2005). A multi-objective PSO technique was developed to solve the problem via a non-stationary multi-stage assignment penalty function. A method based on PSO to solve OPF was proposed by (Zhao et al., 2004). The feasibility of the proposed method was tested on IEEE 30-bus system for three different objective functions to minimize the total real power generation cost, reactive power optimization and the optimal real and reactive dispatches. Results from the proposed technique were compared to those of LP and GA and better results were reported. (Gaing 2005) developed a mixed integer particle swarm optimization technique for constrained OPF with valve point nonlinearities. The technique was tested on 9-bus and 26-bus systems and the effectiveness of the proposed method was compared to that of EP and GA. (AlRashidi & El-Hawary, 2007) developed a hybrid PSO technique that preserved only feasible solutions to solve the discrete OPF. They considered different objectives in their formulation such as emission, real power losses, and the non-convex fuel cost that incorporated the valve point loading effects. A dynamic version of the BFO technique was developed by (Tang et al., 2006) to minimize the quadratic fuel cost function. The OPF was solved considering a dynamic loading environment using the traditional control variables.

Das & Patvardhan (2003) proposed a hybrid evolutionary approach to solve a multi-objective optimization problem with four different objective functions, as fuel cost, emissions, transmission line real power losses, and the security margin index. In this paper GA was combined with SA to form a hybrid solution tool with enhanced capability of handling both discrete and continuous control variables. The shape of the Pareto front was detected which determined the best comprising solution of the multi-objective OPF problem. Abido (2004) presented a strength Pareto evolutionary algorithm to solve the multi-objective OPF problem. In his formulation, Abido considered two objective functions, namely the fuel cost and the voltage stability indicator.

An entire set of Pareto optimal solutions was calculated and the Pareto shape was captured. Ramesh & Li (1997) modeled the OPF problem as a multi-objective optimization problem with the conflicting objectives being the pre-contingency operating cost and post-contingency correction time. The two functions were formulated as fuzzy objectives subject to hard constraints and sequential quadratic programming (SQP) was used to solve the problem.

MATHEMATICAL PROBLEM FORMULATION

The objective of OPF is to minimize the objective function of single objective, multi-objective OPF and Objective functions of ORPD while satisfying all the equality and inequality constraints such as generator constraints, load bus constraints, transmission line constraints, transformer tap constraints, and shunt compensator constraints, have already been discussed in the sections "Motivation of the Present Work" and "Mathematical Problem Formulation" of chapter 3.

EVOLUTIONARY TECHNIQUES APPLIED TO OPF AND ORPD

BBO Technique Applied to OPF and ORPD

In this chapter, BBO technique has been employed to solve constrained OPF and ORPD problem and to find optimal solutions satisfying both equality and inequality constraints. The algorithm of the proposed method is as enumerated below:

Step 1: Initialization of BBO parameters

Step 2: The initial position of *SIV* i.e. independent variables (active powers of all generators except slack bus, generators' voltages, tap settings of regulating transformers, reactive power injections for OPF and generators' voltages, tap settings of regulating transformers, reactive power injections for ORPD) of each habitat should be randomly selected while satisfying different equality and inequality constraints. Several numbers of habitats depending upon the population size are being generated. Each habitat represents a potential solution to the given problem. SIV^{iq} represents the value of the output of the qth independent variable of the ith habitat set H^i. In this work, each habitat is a vector with m independent variables. Each individual habitat within the total of H habitat represents a candidate solution of the problem. For example, if there are m independent variables, then, the ith individual habitat H^i can be defined as follows:

$$H^i = SIV^{iq} = [SIV^{i1}, SIV^{i2}, ..., SIV^{im}]\; i = 1,2,...,S; q = 1,2,...m \tag{1}$$

The matrix represents the total population of habitat set.

Step 3: Perform load flow using Newton-Raphson method and determine all dependent variables such as load voltages, active power of the slack bus, generators' reactive powers etc.
Step 4: Calculate the *HSI* i.e. value of the objective function for each habitat of the population set for given emigration rate μ, immigration rate, λ and species, S.
Step 5: Based on the *HSI* value elite habitats are identified.
Step 6: Each non-elite habitat is modified by performing probabilistically migration operation and *HSI* of each modified set is recalculated. Feasibility of a solution is verified i.e. each *SIV* should satisfy equality and inequality constraints as mentioned in the specific problem.
Step 7: Species count probability of each habitat is updated using equation (217) of chapter 2. Mutation operation is performed on the non-elite habitat and *HSI* value of each new habitat is computed.
Step 8: Feasibility of a problem solution is verified i.e. each *SIV* should satisfy equality and inequality constraints.
Step 9: Go to step 3 for the next iteration.
Step 10: Stop iterations after a predefined number of iterations.

TLBO Algorithm Applied to OPF and ORPD

The main steps of the proposed TLBO approach applied to OPF problem are listed as follows.

Step 1: Depending upon the population size, several numbers of initial solutions are initialized randomly within their specified operating limits using (17) - (23).

Step 2: The independent variables of OPF problem (active power of all generator buses except slack bus, voltage of generator buses, reactive power injections for OPF and generators' voltages, tap settings of regulating transformers, reactive power injections for ORPD) of each initial solution string are updated using the following expression to generate oppositional population.

Step 3: The dependent variables of the problem are evaluated using load flow analysis.

Step 4: The fitness value of the current population (P) is evaluated.

Step 5: n_p number of fittest individuals are selected from $\{P \cup OP\}$.

Step 6: Sort the requisite number of fittest vectors are ed from best to worst.

Step 7: Few elite solutions are selected based on the fitness values.

Step 8: The independent variables (active power of all generator buses except the slack bus, the voltage of generator buses, reactive power injection, tap settings of transformers,) of each non-elite solution are modified by performing teaching and learning phase of TLBO algorithm using (16) - (18) of chapter 2.

Step 9: The dependent variables as mentioned above of based OPF are evaluated. Afterward, the fitness value of the population set is calculated.

Step 10: Based on a jumping rate j_r (i.e. jumping probability), the opposite population of the current population is evaluated and fitness values of the opposite population are calculated.

Step 11: n_p fittest individuals are selected from the union of the current population and the opposite population.

Step 12: Go to Step 8 for the next iteration until the termination criterion is made.

HCRO Algorithm Applied to OPF and ORPD

The HCRO algorithm is applied to seek out the optimal or near-optimal solutions for the OPF and ORPD problem. Its computational procedure can be illustrated using the following steps:

Step 1: Confine the search space, i.e., specify the lower and upper limits of each decision variable such as active powers of all generators except slack bus, generators' voltages, tap settings of regulating transformers, reactive power injections for OPF and generators' voltages, tap settings of regulating transformers, reactive power injections for ORPD.

Step 2: Initialize the input parameters of the proposed HCRO approach.

Step 3: Run the load flow to find the slack bus power, load voltages, generators' reactive powers, etc. In this study, load flow is performed by Newton-Raphson method. Each feasible solution represents the molecular structure of each molecule.

Step 4: Evaluate the fitness value. Based on the objective value, sort the molecules and few are kept as elite molecules.

Step 5: Assign potential energy same as fitness value for each molecule of the population set.

Step 6: Apply on-wall ineffective collision operations, decomposition operation, inter-molecular ineffective collision and synthesis collision operation on non-elite molecules as described in Steps 4, 5, and 6 in the section titled "HCRO Algorithm Applied to ELD" (in chapter 3) to modify the independent variables i.e generators' active power, the voltage of generator buses, reactive power injection, tap settings of transformers.

Step 7: To check the feasibility, run load flow and find the dependent variables. If any dependent variable violates the operating limit, replace that solution set by the new feasible solution.

Step 8: Compute fitness value for each newly generated feasible solution set.

Step 9: If the stopping criteria is satisfied, go to Step 10. Otherwise, go to Step 4.

Step 10: Output optimal settings of the control variables and the corresponding cost.

GWO Algorithm Applied to OPF and ORPD

Step 1: Read line data, bus data, fuel cost co-efficient, load flow parameters, etc. of the power system and specify the upper and lower limits of each control variables of OPF like active power generation, generators' bus voltage, load bus voltages, reactive power generations, tap settings of the tap changing transformers, VAr injection of the shunt compensating devices and line flow through each transmission lines. Afterward, randomly select the independent control variables within their operating limits. Thereafter, Newton Raphson based load flow method is run to evaluate the slack bus power and the other dependent variables. Check whether the control variables are within the limits or not. If any of them violates the limits, the corresponding set is discarded and initialise again. In this fashion, a feasible initial solution is generated. Now several initial solution set depending upon the population size are generated. A feasible solution set (control variables) represents the position of different GWO individuals.

Step 2: The input parameters such as a, A and C of GWO algorithm is initialized.

Step 3: Evaluate the fitness function of the initial solution set and elite solutions are identified.

Step 4: Find the updated best position of alpha, beta and delta wolves and independent variables of OPF are modified. Thereafter, Newton Raphson based load flow is run to obtain the dependent variables. The feasibility of the controlled variables is checked. If any one of them does not satisfy the inequality constraints, the particular set is replaced by the best feasible solution set.

Step 5: Evaluate the fitness value of the feasible population set.

Step 6: Check if the number of iteration is less than the maximum specified iteration or not.

Step 7: When the maximum iteration is reached display the fitness value and the updated final position of the wolves denoted by X(t+1) from the prey which is generally a circle around the prey.

Step 8: If the maximum iteration is not reached, iteration cycle is incremented.

Step 9: Go to Step 4 until a stopping criterion is satisfied.

KHA Algorithm Applied to OPF and ORPD

The procedure for implementing the KHA algorithm in solving optimal power flow problem to get minimum cost may be summarized by the following steps:

Step 1: All independent variables such as all generators' active power except the slack bus, generators' voltages, tap settings of regulating transformers, should be randomly selected between their operating limits

Step 2: Run the Newton-Raphson load to determine the slack bus power, load voltages, power flow through the transmission line, and check whether they satisfy the operating limits or not. If any of these parameters violate the operating limits; discard that population set and re-initialize the corresponding set. Depending upon the population size, several solutions are generated. Each feasible solution set represents the initial position of each krill individuals.

Step 3: Evaluate the fitness value and sort the solutions from best to worst.

Step 4: Select few best solutions for elitism.

Step 5: Evaluate the three motion index, namely, motion induced by other individual; foraging motion and random diffusion.

Step 6: The non-elite krill individuals' position is modified.

Step 7: Applying crossover and mutation to update the position of each non-elite krill individuals.

Step 8: Check whether the independent variables violate the operating limits or not. If any independent variable is less than the minimum level it is made equal to the minimum value and if it is greater than the maximum level it is made equal to maximum level.

Step 9: Run Newton-Raphson based load flow to determine the dependent variables and check whether they satisfy the system operating constraints or not. The infeasible solutions are replaced by the best feasible solutions.

Step 10: Go to Step 4 until the current iteration number reaches the pre-specified maximum iteration number.

INPUT PARAMETERS FOR DIFFERENT ALGORITHM

To successfully implement of BBO and CRO the values of the parameters, such as Habitat modification probability (P_H) = 1, mutation probability (P_m) = 0.005, maximum immigration rate (I_m) = 1, maximum emigration rate (E_m) = 1, elitism parameter (E_P) = 4, step size for numerical integration (*dt*) = 1, etc. have to be determined. Different trials have been performed for the IEEE 30 bus test systems.

BBO

The performance of the BBO algorithm mainly depends on the three parameters P_m, I_m and E_P. The average value of the fuel cost of generating (FC) over 25 different trials of OPF without FACTs of the IEEE 30 bus system for different values of P_m, I_m and E_P are listed in Table 1, and it is found that the best setting of these input since the performance of the algorithms depends on input parameters, they should be carefully chosen. After several runs, the the following input control parameters are found to be best for optimal performance of the proposed algorithm.

Table 1. Effect of control parameters on BBO performance

Input parameters			Objective function (FC)	Input parameters			Objective function(FC)
E_m	I_m	P_m		E_m	I_m	P_m	
1	1	0.0500	822.1201	0.75	1	0.0500	822.6201
1	1	0.0250	822.0120	0.75	1	0.0250	822.4120
1	1	0.0010	821.6121	0.75	1	0.0010	821.8121
1	1	0.0075	821.5612	0.75	1	0.0075	821.9612
1	1	0.0050	821.4153	0.75	1	0.0050	821.7153
1	1	0.0025	821.4954	0.75	1	0.0025	821.6154
1	0.75	0.0500	821.7001	0.75	0.75	0.0500	821.5601
1	0.75	0.0250	821.7821	0.75	0.75	0.0250	821.8921
1	0.75	0.0010	821.9851	0.75	0.75	0.0010	821.7751
1	0.75	0.0075	822.2315	0.75	0.75	0.0075	822.0015
1	0.75	0.0050	822.560	0.75	0.75	0.0050	822.0202
1	0.75	0.0025	821.985	0.75	0.75	0.0025	821.897

CRO

For CRO, the average value of the transmission loss over 25 different trials of IEEE 30-bus system without FACTs for different values of $\mathbf{KE}_{loss\_rate}$ and $\mathbf{KE}_{initial}$ is listed in Table 2. It is clearly observed from Table 2 that the optimal settings of these input parameters for the optimal performance of the proposed CRO algorithm are as follows: $\mathbf{KE}_{loss\_rate} = 0.2$; $\mathbf{KE}_{initial}$ for each molecule = 10,000.

SIMULATION RESULTS AND DISCUSSIONS

IEEE 30 Bus System

The applicability and validity of the different techniques are tested on IEEE 30-bus system for OPF and ORPD problems. IEEE 30-bus system contains 6 generating units connected at bus number. 1, 2, 5, 8, 11 and 13; 30 buses, 4 regulating transformers connected between the line numbers 6-9, 6-10, 4-12 and 27-28; 2 shunt compensators connected at bus number 10 and 24; 41 transmission lines. The system bus data and line data, the generators'

Table 2. Effect of KE loss rate and KE initial on CRO performance

Input parameters		Objective function(FC)	Input parameters		Objective function(FC)
KE_{loss_rate}	$KE_{initial}$		KE_{loss_rate}	$KE_{initial}$	
0.05	5000	820.7798	0.20	5000	820.9852
0.05	7500	821.2301	0.20	7500	821.5621
0.05	10000	821.5621	0.20	10000	821.2310
0.05	12500	821.5623	0.20	12500	821.0122
0.05	15000	820.8956	0.20	15000	820.9985
0.10	5000	820.9998	0.25	5000	820.8912
0.10	7500	820.7895	0.25	7500	820.6833
0.10	10000	820.9856	0.25	10000	820.8956
0.10	12500	821.5421	0.25	12500	821.5623
0.10	15000	821.2315	0.30	15000	821.5677
0.15	5000	820.9856	0.30	5000	820.8822
0.15	7500	820.7784	0.30	7500	820.9809
0.15	10000	821.3322	0.30	10000	821.2301
0.15	12500	821.4458	0.30	12500	821.5623
0.15	15000	821.8956	0.30	15000	821.7654

capacities and cost coefficient data including valve point discontinuities are taken form (Basu, 2008) and are also shown in Tables 12-14 illustrated in Appendix. The generators' active powers (except slack bus-1) and generators' voltages, tap settings of the regulating transformers, VAr injections of the shunt capacitors, are considered as controlled variables. The voltage magnitudes of the generator and load buses are considered within the range of [0.95 *p.u.,* 1.1 *p.u.*] and [0.95 *p.u.,* 1.05 *p.u.*], respectively (Basu, 2008). Tap settings of regulating transformers are within the range of [$0.9p.u., 1.1p.u.$] (Roy et al., 2011). The VAr injections of the shunt capacitors are taken within the interval of [0 MVAr, 30 MVAr] (Roy et al.,2011). The limits of transmission line loading are adopted from (Basu, 2008). The reactive power generation limits of different generator buses are taken from (Roy et al., 2011). The base values of apparent power and voltage used for this test system are taken as 100 MVA and 11 KV, respectively. All the programs are developed in MATLAB language, using MATLAB 7.1 and the system configuration is on core 2 duo processor, 2.0 GHz with 1GB RAM. The feasibility of the proposed BBO, TLBO, KHA, GWO, CRO, and HCRO methods are tested on modified IEEE 30-bus system in seven different cases (incorporating single and multi-objective). The simulation results shown in different Tables are the best solutions obtained in fifty runs.

Simulation Results of Single Objective OPF Problems

Minimization of Fuel Cost

The objective is to minimize the total fuel cost having quadratic cost function and with valve-point discontinuities. The optimal settings of control variables, fuel cost and loss obtained by BBO, TLBO, CRO, KHA, GWO and HCRO are shown in Table 3 and Table 4. The comparative simulation results show the validity and superiority of the HCRO method in terms of the least fuel cost over other. The convergence characteristics of the total fuel cost having quadratic cost function are plotted for all techniques in Figure 1.

Minimization of Voltage Deviation

The simulation results of control variables, the value of the objective function (OF), transmission loss, summation of voltage deviations (SVD), computational time per iteration (CT) for objective function of minimization of voltage deviation are given in Table 5. The results show that SVD (objective function) found by the HCRO method is lower ($0.1251 p.u.$) than other techniques.

Minimization of Transmission Loss

The optimal settings of control variables that are obtained by the different methods for transmission loss minimization objective function are shown in Table 6. The objective value, total transmission loss and sum of voltage deviations that are obtained from different techniques are also shown in Table 6. The results clearly show that the transmission losses obtained by all methods significantly improved in this case. The simulation results also show that HCRO method gives lower transmission loss (2.9784MW) than others. Figure 2 gives the convergence profile of transmission loss for all the techniques.

Table 3. Simulation results using BBO, TLBO, CRO, KHA, GWO and HCRO for fuel cost minimization having quadratic cost function

Control variables	HCRO	GWO	KHA	CRO	TLBO	BBO
$P_{g1(1)} \cdot P_{g1(1)}$ .u.)	0.6111	0.6082	0.6356	0.5975	0.6291	0.6128
$P_{g2(2)} \cdot P_{g2(2)}$ .u.)	0.7802	0.7814	0.7643	0.7858	0.7819	0.7917
$P_{g3(5)} \cdot P_{g3(5)}$ .u.)	0.3466	0.3439	0.3346	0.3623	0.3304	0.3578
$P_{g4(8)} \cdot P_{g4(8)}$ .u.)	0.2919	0.2962	0.2952	0.2814	0.2879	0.2982
$P_{g5(11)} \cdot P_{g5(11)}$ .u.)	0.2995	0.2946	0.3000	0.2983	0.2992	0.2928
$P_{g6(13)} \cdot P_{g6(13)}$ .u.)	0.5431	0.5488	0.5456	0.5496	0.5482	0.5213
$V_{g1(1)} \cdot V_{g1(1)}$ .u.)	1.0711	1.0572	1.0377	1.0126	1.0207	1.0321
$V_{g2(2)} \cdot V_{g2(2)}$ .u.)	1.0636	1.0533	1.0298	1.0077	1.0150	1.0242
$V_{g3(5)} \cdot V_{g3(5)}$ .u.)	1.0417	1.0317	1.0090	0.9881	0.9882	1.0038
$V_{g4(8)} \cdot V_{g4(8)}$ .u.)	1.0532	1.0403	1.0184	0.9956	1.0041	1.0153
$V_{g5(11)} \cdot V_{g5(11)}$ .u.)	1.0997	1.0952	1.0999	1.0934	1.0769	1.0871
$V_{g6(13)} \cdot V_{g6(13)}$ (u.)	1.0987	1.0983	1.0988	1.0995	1.0867	1.0881
$TAP_{1(6-9)} \cdot TAP_{1(6-9)}$ .u.)	1.0256	1.0218	0.9921	0.9677	1.0057	0.9187
$TAP_{2(6-10)} \cdot TAP_{2(6-10)}$ .u.)	0.9329	0.9000	0.9000	0.9000	0.9000	1.0860

continued on following page

Table 3. Continued

Control variables	HCRO	GWO	KHA	CRO	TLBO	BBO
$TAP_{3(4-12)}$. $TAP_{3(4-12)}$ .u.)	0.9387	0.9208	0.9314	0.9000	0.9231	0.9302
$TAP_{4(28-27)}$. $TAP_{4(28-27)}$ .u.)	0.9499	0.9247	0.9000	0.9000	0.9258	0.9376
$Q_{C1(10)}$. $Q_{C1(10)}$ .u.)	0.2700	0.2812	0.1932	0.2818	0.2783	0.2994
$Q_{C2(24)}$. $Q_{C2(24)}$ .u.)	0.1004	0.1068	0.1173	0.1089	0.1267	0.1077
Cost ($/h)	969.8067	969.9656	970.8972	971.1596	971.1690	972.0023
Power Loss (MW)	3.8414	3.9085	4.1266	4.0906	4.2825	4.0642
Voltage Deviation (pu)	1.7295	1.7964	1.5199	1.4277	1.1209	1.1062

Figure 1. Convergence characteristic of fuel cost minimization having quadratic cost function using BBO, TLBO, CRO, KHA, GWO and HCRO

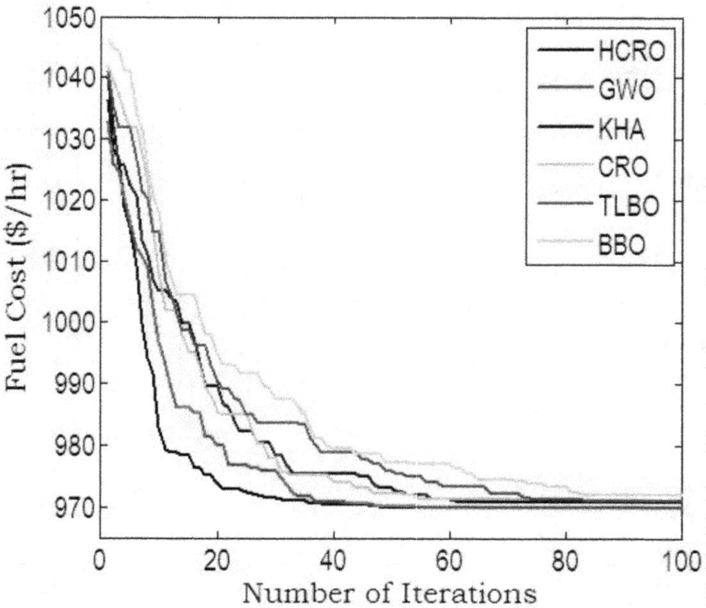

Table 4. Comparative Simulation results obtained using BBO, TLBO, CRO, KHA, GWO and HCRO for fuel cost minimization with valve point discontinuities

Control Variables	HCRO	GWO	KHA	CRO	TLBO	BBO
$P_{g1(1)}$ · $P_{g1(1)}$ .u.)	0.9418	0.7852	0.7834	0.7854	0.9417	0.9425
$P_{g2(2)}$ · $P_{g2(2)}$ .u.)	0.7093	0.7138	0.7176	0.7072	0.7088	0.7147
$P_{g3(5)}$ · $P_{g3(5)}$ .u.)	0.2678	0.3650	0.2725	0.3603	0.2692	0.2741
$P_{g4(8)}$ · $P_{g4(8)}$ .u.)	0.2229	0.3707	0.4475	0.4477	0.3747	0.3727
$P_{g5(11)}$ · $P_{g5(11)}$ .u.)	0.2719	0.2684	0.2684	0.1774	0.0874	0.2668
$P_{g6(13)}$ · $P_{g6(13)}$ .u.)	0.5223	0.4150	0.4226	0.4107	0.5306	0.3203
$V_{g1(1)}$ · $V_{g1(1)}$ .u.)	0.9997	1.0961	1.0358	1.0075	1.0416	1.0744
$V_{g2(2)}$ · $V_{g2(2)}$ .u.)	1.0655	1.0452	1.0222	1.0222	1.0521	1.0588
$V_{g3(5)}$ · $V_{g3(5)}$ .u.)	1.0842	0.9755	1.0885	0.9994	1.0415	1.0704
$V_{g4(8)}$ · $V_{g4(8)}$ .u.)	1.0309	0.9572	1.0484	0.9838	0.9805	1.0508
$V_{g5(11)}$ · $V_{g5(11)}$ .u.)	1.0119	1.0421	1.0010	0.9900	1.0751	0.9914
$V_{g6(13)}$ · $V_{g6(13)}$ (u.)	1.0618	1.0426	1.0261	1.0276	0.9549	1.0300
$TAP_{1(6-9)}$ · $TAP_{1(6-9)}$ .u.)	1.0511	1.0075	0.9000	1.0288	0.9151	1.0191
$TAP_{2(6-10)}$ · $TAP_{2(6-10)}$ .u.)	1.0045	1.0694	0.9751	1.0332	0.9963	0.9630

continued on following page

Table 4. Continued

Control Variables	HCRO	GWO	KHA	CRO	TLBO	BBO
$TAP_{3(4-12)}$ · $TAP_{3(4-12)}$ .u.)	1.0172	1.0927	0.9000	1.0278	0.9408	1.0914
$TAP_{4(28-27)}$ · $TAP_{4(28-27)}$ .u.)	0.9661	1.0886	1.0957	1.0379	0.9074	0.9310
$Q_{C1(10)}$ $Q_{C1(10)}$ .u.)	0.1794	0.0352	0.1579	0.0814	0.1125	0.1026
$Q_{C2(24)}$ $Q_{C2(24)}$ .u.)	0.1799	0.0337	0.0253	0.0462	0.0426	0.0327
Cost ($/h)	1088.7997	1089.9270	1091.8996	1095.6236	1095.6205	1096.8482
Power Loss (MW)	10.1924	8.3992	7.8064	5.4581	7.8525	5.7116
Voltage Deviation(pu)	0.4740	1.4865	0.8142	1.2259	0.3075	0.5485

Table 5. Comparative Simulation results obtained using BBO, TLBO, CRO, KHA, GWO and HCRO for minimization of voltage deviation

Control Variables	HCRO	GWO	KHA	CRO	TLBO	BBO
$P_{g1(1)}$ · $P_{g1(1)}$ .u.)	0.9140	1.9732	1.3507	1.5964	1.5603	2.4570
$P_{g2(2)}$ · $P_{g2(2)}$ .u.)	0.7666	0.7185	0.5145	0.7828	0.0561	0.1003
$P_{g3(5)}$ · $P_{g3(5)}$ .u.)	0.3693	0.0149	0.2366	0.1407	0.3589	0.0643
$P_{g4(8)}$ · $P_{g4(8)}$ .u.)	0.3051	0.1391	0.4963	0.4706	0.4194	0.1480
$P_{g5(11)}$ · $P_{g5(11)}$ .u.)	0.0620	0.0405	0.0153	0.0167	0.2373	0.0115
$P_{g6(13)}$ · $P_{g6(13)}$ .u.)	0.4853	0.1161	0.3621	0.0123	0.2805	0.2788
$V_{g1(1)}$ · $V_{g1(1)}$ .u.)	0.9963	1.0155	0.9712	0.9712	1.0560	0.9821

continued on following page

Table 5. Continued

Control Variables	HCRO	GWO	KHA	CRO	TLBO	BBO
$V_{g2(2)} \cdot V_{g2(2)}$ .u.)	0.9825	0.9643	1.0563	1.0675	1.0034	1.0579
$V_{g3(5)} \cdot V_{g3(5)}$ .u.)	1.0162	1.0120	1.0154	1.0168	1.0188	1.0106
$V_{g4(8)} \cdot V_{g4(8)}$ .u.)	1.0086	1.0152	1.0031	0.9899	1.0062	1.0086
$V_{g5(11)} \cdot V_{g5(11)}$ .u.)	1.0464	1.0657	0.9722	1.0291	1.0022	0.9812
$V_{g6(13)} \cdot V_{g6(13)}$ (u.)	1.0432	1.0594	1.0296	1.0514	0.9957	0.9957
$TAP_{1(6-9)} \cdot TAP_{1(6-9)}$ .u.)	1.0664	1.0836	0.9961	1.0474	1.0153	1.0110
$TAP_{2(6-10)} \cdot TAP_{2(6-10)}$ .u.)	0.9499	0.9633	0.9194	0.9263	0.9299	0.9965
$TAP_{3(4-12)} \cdot TAP_{3(4-12)}$ .u.)	1.0232	1.0288	0.9976	1.0310	0.9271	0.9440
$TAP_{4(28-27)} \cdot TAP_{4(28-27)}$ .u.)	0.9571	0.9534	0.9535	0.9619	0.9521	0.9636
$Q_{C1(10)} . Q_{C1(10)}$ .u.)	0.1966	0.2120	0.1940	0.0934	0.1660	0.2850
$Q_{C2(24)} . Q_{C2(24)}$ .u.)	0.1945	0.1234	0.1404	0.2180	0.1382	0.2189
Cost ($/h)	1725.7718	2180.3075	2188.4050	2122.4834	1942.5813	2630.3732
Power Loss (MW)	6.8375	16.8485	14.1460	18.5737	7.8533	22.5891
Voltage Deviation (pu)	0.1251	0.1288	0.1323	0.1368	0.1372	0.1380

Figure 2. Convergence characteristic of Transmission loss using BBO, TLBO, CRO, KHA, GWO and HCRO

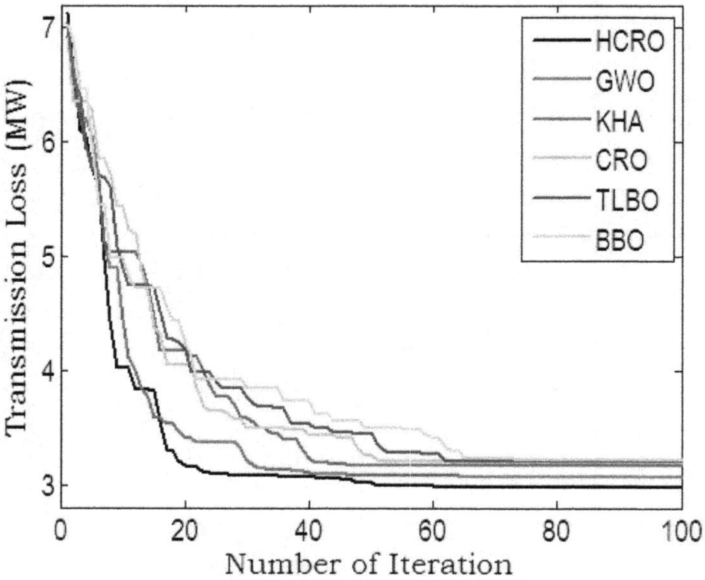

Table 6. Comparative Simulation results obtained using BBO, TLBO, CRO, KHA, GWO and HCRO for minimization of transmission loss

Control Variables	HCRO	GWO	KHA	CRO	TLBO	BBO
$P_{g1(1)} \cdot P_{g1(1)}$ .u.)	0.3238	0.3784	0.3414	0.3593	0.3984	0.4121
$P_{g2(2)} \cdot P_{g2(2)}$ .u.)	0.7978	0.7430	0.7929	0.7920	0.7692	0.7738
$P_{g3(5)} \cdot P_{g3(5)}$ .u.)	0.3996	0.3989	0.3937	0.3911	0.3942	0.3943
$P_{g4(8)} \cdot P_{g4(8)}$ .u.)	0.4976	0.4988	0.4958	0.4870	0.4846	0.4897
$P_{g5(11)} \cdot P_{g5(11)}$ .u.)	0.2987	0.2970	0.2978	0.2990	0.2704	0.2754
$P_{g6(13)} \cdot P_{g6(13)}$ .u.)	0.5462	0.5485	0.5439	0.5375	0.5493	0.5209
$V_{g1(1)} \cdot V_{g1(1)}$ .u.)	1.0585	1.0477	1.0249	1.0275	1.0681	1.0993

continued on following page

Table 6. Continued

Control Variables	HCRO	GWO	KHA	CRO	TLBO	BBO
$V_{g2(2)} \cdot V_{g2(2)}$ .u.)	1.0568	1.0448	1.0214	1.0255	1.0676	1.0892
$V_{g3(5)} \cdot V_{g3(5)}$ .u.)	1.0400	1.0226	0.9963	1.0043	1.0364	1.0794
$V_{g4(8)} \cdot V_{g4(8)}$ .u.)	1.0522	1.0375	1.0145	1.0166	1.0525	1.0882
$V_{g5(11)} \cdot V_{g5(11)}$ .u.)	1.0970	1.0938	1.0893	1.0866	1.0768	1.0349
$V_{g6(13)} \cdot V_{g6(13)}$ (u.)	1.1000	1.0973	1.0991	1.0966	1.1000	1.0491
$TAP_{1(6-9)} \cdot$ $TAP_{1(6-9)}$ .u.)	1.0289	1.0206	0.9489	0.9606	1.0251	1.0704
$TAP_{2(6-10)} \cdot$ $TAP_{2(6-10)}$ .u.)	0.9058	0.9021	0.9407	0.9531	0.9788	1.0545
$TAP_{3(4-12)} \cdot$ $TAP_{3(4-12)}$ .u.)	0.9373	0.9302	0.9184	0.9366	1.0110	1.0713
$TAP_{4(28-27)}$. $TAP_{4(28-27)}$ .u.)	0.9366	0.9165	0.9122	0.9162	0.9766	1.0210
$Q_{C1(10)}$. $Q_{C1(10)}$ .u.)	0.2810	0.2870	0.2698	0.2728	0.1899	0.1642
$Q_{C2(24)}$. $Q_{C2(24)}$ .u.)	0.1026	0.1128	0.1100	0.1021	0.1096	0.1039
Cost ($/h)	1830.4486	1976.5764	1916.7775	1985.9286	1906.2227	1865.0757
Power Loss (MW)	2.9784	3.0719	3.1661	3.1955	3.2066	3.2194
Voltage Deviation (pu)	1.8211	1.7404	1.4698	1.3595	1.1537	0.6079

Simulation Results of Multi-Objective OPF Problems

To test the effectiveness of BBO, TLBO, CRO, KHA, GWO and HCRO techniques, it is tested for multi-objective OPF problems on IEEE 30-bus system over fifty runs. The programming was written in MATLAB 7.1 on core 2 duo processor, 2.0 GHz, 1GB RAM.

Case 1 (Minimization of Fuel Cost and Improvement of Voltage Profile)

The objective in this case, is to minimize of voltage deviation together with fuel cost. The optimal settings of control variables, transmission loss, the summation of voltage deviation and fuel costs obtained by the BBO, TLBO, CRO, KHA, GWO and HCRO are shown in Table 7. The simulation results show that HCRO method has a better value of objective function than other methods.

Case 2 (Minimization of Fuel Cost and Transmission Loss)

The optimal settings of control variables, value of objective function, loss, fuel cost and summation of voltage deviation obtained by BBO, TLBO, CRO, KHA, GWO and HCRO for objective function of minimization of cost as well as transmission loss are given in Table 8. The simulation results show that HCRO outperforms in terms of solution quality (objective function) without violating any system operating constraints.

Case 3 (Minimization of Fuel Cost, Voltage Profile, and Transmission Loss)

The optimal settings of control variables, value of objective function, loss, fuel cost and summation of voltage deviation obtained by BBO, TLBO, CRO, KHA, GWO and HCRO for objective function of minimization of cost, voltage profile as well as transmission loss are given in Table 9. The simulation results show that HCRO outperforms in terms of solution quality (objective function) without violating any system operating constraints.

Table 7. Comparative Simulation results obtained using BBO, TLBO, CRO, KHA, GWO and HCRO for Minimization of fuel cost and improvement of voltage profile

Control Variables	HCRO	GWO	KHA	CRO	TLBO	BBO
$P_{g1(1)} \cdot P_{g1(1)}$ .u.)	0.7852	0.9425	0.9425	1.0995	0.7874	0.7854
$P_{g2(2)} \cdot P_{g2(2)}$ .u.)	0.7167	0.7106	0.7280	0.5740	0.7123	0.5804
$P_{g3(5)} \cdot P_{g3(5)}$ .u.)	0.2662	0.3590	0.1797	0.1796	0.3642	0.3631
$P_{g4(8)} \cdot P_{g4(8)}$ .u.)	0.4470	0.3722	0.2988	0.4488	0.3762	0.3729
$P_{g5(11)} \cdot P_{g5(11)}$ .u.)	0.2688	0.1795	0.2668	0.1817	0.1814	0.2673
$P_{g6(13)} \cdot P_{g6(13)}$ .u.)	0.4020	0.3329	0.5248	0.4292	0.5202	0.5449
$V_{g1(1)} \cdot V_{g1(1)}$ .u.)	1.0091	0.9911	1.0621	1.0581	1.0243	1.0498
$V_{g2(2)} \cdot V_{g2(2)}$ .u.)	0.9987	0.9778	0.9825	0.9980	1.0937	0.9807
$V_{g3(5)} \cdot V_{g3(5)}$ .u.)	1.0005	1.0048	1.0245	0.9992	0.9912	1.0028
$V_{g4(8)} \cdot V_{g4(8)}$ .u.)	1.0072	0.9909	1.0081	1.0013	0.9957	0.9939
$V_{g5(11)} \cdot V_{g5(11)}$ .u.)	0.9807	1.0385	0.9988	1.0591	0.9787	1.0748
$V_{g6(13)} \cdot V_{g6(13)}$ (u.)	1.0593	1.0623	1.0219	1.0407	1.0420	1.0085
$TAP_{1(6-9)} \cdot TAP_{1(6-9)}$ .u.)	0.9301	0.9490	0.9020	0.9457	1.0890	0.9661
$TAP_{2(6-10)} \cdot TAP_{2(6-10)}$ .u.)	1.0134	1.0998	0.9231	1.0425	1.0184	0.9054

continued on following page

Table 7. Continued

Control Variables	HCRO	GWO	KHA	CRO	TLBO	BBO
$TAP_{3(4-12)}$ · $TAP_{3(4-12)}$ .u.)	1.0686	1.0761	1.0513	0.9756	1.0425	1.0814
$TAP_{4(28-27)}$ · $TAP_{4(28-27)}$ .u.)	0.9493	0.9253	0.9222	0.9486	1.0664	0.9424
$Q_{C1(10)}$ · $Q_{C1(10)}$ .u.)	0.1857	0.2548	0.0240	0.1154	0.2963	0.0586
$Q_{C2(24)}$ · $Q_{C2(24)}$ .u.)	0.1465	0.1782	0.0989	0.0982	0.1749	0.2864
Cost ($/h)	1108.1653	1110.2788	1125.6478	1134.2244	1136.3586	1166.4202
Power Loss (MW)	5.1941	6.2629	10.6617	7.8872	10.7703	7.9935
Voltage Deviation (pu)	0.1719	0.2237	0.2510	0.2115	0.7627	0.3381

Table 8. Comparative Simulation results obtained using BBO, TLBO, CRO, KHA, GWO and HCRO for Minimization of fuel cost and transmission loss

Control Variables	HCRO	GWO	KHA	CRO	TLBO	BBO
$P_{g1(1)}$ · $P_{g1(1)}$ .u.)	0.9437	0.9411	0.6283	1.1024	0.7854	0.9443
$P_{g2(2)}$ · $P_{g2(2)}$ .u.)	0.7175	0.5736	0.7037	0.7149	0.6989	0.7104
$P_{g3(5)}$ · $P_{g3(5)}$ .u.)	0.3580	0.3561	0.3623	0.0011	0.2661	0.2702
$P_{g4(8)}$ · $P_{g4(8)}$ .u.)	0.2996	0.2270	0.3760	0.2978	0.4484	0.4481
$P_{g5(11)}$ · $P_{g5(11)}$ .u.)	0.0893	0.2678	0.2715	0.2714	0.2658	0.1802
$P_{g6(13)}$ · $P_{g6(13)}$ .u.)	0.5248	0.5303	0.5389	0.5232	0.4215	0.3344

continued on following page

Table 8. Continued

Control Variables	HCRO	GWO	KHA	CRO	TLBO	BBO
$V_{g1(1)} \cdot V_{g1(1)}$ .u.)	1.0802	1.0642	1.0138	1.0255	1.0529	1.0510
$V_{g2(2)} \cdot V_{g2(2)}$ .u.)	0.9927	1.0176	1.0011	1.0214	1.0204	1.0311
$V_{g3(5)} \cdot V_{g3(5)}$ .u.)	0.9995	1.0015	0.9986	0.9979	0.9997	1.0017
$V_{g4(8)} \cdot V_{g4(8)}$ .u.)	1.0217	1.0026	1.0206	1.0083	1.0001	1.0126
$V_{g5(11)} \cdot V_{g5(11)}$ .u.)	1.0391	0.9904	1.0519	1.0668	1.0815	1.0828
$V_{g6(13)} \cdot V_{g6(13)}$ (u.)	1.0322	1.0590	0.9952	1.0640	1.0255	1.0339
$TAP_{1(6-9)} \cdot TAP_{1(6-9)}$ .u.)	1.0764	0.9855	0.9593	1.0578	1.0921	0.9762
$TAP_{2(6-10)} \cdot TAP_{2(6-10)}$ .u.)	0.9380	1.0261	0.9575	0.9521	0.9457	1.0430
$TAP_{3(4-12)} \cdot TAP_{3(4-12)}$ .u.)	0.9926	1.0017	0.9611	1.0412	0.9821	0.9764
$TAP_{4(28-27)} \cdot TAP_{4(28-27)}$ .u.)	0.9376	0.9416	0.9335	0.9268	0.9375	0.9480
$Q_{C1(10)} \cdot Q_{C1(10)}$ .u.)	0.2261	0.0147	0.0295	0.0389	0.1575	0.0357
$Q_{C2(24)} \cdot Q_{C2(24)}$ .u.)	0.0695	0.0820	0.1122	0.0987	0.1077	0.1492
Cost ($/h)	1113.3380	1132.2988	1145.7108	1147.4855	1153.2874	1159.7496
Power Loss (MW)	9.8873	6.1835	4.6652	7.6817	5.2135	5.3762
Voltage Deviation (pu)	0.2156	0.3781	0.1910	0.2280	0.1813	0.2808

Table 9. Comparative Simulation results obtained using BBO, TLBO, CRO, KHA, GWO and HCRO for Minimization of fuel cost, voltage profile and transmission loss

Control Variables	HCRO	GWO	KHA	CRO	TLBO	BBO
$P_{g1(1)} \cdot P_{g1(1)}$ .u.)	0.6310	1.0996	0.7854	0.7886	0.7854	0.9444
$P_{g2(2)} \cdot P_{g2(2)}$ .u.)	0.7135	0.7111	0.7053	0.7107	0.7037	0.5681
$P_{g3(5)} \cdot P_{g3(5)}$ .u.)	0.3599	0.1792	0.2688	0.3582	0.3583	0.3603
$P_{g4(8)} \cdot P_{g4(8)}$ .u.)	0.3705	0.2241	0.3748	0.2360	0.2295	0.2233
$P_{g5(11)} \cdot P_{g5(11)}$ .u.)	0.2746	0.1813	0.2701	0.2688	0.2711	0.2788
$P_{g6(13)} \cdot P_{g6(13)}$ .u.)	0.5312	0.5241	0.5121	0.5161	0.5389	0.5213
$V_{g1(1)} \cdot V_{g1(1)}$ .u.)	1.0457	0.9982	1.0660	1.0613	1.0087	1.0498
$V_{g2(2)} \cdot V_{g2(2)}$ .u.)	1.0446	0.9721	1.0553	1.0525	1.0057	1.0199
$V_{g3(5)} \cdot V_{g3(5)}$ .u.)	1.0241	0.9813	0.9543	1.0229	0.9547	0.9823
$V_{g4(8)} \cdot V_{g4(8)}$ .u.)	1.0771	0.9949	1.0802	1.0285	0.9826	1.0413
$V_{g5(11)} \cdot V_{g5(11)}$ .u.)	1.0773	0.9712	1.0477	1.0735	0.9588	0.9644
$V_{g6(13)} \cdot V_{g6(13)}$ (u.)	1.0674	1.0870	0.9724	1.0970	1.0138	0.9534
$TAP_{1(6-9)} \cdot TAP_{1(6-9)}$ .u.)	0.9782	1.0595	0.9183	0.9689	1.0474	0.9708
$TAP_{2(6-10)} \cdot TAP_{2(6-10)}$ .u.)	1.0947	0.9610	0.9264	1.0355	0.9000	0.9955

continued on following page

Table 9. Continued

Control Variables	HCRO	GWO	KHA	CRO	TLBO	BBO
$TAP_{3(4-12)}$. $TAP_{3(4-12)}$ .u.)	0.9628	0.9000	0.9799	0.9865	1.0836	1.0074
$TAP_{3(4-12)}$. $TAP_{3(4-12)}$ .u.)	1.0487	0.9338	1.0510	0.9534	0.9778	0.9791
$TAP_{3(4-12)}$. $TAP_{3(4-12)}$ .u.)	0.2216	0.1300	0.2058	0.2939	0.2357	0.0968
$TAP_{3(4-12)}$. $TAP_{3(4-12)}$ .u.)	0.1323	0.2821	0.0622	0.0998	0.1808	0.1553
Cost ($/h)	1104.1490	1109.6987	1120.2385	1134.4375	1144.8037	1150.3411
Power Loss (MW)	4.6540	8.5397	8.2532	4.4388	5.2840	6.2301
Voltage Deviation (pu)	0.9077	0.4675	0.6320	1.1364	0.7714	0.6646

Simulation Results of ORPD Problems

To test the effectiveness of the proposed techniques, IEEE 30-bus system is tested for multi-objective ORPD problem

Test System (IEEE 30-Bus System)

This system has been described in this chapter in the section titled "IEEE 30 bus System." The feasibility of the proposed methods is tested on IEEE 30-bus system with two different objective functions given below.

Case 1 (Minimization of Voltage Deviation)

The objective function of minimization of voltage deviation is considered in this case. The optimal settings of control variables that are obtained by the different methods are shown in Table 10. The summation of voltage deviation (SVD) and total computational time that are found from different techniques are also shown in Table 10. It can be seen from Table 10 that SVDs $TAP_{3(4-12)}$ obtained by the proposed HCRO and GWO methods are 0.1648 p.u .and 0.1656 p.u, however with the same system data and control variable limits, SVD found by conventional CRO, KHA, TLBO and BBO are 0.1712 $p.u.$,0.1683 $p.u.$, 0.1764 $p.u.$, and 0.1862 $p.u.$ respectively. The results clearly show that HCRO and GWO outperform other methods in terms of solution quality. The convergence characteristics of the objective function are plotted in Figure 3 for BBO, TLBO, CRO, KHA, GWO and HCRO techniques.

Table 10. Comparative Simulation results obtained using BBO, TLBO, CRO, KHA, GWO and HCRO for Minimization of voltage profile

Control Variables	HCRO	GWO	KHA	CRO	TLBO	BBO
$TAP_{3(4-12)}$. $TAP_{3(4-12)}$ .u.)	0.9872	0.9840	0.9754	0.9946	0.9814	0.9937
$TAP_{3(4-12)}$. $TAP_{3(4-12)}$ .u.)	1.0002	1.0252	1.0112	0.9622	0.9993	0.9384
$TAP_{3(4-12)}$. $TAP_{3(4-12)}$ .u.)	1.0149	1.0118	1.0142	1.0089	1.0039	0.9987
$TAP_{3(4-12)}$. $TAP_{3(4-12)}$ .u.)	1.0123	1.0130	1.0097	1.0327	1.0277	1.0598
$TAP_{3(4-12)}$. $TAP_{3(4-12)}$ .u.)	1.0675	1.0418	1.0604	1.0712	1.0739	1.0246

continued on following page

Table 10. Continued

Control Variables	HCRO	GWO	KHA	CRO	TLBO	BBO
$TAP_{3(4-12)}$ · $TAP_{3(4-12)}$ (u.)	1.0613	1.0404	1.0737	1.0665	1.0451	1.0698
$TAP_{3(4-12)}$ · $TAP_{3(4-12)}$ .u.)	0.9959	0.9744	0.9937	1.0093	1.0158	0.9767
$TAP_{3(4-12)}$ · $TAP_{3(4-12)}$ .u.)	0.9516	0.9505	0.9504	0.9506	0.9523	0.9502
$TAP_{3(4-12)}$ · $TAP_{3(4-12)}$ .u.)	1.0195	0.9819	1.0465	1.0304	0.9912	1.0369
$TAP_{3(4-12)}$. $TAP_{3(4-12)}$ .u.)	0.9503	0.9507	0.9513	0.9550	0.9540	0.9648
$TAP_{3(4-12)}$. $TAP_{3(4-12)}$ .u.)	0.0447	0.0495	0.0496	0.0454	0.0489	0.0490
$TAP_{3(4-12)}$. $TAP_{3(4-12)}$ .u.)	0.0500	0.0500	0.0498	0.0499	0.0499	0.0497
Power Loss (MW)	6.5213	7.4160	7.3290	8.7510	7.0336	13.0777
Voltage Deviation (pu)	0.1648	0.1656	0.1683	0.1712	0.1764	0.1862

Figure 3. Convergence characteristic of voltage deviation using BBO, TLBO, CRO, KHA, GWO and HCRO

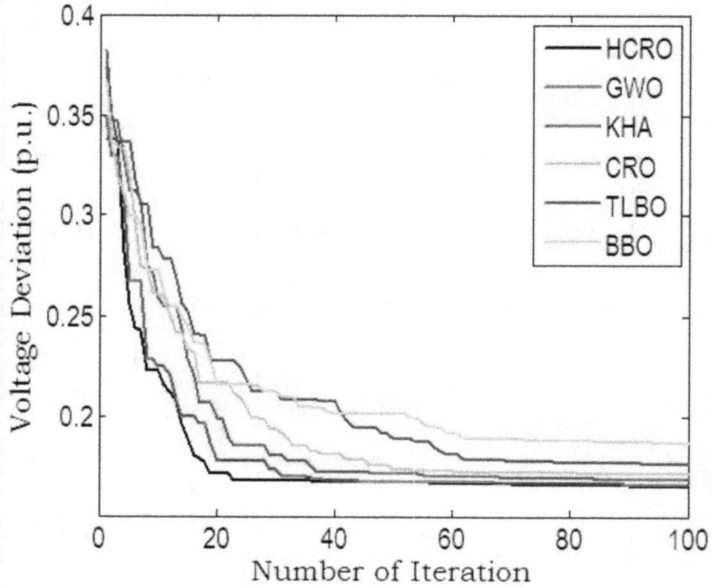

Case 2 (Minimization of Transmission Loss)

The objective in this case, is to minimize the total transmission loss. The optimal settings of control variables, transmission loss and computational time obtained by the different methods are shown in Table 11. The simulation results clearly show that HCRO and GWO methods depict lower transmission losses of 5.4342MW and 5.4893MW amongst all the techniques without violating any system constraints. The computational times required for BBO and TRPSO methods are very much comparable with those of other methods.

It is clear from Table and also Fig. that, the outcomes of the HCRO algorithm are too close to the global optimal solution. Therefore, the proposed algorithm of this chapter has small standard deviation for different independent runs and this means that the HCRO algorithm is stable and robust. In other word, most of times the proposed algorithm convergence to global optimal solution and the small deviation from desired results is reason-ably acceptable.

Table 11. Comparative Simulation results obtained using BBO, TLBO, CRO, KHA, GWO and HCRO for Minimization of transmission loss.

Control Variables	HCRO	GWO	KHA	CRO	TLBO	BBO
$TAP_{3(4-12)} \cdot TAP_{3(4-12)}$ .u.)	1.0319	1.0636	1.0115	1.0670	1.0068	1.0068
$TAP_{3(4-12)} \cdot TAP_{3(4-12)}$ .u.)	1.0234	1.0471	0.9993	1.0582	0.9988	0.9939
$TAP_{3(4-12)} \cdot TAP_{3(4-12)}$ .u.)	1.0078	1.0384	0.9805	1.0247	0.9748	0.9734
$TAP_{3(4-12)} \cdot TAP_{3(4-12)}$ .u.)	1.0037	1.0173	0.9838	1.0350	0.9772	0.9772
$TAP_{3(4-12)} \cdot TAP_{3(4-12)}$ .u.)	1.0602	1.0260	1.0999	1.0705	1.0999	1.0961
$TAP_{3(4-12)} \cdot TAP_{3(4-12)}$ (u.)	1.0787	1.0348	1.0684	0.9987	1.0657	1.0684
$TAP_{3(4-12)} \cdot TAP_{3(4-12)}$ .u.)	0.9526	1.0019	0.9524	0.9565	0.9524	0.9602
$TAP_{3(4-12)} \cdot TAP_{3(4-12)}$ .u.)	1.0215	0.9845	0.9506	0.9808	0.9506	0.9506
$TAP_{3(4-12)} \cdot TAP_{3(4-12)}$ .u.)	1.0114	1.0206	0.9749	0.9541	0.9779	0.9779
$TAP_{3(4-12)} \cdot TAP_{3(4-12)}$ .u.)	0.9784	1.0413	0.9538	1.0456	0.9538	0.9538
$TAP_{3(4-12)} \cdot TAP_{3(4-12)}$ .u.)	0.0046	0.0091	0.0488	0.0314	0.0488	0.0488
$TAP_{3(4-12)} \cdot TAP_{3(4-12)}$ .u.)	0.0474	0.0473	0.0499	0.0172	0.0499	0.0499
Power Loss (MW)	5.4342	5.4893	5.5434	5.5514	5.5758	5.6124
Voltage Deviation (pu)	0.2944	0.6968	0.3734	0.5325	0.3617	0.3546

CONCLUSION

In this chapter BBO, TLBO, CRO, KHA, GWO and HCRO techniques have been proposed, developed and successfully implemented to solve OPF and ORPD problems considering the valve point discontinuities of the generating units. The proposed methods have been tested on the modified IEEE 30-bus test system for single and multi-objective different cases with the statistical comparison. The simulation results demonstrate the effectiveness and robustness of the proposed techniques to solve OPF and ORPD problem. Hence, the results presented explicitly show the superiority of new techniques in order to achieve better performances of finding optimal solutions.

REFERENCES

Abido, M. A. (2002). Optimal power flow using particle swarm optimization. *International Journal of Electrical Power & Energy Systems*, *24*(7), 563–571. doi:10.1016/S0142-0615(01)00067-9

Abido, M. A. (2002). Optimal power flow using tabu search algorithm. *Electric Power Components and Systems*, *30*(5), 469–483. doi:10.1080/15325000252888425

Abido, M. A. (2004). Multi-objective optimal power flow using strength pareto evolutionary algorithm. *The 39th International Universities Power Engineering Conference*, 457–461.

AlRashidi, M. R., & El-Hawary, M. E. (2007). Hybrid particle swarm optimization approach for solving the discrete OPF problem considering the valve point effect. *IEEE Transactions on Power Systems*, *22*(4), 2030–2038. doi:10.1109/TPWRS.2007.907375

Basu, M. (2008). Optimal power flow with FACTS devices using differential evolution. *International Journal of Electrical Power & Energy Systems*, *30*(2), 150–156. doi:10.1016/j.ijepes.2007.06.011

Bechert, T., & Chen, N. (1997). Area automatic generation control by multi-pass dynamic programming. *IEEE Transactions on Power Apparatus and Systems*, *96*(5), 1460–1469. doi:10.1109/T-PAS.1977.32474

Burchett, B. T. (2013). Aerodynamic parameter identification for symmetric projectiles: An improved gradient based method. *Aerospace Science and Technology*, *30*(1), 119–127. doi:10.1016/j.ast.2013.07.010

Capitanescu, F., & Wehenkel, L. (2013). Experiments with the interior-point method for solving large scale optimal power flow problems. *Electric Power Systems Research*, *95*, 276–283. doi:10.1016/j.epsr.2012.10.001

Chan, K. Y., Ling, S. H., Chan, K. W., Iu, H. H. C., & Pong, G. T. Y. (2007). Solving multi-contingency transient stability constrained optimal power flow problems with an improved GA. *IEEE Congress on Evolutionary Computation*, 2901–2908. 10.1109/CEC.2007.4424840

Chaohua, D., Weirong, C., Yunfang, Z., & Xuexia, Z. (2009). Seeker optimization algorithm for optimal reactive power dispatch. *IEEE Transactions on Power Systems*, *24*(3), 1218–1231. doi:10.1109/TPWRS.2009.2021226

Das, D. B., & Patvardhan, C. (2003). Useful multi-objective hybrid evolutionary approach to optimal power flow. *IEE Proceedings. Generation, Transmission and Distribution*, *150*(3), 275–282. doi:10.1049/ip-gtd:20030188

Dommel, H. W., & Tinney, W. F. (1968). Optimal power flow solutions. *IEEE Transactions on Power Apparatus and Systems*, *87*(10), 1866–1876. doi:10.1109/TPAS.1968.292150

El-Fergany, A. A., & Hasanien, H. M. (2018). Tree-seed algorithm for solving optimal power flow problem in large-scale power systems incorporating validations and comparisons. *Applied Soft Computing*, *64*, 307–316. doi:10.1016/j.asoc.2017.12.026

Gaing, Z., & Chang, R. (2006). Security-constrained optimal power flow by mixed-integer genetic algorithm with arithmetic operators. *2006 IEEE Power Engineering Society General Meeting*, 8 10.1109/PES.2006.1709334

Gaing, Z., & Huang, H. S. (2004). Real-coded mixed integer genetic algorithm for constrained optimal power flow. *Proceedings of IEEE Region 10 Conference TENCON, 3*, 323-326.

Gaing, Z. L. (2005). Constrained optimal power flow by mixed-integer particle swarm optimization. *IEEE Power Engineering Society General Meeting, 1*, 243-250.

Granada, M., Rider, M. J., Mantovani, J. R. S., & Shahidehpour, M. A. (2012). Decentralized approach for optimal reactive power dispatch using a Lagrangian decomposition method. *Electric Power Systems Research, 89*, 148–156. doi:10.1016/j.epsr.2012.02.015

Granville, S. (1995). Optimal reactive power dispatch through interior point methods. *IEEE Transactions on Power Systems, 9*(1), 136–146. doi:10.1109/59.317548

Grudinin, N. (1998). Reactive power optimization using successive quadratic programming method. *IEEE Transactions on Power Systems, 13*(4), 1219–1225. doi:10.1109/59.736232

Habiabollahzadeh, H., Luo, X., & Semlyen, A. (1989). Hydrothermal optimal power flow based on a combined linear and non-linear programming methodology. *IEEE Transactions on Power Systems, 4*(2), 530–537. doi:10.1109/59.193826

Hadi, N., Mahdi, N., & Patrick, S. (2012). A multi objective gravitational search algorithm based on non-dominated sorting. *International Journal of Swarm Intelligence Research, 3*(3), 32–49. doi:10.4018/jsir.2012070103

Khazali, A. H., & Kalantar, M. (2011). Optimal reactive power dispatch based on harmony search algorithm. *International Journal of Electrical Power & Energy Systems, 33*(3), 684–692. doi:10.1016/j.ijepes.2010.11.018

Lai, L. L., & Ma, J. T. (1997). Application of evolutionary programming to reactive power planning-comparison with nonlinear programming approach. *IEEE Transactions on Power Systems, 12*(1), 198–206. doi:10.1109/59.574940

Lai, L. L., Ma, J. T., Yokoyama, R., & Zhao, M. (1997). Improved genetic algorithm for optimal power flow under both normal and contingent operation states. *International Journal of Electrical Power & Energy Systems, 19*(5), 287–292. doi:10.1016/S0142-0615(96)00051-8

Lo, K. L., & Zhu, S. P. (1991). A decoupled quadratic programming approach for optimal power dispatch. *Electric Power Systems Research, 22*(1), 4760. doi:10.1016/0378-7796(91)90079-3

Lobato, E., Rouco, L., Navarrete, M. I., Casanova, R., & Lopez, G. (2001). An LP-based optimal power flow for transmission losses and generator reactive margins minimization. *Proceedings of IEEE Porto Power Tech Conference*. 10.1109/PTC.2001.964894

Lu, F. C., & Hsu, Y. Y. (1995). Reactive power/voltage control in a distribution substation using dynamic programming. *IEE Proceedings. Generation, Transmission and Distribution, 142*(6), 639–645. doi:10.1049/ip-gtd:19952210

Ma, R., Wang, P., Yang, H., & Hu, G. (2005). Environmental/economic transaction planning using multiobjective particle swarm optimization and non-stationary multi-stage assignment penalty function. Proceedings of 2005 IEEE/PES Transmission and Distribution Conference and Exhibition: Asia and Pacific, 1–6.

Mao, Y., & Li, M. (2008). Optimal reactive power planning based on simulated annealing particle swarm algorithm considering static voltage stability. *Proceeding of the International Conference on Intelligent Computation Technology and Automation (ICICTA)*, 1, 106-110. 10.1109/ICICTA.2008.427

Mei, R. N. S., Sulaiman, M. H., Mustaffa, Z., & Daniyal, H. (2017). optimal reactive power dispatch. *Applied Soft Computing, 59*, 210–222.

Mirjalili, S. (2015). Moth-flame optimization algorithm: A novel nature-inspired heuristic paradigm. *Knowledge-Based Systems, 89*, 228–249. doi:10.1016/j.knosys.2015.07.006

Momoh, J. A. (1989). A generalized quadratic-based model for optimal power flow. *IEEE International Conference on Systems, Man and Cybernetics, 1*, 261-267. 10.1109/ICSMC.1989.71294

Mouassa, S., Bouktir, T., & Salhi, A., (2017). Ant lion optimizer for solving optimal reactive power dispatch problem in power systems. *Engineering Science and Technology, an International Journal, 20*(3), 885-895.

Nanda, J., Mishra, S., & Saikia, L. C. (2009). Maiden Application of Bacterial Foraging-Based Optimization Technique in Multiarea Automatic Generation Control. *IEEE Transactions on Power Systems, 24*(2), 602–609. doi:10.1109/TPWRS.2009.2016588

Nuaekaew, K., Artrit, P., Pholdee, N., & Bureerat, S. (2017). Optimal reactive power dispatch problem using a two-archive multi-objective grey wolf optimizer. *Expert Systems with Applications, 87*, 79–89. doi:10.1016/j.eswa.2017.06.009

Paranjothi, S. R., & Anburaja, K. (2002). Optimal power flow using refined genetic algorithm. *Electric Power Components and Systems, 30*(10), 1055–1063. doi:10.1080/15325000290085343

Pudjianto, D., Ahmed, S., & Strbac, G. (2002). Allocation of VAR support using LP and NLP based optimal power flows. *IEE Proceedings. Generation, Transmission and Distribution*, *149*(4), 377–383. doi:10.1049/ip-gtd:20020200

Quintana, V. H., & Santos-Nieto, M. (1989). Reactive-power dispatch by successive quadratic programming. *IEEE Transactions on Energy Conversion*, *4*(3), 425–435. doi:10.1109/60.43245

Ramesh, V. C., & Li, X. (1997). A fuzzy multi-objective approach to contingency constrained OPF. *IEEE Transactions on Power Systems*, *12*(3), 1348–1354. doi:10.1109/59.630480

Rao, R. V., Savsani, V. J., & Vakharia, D. P. (2011). Teaching–learning-based optimization: A novel method for constrained mechanical design optimization problems. *Computer Aided Design*, *43*(3), 303–315. doi:10.1016/j.cad.2010.12.015

Roa-Sepulveda, C. A., & Pavez-Lazo, B. J., (2001). A solution to the optimal power flow using simulated annealing. *IEEE Porto Power Tech Proceedings, 2*.

Roa-Sepulveda, C. A., & Pavez-Lazo, B. J. (2003). A solution to the optimal power flow using simulated annealing. *International Journal of Electrical Power & Energy Systems*, *25*(1), 47–57. doi:10.1016/S0142-0615(02)00020-0

Roy, P. K., Ghoshal, S. P., & Thakur, S. S. (2011). Optimal reactive power dispatch considering flexible AC transmission system devices using biogeography based optimization. *Electric Power Components and Systems*, *39*(8), 733–750. doi:10.1080/15325008.2010.541410

Roy, P. K., Ghoshal, S. P., & Thakur S. S. (2011). Optimal power flow using biogeography based optimization. *International Journal of Power and Energy Conversion*, *2*(3).

Roy, P. K., Mandal, B., & Bhattacharya, K. (2012). Gravitational search algorithm based optimal reactive power dispatch for voltage stability enhancement. *Electric Power Components and Systems*, *40*(9), 956–976. doi:10.1080/15325008.2012.675405

Santos, A., & Costa, G. R. D. (1995). Optimal power flow solution by Newton's method applied to an augmented Lagrangian function. *IEE Proceedings. Generation, Transmission and Distribution*, *142*(1), 33–36. doi:10.1049/ip-gtd:19951586

Shouzheng, W., Lixin, M., & Dashuai, S. (2010). Hybrid differential evolution particle swarm optimization algorithm for reactive power optimization. *Asia-Pacific Power and Energy Engineering Conference (APPEEC).*

Sivasubramani, S., & Swarup, K. S. (2011). Multi-objective harmony search algorithm for optimal power flow problem. *International Journal of Electrical Power & Energy Systems, 33*(3), 745–752. doi:10.1016/j.ijepes.2010.12.031

Sun, I., Ashley, B., Brewer, B., Hughes, A., & Tinney, W. F. (1984). Optimal power flow by Newton approach. *IEEE Transactions on Power Apparatus and Systems, 103*(10), 2864–2880. doi:10.1109/TPAS.1984.318284

Tang, W. J., Li, M. S., He, S., Wu, Q. H., & Saunders, J. R. (2006). Optimal power flow with dynamic loads using bacterial foraging algorithm. *Proceedings of the International Conference on Power System Technology, Powercon*, 1–5. 10.1109/ICPST.2006.321538

Tripathy, M., & Mishra, S. (2007). Bacteria foraging-based solution to optimize both real power loss and voltage stability limit. *IEEE Transactions on Power Systems, 22*(1), 240–248. doi:10.1109/TPWRS.2006.887968

Urdaneta, A. J., Gomez, J. F., Sorrentino, E., Flores, L., & Diaz, R. (1999). A hybrid genetic algorithm for optimal reactive power planning based upon successive linear programming. *IEEE Transactions on Power Systems, 14*(4), 1292–1298. doi:10.1109/59.801887

Wei, Y., Fang, L., Chung, C. Y., & Wong, K. P. (2006). A hybrid genetic algorithm interior point method for optimal reactive power flow. *IEEE Transactions on Power Systems, 21*(3), 1163–1169. doi:10.1109/TPWRS.2006.879262

Wennan, L., Yihua Xingtao, L. X., & Maojun, I. (2008). Reactive power optimization in area power grid based on improved Tabu search algorithm. *Proceedings of the Third International Conference on Electric Utility Deregulation and Restructuring and Power Technologies*, 1472–1477. 10.1109/DRPT.2008.4523637

Yan, W., Lu, S., & Yu, D. C. (2004). A novel optimal reactive power dispatch method based on an improved hybrid evolutionary programming technique. *IEEE Transactions on Power Systems, 19*(2), 913–918. doi:10.1109/TPWRS.2004.826716

Yan, X., & Quintana, V. H. (1999). Improving an interior point based OPF by dynamic adjustments of step sizes and tolerances. *IEEE Transactions on Power Systems*, *14*(2), 709–717. doi:10.1109/59.761902

Yu, D. C., Fagan, J. E., Foote, B., & Aly, A. A. (1986). An optimal load flow study by the generalized reduced gradient approach. *Electric Power Systems Research*, *10*(1), 47–53. doi:10.1016/0378-7796(86)90048-9

Zhao, B., Guo, C. X., & Cao, Y. J. (2004). An improved particle swarm optimization for OPF Problems. *Proceedings of IEEE/PES Power System Conference and Exposition 2004*, 233-238.

Zhu, J. Z., & Xiong, X. F. (2003). Optimal reactive power control using modified interior point method. *Electric Power Systems Research*, *66*(2), 187–192. doi:10.1016/S0378-7796(03)00078-6

APPENDIX 1: IEEE 30-BUS SYSTEM

Table 12. Generator data of IEEE-30 bus system

Unit	Bus No.	Unit Cost Coefficients			Pg^{max} (MW)	Pg^{min} (MW)	Qg^{max} (MVAr)	Qg^{min} (MVAr)
		a ($/MW²hr)	b ($/ MWhr)	c ($/hr)				
1	1	0.00375	2	0	2	0.5	0.1	-0
2	2	0.0175	1.75	0	0.8	0.2	0.50	-0.4
3	5	0.06250	1	0	0.3	0.1	0.4	-0.4
4	8	0.00834	3.25	0	1.1	0.95	0.4	-0.1
5	11	0.0250	3	0	1.1	0.95	0.24	-0.06
6	13	0.025	3	0	1.1	0.9	0.24	-0-06

APPENDIX 2

Table 13. Load data of IEEE-30 bus system

Bus no.	PD(MW)	QD(MVA)	Bus no.	PD(MW)	QD(MVA)
1	00.00	00.00	16	3.50	01.80
2	21.70	12.70	17	9.00	05.80
3	02.40	01.20	18	3.20	0.90
4	07.60	01.60	19	9.50	03.40
5	94.20	19.00	20	2.20	00.70
6	00.00	00.00	21	17.50	11.20
7	22.80	10.90	22	00.00	0.00
8	30.00	30.00	23	03.20	01.60
9	00.00	00.00	24	08.70	06.70
10	05.80	02.00	25	00.00	00.00
11	00.00	00.00	26	03.50	02.30
12	11.20	07.50	27	00.00	00.00
13	00.00	00.00	28	00.00	00.00
14	06.20	01.60	29	02.40	00.90
15	08.20	02.50	30	10.60	01.90

Table 14. Line data of IEEE-30 bus system

Line	From Bus	To Bus	Resistance (*p.u.*)	Reactance (*p.u.*)	Susceptance (*p.u.*)	MVA rating
1	1	2	0.0192	0.0575	0.0264	130
2	1	3	0.0452	0.1852	0.0204	130
3	2	4	0.057	0.1737	0.0184	65
4	3	4	0.0132	0.0379	0.0042	130
5	2	5	0.0472	0.1983	0.0209	130
6	2	6	0.0581	0.1763	0.0187	65
7	4	6	0.0119	0.0414	0.0045	90
8	5	7	0.0460	0.1160	0.0102	70
9	6	7	0.0267	0.0820	0.0085	130
10	6	8	0.0120	0.0420	0.0045	32
11	6	9	0.0000	0.2080	0.0000	65
12	6	10	0.0000	0.5560	0.0000	32

continued on following page

Table 14. Continued

Line	From Bus	To Bus	Resistance (*p.u.*)	Reactance (*p.u.*)	Susceptance (*p.u.*)	MVA rating
13	9	11	0.0000	0.2080	0.0000	65
14	9	10	0.0000	0.1100	0.0000	65
15	4	12	0.0000	0.2560	0.0000	65
16	12	13	0.0000	0.1400	0.0000	65
17	12	14	0.1231	0.2559	0.0000	32
18	12	15	0.0662	0.1304	0.0000	32
19	12	16	0.0945	0.1987	0.0000	32
20	14	15	0.2210	0.1997	0.0000	16
21	16	17	0.0824	0.1923	0.0000	16
22	15	18	0.1070	0.2185	0.0000	16
23	18	19	0.0639	0.1292	0.0000	16
24	19	20	0.0340	0.0680	0.0000	32
25	10	20	0.0936	0.2090	0.0000	32
26	10	17	0.0324	0.0845	0.0000	32
27	10	21	0.0348	0.0749	0.0000	32
28	10	22	0.0727	0.1499	0.0000	32
29	21	22	0.0116	0.0236	0.0000	32
30	15	23	0.1000	0.2020	0.0000	16
31	22	24	0.1150	0.1790	0.0000	16
32	23	24	0.1320	0.2700	0.0000	16
33	24	25	0.1885	0.3292	0.0000	16
34	25	26	0.2544	0.3800	0.0000	16
35	25	27	0.1093	0.2087	0.0000	16
36	28	27	0.0000	0.3960	0.0000	65
37	27	29	0.2198	0.4153	0.0000	16
38	27	30	0.3202	0.6027	0.0000	16
39	29	30	0.2399	0.4533	0.0000	16
40	8	28	0.0636	0.2000	0.0214	32
41	6	28	0.0169	0.0599	0.0065	32

Chapter 5
Flexible AC Transmission Systems (FACTS) Controller in Power Transmission

ABSTRACT

Secure and reliable operation of the power system is a critical issue for large, complicated, and interconnected power system networks. Security constraints such as thermal limits of transmission lines and bus voltage limits must be satisfied under any operating point in order to deliver reliable power to the consumers. One of the best alternative solutions of improvement of the security of power system is the use of flexible AC transmission systems (FACTS) devices. FACTS devices can be used to limit the power flow on the overloaded line and to increase the use of alternative paths to improve power transmission capacity. This chapter briefly describes all three categories of FACTS devices, namely shunt controllers (static synchronous compensator, static var compensator, thyristor-controlled reactor, thyristor switched reactor, thyristor switched capacitor), series controllers (static synchronous series compensator, thyristor controlled series capacitor, thyristor-controlled series reactor), and combined series-shunt controllers (unified power flow controller).

DOI: 10.4018/978-1-5225-6971-8.ch005

INTRODUCTION

The Flexible AC transmission system provides voltage support at critical buses in the system and regulates power flow in critical lines. It is defined as a power electronic based system and other static equipment that provide control of one or more AC transmission system parameter. The objective of power system operation is the electric power supply from supply end to load end stably and most economically, so that the frequency and voltage at any point of the network which remain constant while the voltage waveform at any bus is free from harmonics and the phases are balanced. However, in order to achieve optimum performance of the power system it is required to control the reactive power flow in the network and is possible by applying reactive power management at the different nodes in the power systems. The reactive power of load is to achieve its optimum performance is called load compensation while the reactive power compensation of the transmission network is called line compensation. The rapid growth of electrical energy consumption with demand for cost-effective energy has led to the development of generation. Modern power systems are designed to operate efficiently to supply power on demand to various load centres with high reliability. The generating stations are located at distant locations for economic and safety reasons. There is ac system interconnection the security can be adversely affected as the disturbances are initiated in a particular area and propagate over the entire system resulting in major blackouts caused by cascading outages. The purpose of the transmission network is to pool up the power plants and load centres in order to minimize the total power generation capacity and fuel cost. A power delivery system made up of radial lines from individual. Generators which are local without being part of a grid system, many more generation resources are needed to serve the load with the same reliability, and the cost of electricity would be too much higher. Transmission is often an alternative to a new generation resource. If transmission capability is less it means that more generation resources would be required inspite of taking into account of whether the system is made up of large or small power plants. The cost of transmission lines and losses, as well as difficulties in building new transmission lines, would often limit the available transmission capacity. The power systems of today and large, are controlled mechanically. There is a widespread use of microelectronics, computers and high-speed communications for control and protection of transmission systems; however, when operating signals are sent to the power circuits, where the final power control action is taken, the switching devices are mechanical and there is little high-speed control. Another problem with mechanical devices is that control which cannot be initiated frequently, because these mechanical devices tend to wear out very quickly as compared to static devices. In effect, from the point of view of both dynamic and steady-state operation, the system is really uncontrolled. These represent an asset that can be effectively utilized with prudent use of FACTS technology on a selective, as needed basis. The FACTS (Hingorani, 1988) technology is essential to some but not all of these difficulties by enabling utilities to get the service from their transmission facilities and enhance grid reliability. The FACTS devices is not a single high-power Controller, but rather a collection of Controllers, which can be applied individually or in coordination with others to control one or more of the interrelated system parameters. FACTS technology also lends itself by extending usable transmission limits in a step-by-step manner with incremental investment as and when we required. The concept of FACTS (Othman & Gabbar.,2017 *;* Peng., 2017; *Wang & Crow., 2013;* Watanabe et al., 2018) was developed originally for transmission network for improvement of Power Quality (PQ) in distribution systems operating at low or medium voltages. The power quality referred primarily to the continuity of power supply at acceptable voltage and frequency. The nonlinear loads not only cause PQ problems but are also very sensitive to the voltage deviations. Any problem manifested in voltage, current or frequency deviations that result in failure or disoperation of equipment.

BRIEF DESCRIPTION AND DEFINATIONS OF FACTS CONTROLLERS

FACTS Controllers (Hingorani & Gyugyi, 2000) may be based on thyristor devices with no gate tum-off (only with gate turn-on), or with power devices with gate turn-off capability. the principal Controllers with gate turn-off devices are based on the dc to ac converters, which can exchange active and/ or reactive power with the ac system. When the exchange involves reactive power only, they are provided with a minimal storage on the dc side. However, if the generated ac voltage or current is required to deviate from 90 degrees with respect to the line current or voltage, respectively, the converter dc storage can be augmented beyond the minimum required for the converter operation as a source of reactive power only. For the current-sourced converter, the dc current is presented to the ac side through the sequential switching of devices, as ac current, variable in amplitude and also in any phase relationship to the ac system voltage. The power reversal involves a reversal of voltage and not current. The current-sourced converter is represented symbolically by a box with a power device, and a dc inductor as its current source, the voltage-sourced converter is represented in symbolic form by a box with a gate turn-off device paralleled by a reverse diode, and a dc capacitor as its voltage source.

- **Flexibility of Electric Power Transmission:** The ability to accommodate changes in the electric transmission system or operating conditions while maintaining sufficient steady state and transient margins.
- **Flexible AC Transmission System (FACTS):** Alternating current transmission systems incorporating power electronic-based and other static controllers to enhance controllability and increase power transfer capability.
- **FACTS Controller:** A power electronic-based system and other static equipment that provide control of one or more A C transmission system parameter. FACTS Controllers can be divided into four categories:
    - Series Controllers
    - Shunt Controllers
    - Combined series-series Controllers
    - Combined series-shunt Controllers

Series Controllers

The series Controller could be a variable impedance, such as capacitor, reactor, etc., or a power electronics based variable source of main frequency, sub synchronous and harmonic frequencies (or a combination) to serve the desired need. In general, all series Controllers inject voltage in series with the line. Even a variable impedance multiplied by the current flow through it, represents an injected series voltage in the line. As long as the voltage is in phase quadrature with the line current, the series Controller only supplies or consumes variable reactive power. Any other phase relationship will involve handling of real power as well. Series-connected Controller impacts the driving voltage and hence the current and power flow directly. So, if the purpose is to control the current/power flow and damp oscillations, the series Controller for a given MVA size is several times more powerful than the shunt Controller. It is a compensation which is connected in series with power system and works as controllable voltage source. In long lines, when current flows this causes a large voltage drop. So, to compensate this series capacitors are connected, thereby decreasing the effect of inductance. Also, by connecting series capacitors in series with line, the inductance reactance between the receiving end and the sending end will be reduced due to which power-factor of system can be improved. FACTS (Basu, 2008) for series compensation modify the line impedance which is decreased so as to increase transmittable active power. Series-connected Controllers have to be designed to ride through contingency and dynamic overloads, and bypass short circuit currents. They can be protected by metal-oxide arresters or temporarily bypassed by solid-state devices when the fault current is too high, but they have to be rated to handle dynamic and contingency overload.

Shunt Controllers

As in the case of series Controllers, the shunt Controllers may be variable impedance, variable source, or a combination of these. In general, all shunt Controllers inject current into the system at the point of connection and acts as controllable current source which is used either when charging the transmission line or when there is very low load at the receiving end. Due to no-load very low current flows through transmission line. Shunt capacitance in transmission line causes voltage amplification. Receiving end voltage may become double the sending end voltage. To compensate this shunt inductor are connected across the transmission line. Even a variable shunt impedance connected to the line voltage causes a variable current flow and hence represents injection of current into the line. As long as the injected current is in phase quadrature with the line voltage, the shunt Controller only supplies or consumes variable reactive power and other phase relationship will involve handling of real power as well. The shunt Controller serve as an effective means to control voltage at and around the point of connection through injection of reactive current (leading or lagging), alone or a combination of active and reactive current for a more effective voltage control and damping of voltage oscillations at a substation bus. An important advantage of the shunt Controller is that it serves the bus node independently of the individual lines connected to the bus.

Combined Series-Series Controllers

This could be a combination of separate series controllers, which are controlled in a coordinated manner, in a multiline transmission system. Or it could be a unified Controller, in which series Controllers provide independent series reactive compensation for each line but also transfer real power among the lines via the power link. The real power transfer capability of the unified series-series Controller, referred to as *Interline Power Flow Controller,* makes it possible to balance both the real and reactive power flow in the lines and thereby maximize the utilization of the transmission system. The term "unified" here means that the dc terminals of all Controller converters are all connected together for real power transfer.

Combined Series-Shunt Controllers

This could be a combination of separate shunt and series Controllers, which are controlled in a coordinated manner or a *Unified Power Flow Controller* with series and shunt elements. In principle, combined shunt and series Controllers inject current into the system with the shunt part of the Controller and voltage in series in the line with the series part of the Controller. However, when the shunt and series Controllers are unified, there can be a real power exchange between the series and shunt Controllers via the power link.

SHUNT CONNECTED CONTROLLERS

Static Synchronous Compensator (STATCOM)

A Static synchronous generator operated as a shunt-connected static Var compensator whose capacitive or inductive output current can be controlled independent of the ac system voltage.

STATCOM is an important FACTS Controllers based on a voltage sourced or current-sourced converter. For the voltage-source converter, its ac output voltage is controlled such that it is just for the required reactive current flow for any ac bus voltage dc capacitor voltage is automatically adjusted as required in order to serve as a voltage source for the converter. STATCOM can be designed to act as an active filter to absorb system harmonics. STATCOM as defined above by IEEE is a subset of the broad-based shunt connected

Controller which includes then possibility of an active power source or storage on the dc side so that the injected current may include active power.

Static Synchronous Generator (SSG)

A static self-commutated switching power converter supplied from an appropriate electric energy source are operated to produce a set of adjustable multiphase output voltages, which may be coupled to an AC power system for the purpose of exchanging independently controllable real and reactive power.

An electronic interface known as a "chopper" is generally needed between the energy source and the converter. For a voltage-source converter, the energy source serves to appropriately compensate the capacitor charge through the electronic interface and maintain the required capacitor voltage.

Battery Energy Storage System (BESS)

A chemical-based energy storage system using shunt connected, voltage-source converters capable of rapidly adjusting the amount of energy which is supplied to or absorbed from an ac system. For transmission applications, BESS storage unit sizes would tend to be small (a few tens of MWHs), and if the short-time converter rating was large enough, it could deliver MWs with a high MW/MWH ratio for transient stability. The converter can also simultaneously absorb or deliver reactive power within the converter's MVA capacity. When not supplying active power to the system, the converter is used to charge the battery at an acceptable rate. Subset of SSG, suitable for transmission applications, is the Superconducting Magnetic Energy Storage (SMES).

Superconducting Magnetic Energy Storage (SMES)

A Superconducting electromagnetic energy storage device containing electronic converters that rapidly injects and/or absorbs real and/or reactive power or dynamically controls power flow in an ac system. Since the dc current in the magnet does not change rapidly, the power input or output of the magnet is changed by controlling the voltage across the magnet with a suitable electronics interface for connection to a STATCOM.

Static Var Compensator (SVC)

A shunt-connected static Var generator or absorber whose output is adjusted to exchange capacitive or inductive current so as to maintain or control specific parameters of an electrical power system. This is a general term for a Thyristor-controlled or Thyristor-switched reactor, and/or Thyristor-switched capacitor or combination. Static Var Compensator (SVC) is based on without the gate turn-off capability. It includes separate equipment for leading Thyristor and lagging vars; the Thyristor-controlled or Thyristor-switched reactor for absorbing reactive power and Thyristor-switched capacitor for supplying the reactive power. SVC is considered by some as a lower cost alternative to STATCOM.

Thyristor Controlled Reactor (TCR)

A shunt-connected, Thyristor-controlled inductor whose effective reactance is varied in a continuous manner by partial-conduction control of the Thyristor valve. TCR is a subset of SVC in which conduction time and current in a shunt reactor is controlled by a Thyristor-based ac switch with firing angle control.

Thyristor Switched Reactor (TSR)

A shunt-connected, Thyristor-switched inductor whose effective reactance is varied in a stepwise manner by full- or zero conduction operation of the Thyristor valve. TSR is made up of several shunt connected inductors which are switched in and out by Thyristor switches without any firing angle controls in order to achieve the required step changes in the reactive power consumed from the system. Use of Thyristor switches without firing angle control results in lower cost and losses, but without a continuous control.

Thyristor Switched Capacitor (TSC)

A shunt-connected, Thyristor-switched capacitor whose effective reactance is varied in a stepwise manner by full- or zero-conduction operation of the Thyristor valve. TSC is also a subset of SVC in which Thyristor based ac switches are used to switch in and out (without firing angle control) shunt capacitors units, in order to achieve the required step change in the reactive power supplied to the system. Unlike shunt reactors, shunt capacitors cannot be switched continuously with variable firing angle control.

Static Var Generator or Absorber (SVG)

A static electrical device, equipment, or system that is capable of drawing controlled capacitive and/or inductive current from an electrical power system and thereby generating or absorbing reactive power. Thus, both the SVC and the STATCOM are static Var generators equipped with appropriate control loops to vary the Var output so as to meet specific compensation objectives.

Static Var System (SVS)

A combination of different static and mechanically switched Var compensators whose outputs are coordinated are known as static Var system.

Thyristor Controlled Braking Resistor (TCBR)

A shunt-connected Thyristor-switched resistor, which is controlled to aid stabilization of a power system or to minimize power acceleration of a generating unit during a disturbance.TCBR involves cycle-by-cycle switching of a resistor (usually a linear resistor) with a Thyristor-based ac switch with firing angle control. For lower cost, TCBR may be Thyristor switched, i.e., without firing angle control. However, with firing control, half-cycle by half-cycle firing control can be utilized to selectively damp low-frequency oscillations.

SERIES CONNECTED CONTROLLERS

Static Synchronous Series Compensator (SSSC)

A static synchronous generator operated without an external electric energy source as a series compensator whose output voltage is in quadrature with, and controllable independently of, the line current for the purpose of increasing or decreasing the overall reactive voltage drop across the line and thereby controlling the transmitted electric power. The SSSC (Ghadir & Reshma, 2007) may include transiently rated energy storage or energy absorbing devices to enhance the dynamic behaviour of the power system by additional temporary real power compensation, to increase or decrease momentarily, the overall real (resistive) voltage drop across the line. SSSC is one the most important FACTS Controllers. It is like a STATCOM, except that the output ac voltage is in series with the line. Usually the injected voltage in series would be quite small compared to the line voltage, and the insulation to ground would be quite high. With an appropriate insulation between the primary and the secondary of the transformer, the converter equipment is located at the ground potential unless the entire converter equipment is located on a platform duly insulated from ground. The transformer ratio is designed to the most economical converter design. Without an extra energy source, SSSC can only inject a variable voltage, which is 90 degrees leading or lagging the current. The primary of the transformer and hence the secondary as well as the converter has to carry full line current including the fault current unless the converter is temporarily bypassed during severe line faults. Battery-storage or superconducting magnetic storage can also be connected to a series Controller to inject a voltage vector of variable angle in series with the line.

Interline Power Flow Controller (IPFC)

The IPFC is a recently introduced Controller and is the combination of two or more Static Synchronous Series Compensators which are coupled via a common dc link to facilitate bi-directional flow of real power between the ac terminals of the SSSCs, and are controlled to provide independent reactive compensation for the adjustment of real power flow in each line and maintain the desired distribution of reactive power flow among the lines.

The IPFC structure may also include a STATCOM, coupled to the IPFC's common dc link, to provide shunt reactive compensation and supply or absorb the overall real power deficit of the combined SSSCs.

Thyristor Controlled Series Capacitor (TCSC)

A capacitive reactance compensator which consists of a series capacitor bank shunted by a Thyristor-controlled reactor (Acha & Ambriz-Perez, 2000) in order to provide a smoothly variable series capacitive reactance. The TCSC is based on Thyristor without the gate turn-off capability. It is an alternative to SSSC but unlike SSSC, it is a very important FACTS Controller. A variable reactor such as a Thyristor-Controlled Reactor (TCR) is connected across a series capacitor. When the TCR firing angle is 180 degrees, the reactor becomes non-conducting and the series capacitor has its normal impedance. As the firing angle is advanced from 180 degrees, the capacitive impedance increases. On the other end, when the TCR firing angle is 90 degrees, the reactor becomes fully conducting, and the total impedance becomes inductive, because the reactor impedance is designed to be much lower than the series capacitor impedance. With 90 degrees firing angle, the TCSC helps in limiting fault current. The TCSC may be a single, large unit, or may consist of several equal or different-sized smaller capacitors in order to achieve a superior performance.

Thyristor-Switched Series Capacitor (TSSC)

A capacitive reactance compensator which consists of a series capacitor bank shunted by a Thyristor-switched reactor to provide a stepwise control of series capacitive reactance. Instead of continuous control of capacitive impedance, this approach of switching inductors at firing angle of 90 degrees or 180 degrees but without firing angle control, could reduce cost and losses of the Controller.

Thyristor-Controlled Series Reactor (TCSR)

An inductive reactance compensator which consists of a series reactor shunted by a Thyristor controlled reactor in order to provide a smoothly variable series inductive reactance. When the firing angle of the Thyristor controlled reactor is 180 degrees, it stops conducting, and the uncontrolled reactor acts as a fault current limiter. As the angle decreases below 180 degrees, the net inductance decreases until firing angle of 90 degrees, when the net inductance is the parallel combination of the two reactors. As for the TCSC, the TCSR may be a single large unit or several smaller series units.

Thyristor-Switched Series Reactor (TSSR)

An inductive reactance compensator which consists of a series reactor shunted by a Thyristor-controlled switched reactor in order to provide a stepwise control of series inductive reactance. This is a complement of TCSR, but with thyristor switches fully on or off (without firing angle control) to achieve a combination of stepped series inductance.

COMBINED SHUNTAND SERIES CONNECTED CONTROLLERS

Unified Power Flow Controller (UPFC)

A combination of static synchronous compensator (STATCOM) and a static series compensator (SSSC) which are coupled via a common dc link, to allow bidirectional flow of real power between the series output terminals of the SSSC and the shunt output terminals of the STATCOM and are controlled to provide concurrent real and reactive series line compensation without an external electric energy source. The UPFC, by means of angularly unconstrained series voltage injection, is able to control, concurrently or selectively, the transmission line voltage, impedance, and angle or, alternatively, the real and reactive power flow in the line. The UPFC may also provide independently controllable shunt reactive compensation. In UPFC which combines a STATCOM and an SSSC the active power for the series unit (SSSC) is obtained from the line itself via the shunt unit STATCOM the latter is also used for voltage control with control of its reactive power. This is a complete Controller for controlling active and reactive power control through the line, as well as line voltage control. Additional storage such as a superconducting magnet connected to the dc link via an electronic interface would provide the means of further enhancing the effectiveness of the UPFC.

Thyristor-Controlled Phase Shifting Transformer (TCPST)

A phase-shifting transformer adjusted by Thyristor switches to provide a rapidly variable phase angle (Mathur & Basati, 1981).

Phase shifting is obtained by adding a perpendicular voltage vector in series with a phase. This vector is derived from the other two phases via shunt connected transformers. A circuit concept that can handle voltage reversal can provide phase shift in either direction. This Controller is also referred to as *Thyristor-Controlled Phase Angle Regulator* (TCPAR).

Interphase Power Controller (IPC)

A series-connected controller of active and reactive power consisting, in each phase, of inductive and capacitive branches subjected to separately phase-shifted voltages. The active and reactive power can be set independently by adjusting the phase shifts and/or the branch impedances, using mechanical or electronic switches. In the particular case where the inductive and capacitive impedance form a conjugate pair, each terminal of the IPC is a passive current source dependent on the voltage at the other terminal.

STATIC SHUNT COMPENSATORS: SVC AND STATCOM

Objective of Shunt Compensators

The steady-state transmittable power can be increased and the voltage profile along the line controlled by appropriate reactive shunt compensation. The purpose of this reactive compensation is to change the natural electrical characteristics of the transmission line to make it more compatible with the prevailing load demand. Thus, shunt connected, fixed or mechanically switched reactors are applied to minimize line overvoltage under light load conditions, and shunt connected, fixed or mechanically switched capacitors are applied to maintain voltage levels under heavy load conditions.

The ultimate objective of applying reactive shunt compensation in a transmission system is to increase the transmittable power. This may be required to improve the steady-state transmission characteristics as well as the stability of the system. Var compensation is thus used for voltage regulation at the midpoint (or some intermediate) to segment the transmission line and at the end of the (radial) line to prevent voltage instability, as well as for dynamic voltage control to increase transient stability and damp power oscillations.

Midpoint Voltage Regulation for Line Segmentation

Compensator segments the transmission line into two independent parts and exchanges only the reactive power with the transmission line and significantly increase the transmittable power (doubling its maximum value) at the expense of a rapidly increasing reactive power demand on the midpoint compensator. The midpoint of the transmission line is the best location for the compensator. This is because the voltage sag along the uncompensated transmission line is the largest at the midpoint. Also, the compensation at the midpoint breaks the transmission line into two equal segments for each of which the maximum transmittable power is the same. For unequal segments, the transmittable power of the longer segment would clearly determine the overall transmission limit. A distributed compensation hinges on the instantaneous response and unlimited Var generation and absorption capability of the shunt compensators employed, which would have to stay in synchronism with the prevailing phase of the segment voltages and maintain the predefined amplitude of the transmission voltage, independently of load variation. Such a system, however, would tend to be too complex and probably too expensive, to be practical, particularly if stability and reliability requirements under appropriate contingency conditions are also considered.

Line Voltage Support to Prevent Voltage Instability

The voltage stability limit decreases with inductive loads and increases with capacitive loads. Reactive shunt compensation is often used in practical applications to regulate the voltage at a given bus against load variations, or to provide voltage support for the load when, due to generation or line outages, the capacity of the sending-end system becomes impaired. A frequently encountered example is when a large load area is supplied from two or more generation plants with independent transmission lines. The loss of one of the power sources could suddenly increase the load demand on the remaining part of the system, causing severe voltage depression that could result in an ultimate voltage collapse.

Improvement of Transient Stability

Shunt compensation will be able to change the power flow in the system during and following dynamic disturbances so as to increase the transient stability limit and provide effective power oscillation damping. Controlled shunt compensation can provide effective voltage support, it can increase the transmission capability of the post-fault system and thereby enhance transient stability. A substantial increase in the transient stability margin the ideal midpoint compensation with unconstrained Var output can provide by the effective segmentation of the transmission line. In an alternative way if the uncompensated system has a sufficient transient stability margin, shunt compensation can considerably increase the transmittable power without decreasing the margin. The ideal shunt compensator means that the amplitude of the midpoint voltage remains constant all the time, except possibly during faults, and its phase angle follows the generator angle swings so that the compensator would not be involved in real power exchange, but it would continuously provide the necessary reactive power. Thus, a practical compensator can perform as an ideal one only as long as the compensation Var demand does not exceed its rating. Above its maximum rating the compensator either becomes a constant reactive impedance or a constant reactive current source, depending on the power circuit employed for reactive power generation. The necessary rating of the compensator is usually determined by planning studies to meet the desired power flow objectives with defined stability margins under specific system contingency conditions.

Power Oscillation Damping

In the case of an under-damped (Mishra et al., 2002) power system, any minor disturbance can cause the machine angle to oscillate around its steady-state value at the natural frequency of the total electromechanical system. The angle oscillation, of course, results in a corresponding power oscillation around the steady-state power transmitted. The lack of sufficient damping can be a major problem in some power systems and, in some cases, it may be the limiting factor for the transmittable power. Since power oscillation is a sustained dynamic event, it is necessary to vary the applied shunt compensation, and thereby the (midpoint) voltage of the transmission line, to counteract the accelerating and decelerating swings of the disturbed machine(s). For large oscillations the Var output is controlled in a "bang-bang" manner (output is varied between the minimum and maximum values). However, for damping relatively small power oscillations, a strategy that varies the controlled output of the compensator continuously, in sympathy with the generator angle or power, may be preferred.

Methods of Controllable Var Generator

Capacitors generate and reactors (inductors) absorb reactive power when connected to an AC power source. They have been used with mechanical switches for (coarsely) controlled Var generation and absorption since the early days of ac power transmission. Continuously variable Var generation or absorption for dynamic system compensation was originally provided by over- or under-excited rotating synchronous machines and, later, by saturating reactors in conjunction with fixed capacitors.

Earlier, high power, line-commutated Thyristor in conjunction with capacitors and reactors were employed in various circuit configurations to produce variable reactive output. These in turn provide a variable shunt impedance by synchronously switching shunt capacitors and/or reactors "in" and "out" of the network. Using appropriate switch control, the Var output can be controlled continuously from maximum capacitive to maximum inductive output at a given bus voltage. More recently gate turn off Thyristor and other power semiconductors with internal turnoff capability have been used in switching converter circuits to generate and absorb reactive power without the use of ac capacitors or reactors. These perform as an ideal synchronous compensators (condensers), in which the magnitude of the internally generated ac voltage is varied to control the Var output. All of the different semiconductor power circuits, with their internal control enabling them to produce Var output proportional to an input reference. A static Var generator whose output is varied so as to maintain or control specific parameters (e.g., voltage, frequency) of the electric power system. Thus, the static Var generator is a self-sufficiently functioning device that draws controllable reactive current from an alternating power source. The control input to the Var generator can be an arbitrary (within the operating range) reactive current, impedance, or power reference signal that the SVG is to establish at its output. Thus, the static Var generator can be viewed as a power amplifier that faithfully reproduces the reference signal at the desired power level. Evidently, the type and structure of the Var generator will ultimately determine the basic operating characteristics (e.g., voltage vs. Var output, response time, harmonic generation), whereas the external characteristics control the functional capabilities (e.g., voltage regulation, power factor control, power oscillation damping), of the static Var compensator. Modern static Var generators are based on high-power semiconductor switching circuits. These switching circuits inherently determine some of the important operating characteristics, such as the applied voltage versus obtainable reactive output current, harmonic generation, loss versus Var output, and attainable response time, setting limits for the achievable performance of the Var generator and, independent of the external controls used, ultimately also that of the static Var compensator.

Static Var Compensators (SVC)

The Static Var Compensator (SVC), a first-generation FACTS controller which is a variable impedance device where the current through a reactor is controlled using back to back connected Thyristor valves. The application of Thyristor valve technology to SVC is an offshoot of the developments in HVDC technology. The major difference is that Thyristor valves used in SVC are rated for lower voltages as the SVC is connected to an EHV line through a step-down transformer or connected to the tertiary winding of a power transformer.

The application of SVC was initially for load compensation of fast changing loads such as steel mills and arc furnaces. Here the objective is to provide dynamic power factor improvement and also balance the currents on the source side whenever required. Here the objectives are:

1. Increase power transfer in long lines.
2. Improve stability with fast acting voltage regulation.
3. Damp low frequency oscillations due to swing (rotor) modes.
4. Damp sub-synchronous frequency oscillations due to torsional modes.
5. Control dynamic over voltages.

A SVC has no inertia compared to synchronous condensers and can be extremely fast in response (2-3 cycles). This enables the fast control of reactive power in the control range. There are two types of SVC:

1. Fixed Capacitor-Thyristor Controlled Reactor (FC-TCR).
2. Thyristor Switched Capacitor - Thyristor Controlled Reactor (TSC-TCR).

The second type is more flexible than the first one and requires smaller rating of the reactor and consequently generates less harmonics. The TCR and TSC are connected on the secondary side of a step-down transformer. Tuned and high pass filters are also connected in parallel which provide capacitive reactive power at fundamental frequency. The voltage signal is taken from the high voltage SVC bus using a potential transformer. The TSC is switched in using two Thyristor switches (connected back to back) at the instant in a cycle when the voltage across valve is minimum and positive. This results in minimum switching transients. In steady state, TSC does not generate any harmonics. To switch off a TSC, the gate pulses are blocked, and the Thyristor turns off when the current through them fall below the holding currents. Several pairs of Thyristor are connected in series as the voltage rating of a thyristor is not adequate for the voltage level required. However, the voltage ratings of valves for a SVC are much less than the voltage ratings of a HVDC valve as a step-down transformer is used in the case of SVC.

The advantages of TSC-TCR type SVC over FC-TCR type are:

1. The reduction in the reactor size and consequently the harmonics generated.
2. Greater flexibility in control.
3. Better performance under system fault conditions. The power losses in the quiescent operating condition (with SVC output current close to zero) also tend to be less with TSC-TCR type SVC.

The major application of SVC is for rapid voltage regulation and control of dynamic (temporary) over-voltages caused by load throw off faults or other transient disturbances. The dynamic reactive control at the load bus increases power transfer and can solve the problem of voltage instability (collapse) caused by contingency conditions. steady state voltage regulation can be achieved by mechanically switched capacitors and reactors (MSC and MSR). However, fast voltage regulation is required to prevent instability under transient conditions. Thus, generally, a SVC is operated with minimum reactive power output under normal conditions which is achieved by the Susceptance Regulator which ensures that full dynamic range is available for control under contingency conditions.

The fast controllability provided by the Thyristor switches can be also utilized to improve system stability (both transient and small signal). The use of auxiliary damping controllers can help damp low frequency, inter-area power oscillations that can appear at stressed operating conditions. SVC does not raise the fault level which can be a blessing as the requirement of the fault current interruption capability of circuit breakers does not go up. On the other hand, for reactive power control at HVDC converter stations with low Short Circuit Ratios (SCR), the synchronous condenser improves voltage regulation and system performance.

SVC has several advantages over SC (synchronous condenser):

1. Faster response under transient conditions
2. There are no moving parts, hence requires less maintenance
3. There are no problems of loss of synchronism
4. A SVC does not contribute to short circuit currents.

The drawbacks of a SVC are mainly related to the injection of current harmonics. These can be minimized by segmented TCR (with several segments/ modules of TCR connected in parallel) and operating all except one module in the TSR (Thyristor Switched Reactor) mode. The schematic diagram of SVC is shown in the Figure 1

Figure 1. Circuit structure of SVC device

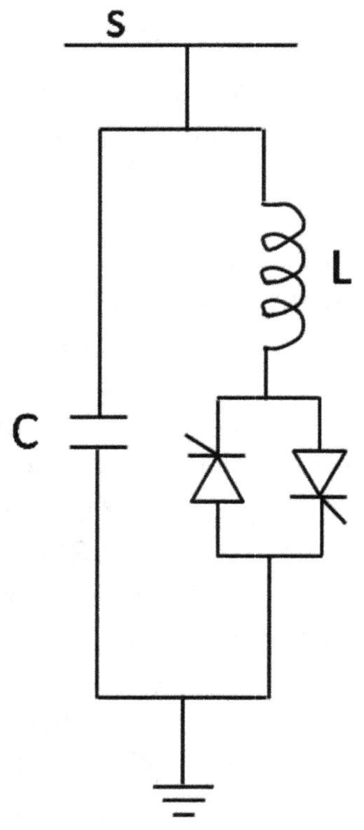

A shunt connected Static Var generator whose output is adjusted to exchange capacitive or inductive current in order to maintain or control specific parameters of electrical power system. The purpose of reactive compensation is to change the natural electrical characteristics of transmission line to make it more compatible with load demand. Shunt connected fixed or mechanically Switched Reactors are applied to minimize line over-voltage under light load and shunt connected fixed or mechanically Switched Capacitors are applied to maintain voltage levels under heavy load conditions. The ultimate objective of applying reactive shunt compensation in a transmission line is to increase transmittable power to improve steady state transmission characteristics and stability of the system. Var compensation is used for voltage regulation at mid-point to segment the transmission line and at the end of radial line to prevent voltage instability as well as dynamic voltage control to increase transient stability and damp power oscillations.

Static Synchronous Compensators (STATCOM)

The shunt connected static compensator was developed as an advanced static VAR compensator where a voltage source convertor (VSC) is used instead of the controllable reactors and switched capacitors. Although VSCs require self-commutated power semiconductor devices such as GTO, IGBT, IGCT, MCT, etc (with higher costs and losses) unlike in the case of variable impedance type SVC which uses Thyristor devices, there are many technical advantages of a STATCOM over a SVC. These are primarily:

1. Faster response
2. Requires less space as bulky passive components (such as reactors) are eliminated
3. Inherently modular and relocatable
4. It can be interfaced with real power sources such as battery, fuel cell or SMES (superconducting magnetic energy storage)
5. A STATCOM has superior performance during low voltage condition as the reactive current can be maintained constant (In a SVC, the capacitive reactive current drops linearly with the voltage at the limit of capacitive susceptance). It is even possible to increase the reactive current in a STATCOM under transient conditions if the devices are rated for the transient overload. In a SVC, the maximum reactive current is determined by the rating of the passive components – reactors and capacitors.

The operation of a STATCOM is comparable to a Synchronous Condenser (or Compensator) which can supply variable reactive power and regulate the voltage of the bus where it is connected. The equivalent circuit of a Synchronous Condenser (SC) is shown in Figure 2. One component of this current is I_p which is automatically determined by the requirement to balance the real power supplied to the series converter through the DC link. This power balance is enforced by regulating the DC capacitor voltage by feedback control.

A STATCOM (previously called as static condenser (STATCOM) has a similar equivalent circuit as that of a SC. The AC voltage is directly proportional to the DC voltage (V_{dc}) across the capacitor. If an energy source (a battery or a rectifier) is present on the DC side, the voltage V_{dc} can be held constant. On comparing SC with STATCOM it is found that while rotation of the DC field winding on the rotor results in the generation of AC voltages in the stator windings through magnetic induction, the synchronous operation of

Figure 2. Circuit structure of STATCOM device

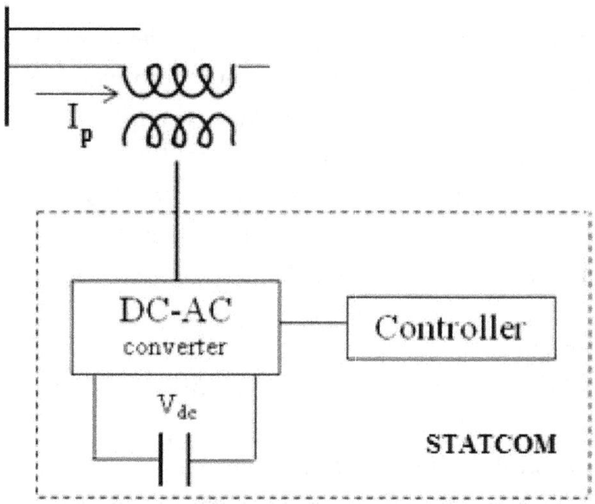

the switches in a STATCOM results in the AC voltage at the output. Unlike in a SC, this output voltage also contains many harmonics and some solution has to be found to eliminate them. In the case of a SC, the capacitors can be charged from the AC side and there is no need of an energy source on the DC side if only reactive current is to be provided in steady state. The losses in the STATCOM can be met from the AC source. The advantages of a STATCOM over a SC are:

1. The response is much faster to changing system conditions.
2. It does not contribute to short circuit current.
3. It has a symmetric lead-lag capability.
4. It has no moving parts and hence the maintenance is easier.
5. It has no problems of loss of synchronism under a major disturbance.

There are two operating (control) modes for a STATCOM or the shunt converter. These are as follows:

1. VAR control mode where the reactive current reference is determined by the inductive or capacitive VAR command. The feedback signals are obtained from current transformers (CT) typically located on the bushings of the coupling (step down) transformer.

2. Automatic voltage control mode where the reactive current reference is determined by the output of the feedback voltage controller which incorporates a droop characteristic (as in the case of a SVC or a STATCOM). The voltage feedback signals are obtained from potential transformers (PT) measuring the voltage at the substation feeding the coupling transformer.

Since, a STATCOM is an advanced type of SVC (based on VSC), the motivations for the applications are similar to that of a SVC. A STATCOM has several advantages over a SVC, namely (i) better performance under low voltage conditions, (ii) faster response that is independent of the system conditions and (iii) reduced space requirement. Also, a STATCOM can be built up in a modular fashion which will reduce the engineering costs.

The primary objective of applying a STATCOM in transmission networks is the fast regulation of voltage at a load or an intermediate bus. This will also increase the power transfer capacity of the network and thus enhance ATC (Available Transfer Capacity). The design of a voltage regulator is similar to that in a SVC. The problems caused by interactions (electromagnetic) with the network also affect the design of voltage regulator in a STATCOM. The smaller footprint and faster response of a STATCOM provides an opportunity to import power (for an urban area) from distant economic generators while the voltage is stabilized with a local STATCOM (older, uneconomic generators have been used in urban areas primarily for voltage-regulation). The use of multi-pulse and/or multilevel converters eliminates the need for harmonic filters in the case of a STATCOM. However, the costs increase not only due to the increased costs of magnetics and self-commutated devices (such as GTO, Thyristor), but also resulting from increased losses.

The new developments in power semiconductor technology are expected to bring down the costs and losses. The voltage and power ratings are expected to increase. At present, the use of STATCOM in distribution systems has become attractive, not only for voltage regulation, but also for eliminating harmonics and improving power quality.

STATIC SERIES COMPENSATORS: TCSC and SSSC

Objective of Series Compensators

Series Capacitors have been used in long distance EHV transmission lines for increasing power transfer. The use of series capacitors is generally the most economical solution for enhancing power flow. However, the problem of SSR has deterred system planners from going in a big way for series compensation. While the use of shunt capacitors doesn't have the problem of SSR, they have drawbacks of their effectiveness being dependent largely on their location. Even when a shunt capacitor is located at the midpoint of a long line, it requires much larger rating to achieve the same level of increase in power transfer as a series capacitor.

Thyristor and GTO Controlled Series Capacitor

By controlled series compensation, it implies dynamic control of the degree of series compensation in a long line. This can be achieved in two ways:

1. Discrete control using TSSC (Thyristor Switched Series Capacitor)
2. Continuous control using TCSC or GTO

Here, a TCR is connected in parallel with a fixed capacitor to enable continuous control over the series compensation. The schematic diagram of TCR is shown in Figure 3. Although harmonics are present in steady state with partial conduction of Thyristor switches, the TCSC can be used to mitigate SSR. In addition, TCSC provides inherent protection against over voltages.

Figure 3. Circuit structure of TCSC device

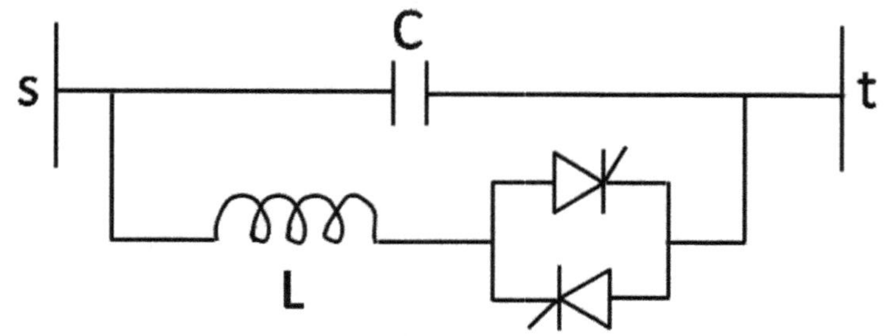

The configuration of GCSC in similar to TCSC except that the switch is made of GTO Thyristor to permit turn off capability when current is flowing through the switch. This enables, phase control to be applied to the capacitor in a similar fashion as in a TCR. GCSC is a dual of the TCR in that the voltage.

Although continuous control of the capacitor is possible in both TCSC and GCSC, the latter is yet to be applied in practice. In comparing TCSC with GCSC, the former has the advantages of (a) lower costs due to the requirements of conventional thyristors as opposed to GTO and (b) boost in the capacitor voltage provided by the discontinuous conduction of the TCR.

TCSC is operated in following manner-

A capacitive reactance compensator which consists of a series capacitor bank shunted by a Thyristor-controlled reactor in order to provide a smoothly variable series capacitive reactance A variable reactor such as a Thyristor-Controlled Reactor (TCR) is connected across a series capacitor. When the TCR firing angle is 180 degrees, the reactor becomes non conducting and the series capacitor has its normal impedance. As the firing angle is advanced from 180 degrees to less than 180 degrees, the capacitive impedance increases. On the other end, when the TCR firing angle is 90 degrees, the reactor becomes fully conducting, and the total impedance becomes inductive, because the reactor impedance is designed to be much lower than the series capacitor impedance. With 90 degrees firing angle, the TCSC helps in limiting fault current. The TCSC may be a single, large unit, or may consist of several equal or different-sized smaller capacitors in order to achieve a superior performance.

A single line diagram of a TCSC is shown in Figure 4. which shows two modules connected in series. There can be one or more modules depending on the requirement. To reduce the costs, TCSC may be used in conjunction with fixed series capacitors.

Figure 4. A single line diagram of a TCSC

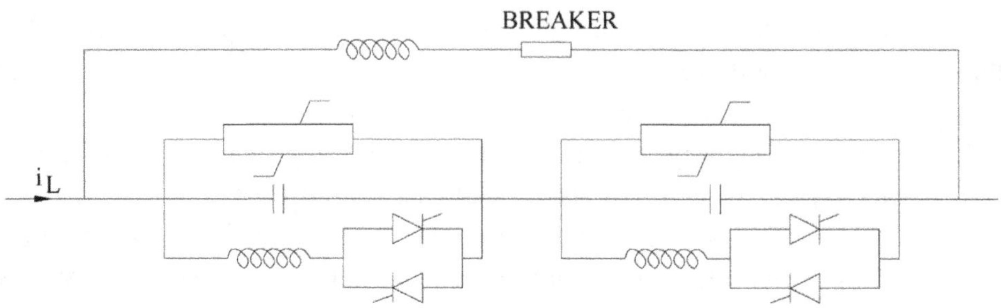

Bypassed

Here the Thyristor valves are gated for 180degree conduction (in each direction) and the current flow in the reactor is continuous and sinusoidal. The net reactance of the module is slightly inductive as the susceptance of the reactor is larger than that of the capacitor. During this mode, most of the line current is flowing through the reactor and Thyristor valves with some current flowing through the capacitor. This mode is used mainly for protecting the capacitor against over voltages. This mode is also termed as TSR (Thyristor Switched Reactor) mode.

Inserted With Thyristor Valve Blocked

In this operating mode no current flows through the valves with the blocking of gate pulses. Here, the TCSC reactance is same as that of the fixed capacitor and there is no difference in the performance of TCSC in this mode with that of a fixed capacitor. Hence this operating mode is generally avoided. This mode is also termed as waiting mode.

Inserted With Vernier Control

In this operating mode, the Thyristor valves are gated such that they conduct for the part of a cycle. The effective value of TCSC reactance (in the capacitive region) increases as the conduction angle increases corresponding to the parallel resonance of TCR and the capacitor (at fundamental frequency). In the inductive Vernier mode, the TCSC (inductive) reactance increases as the conduction angle reduced from 180 degree Generally, Vernier control is used only in the capacitive region and not in the inductive region.

The major objective in applying TCSC is to increase power transfer capacity in critical transmission lines (typically tie lines) under contingency conditions. Under normal steady state conditions, series compensation using fixed capacitors may be adequate unless SSR becomes a problem. A TCSC may be used in such cases to damp SSR by converting a part of the fixed compensation to controllable series compensation.

Often, the contingency conditions are also accompanied by low frequency oscillations that can threaten dynamic security. Sometimes, transient stability may be affected. Thus, it becomes necessary to provide Power Oscillation Damping (POD) using appropriate control signals synthesized from local measurements. Typically, active power or line current have been suggested as input to POD. The optimal size of a FACTS controller such as a TCSC is defined as that which results in the minimum cost of enhancing ATC.

Static Synchronous Series Compensators (SSSC)

The Static Synchronous Series Compensator (SSSC) is a series connected FACTS controller based on VSC and can be viewed as an advanced type of controlled series compensation, just as a STATCOM is an advanced SVC. The equivalent circuit of a SSSC is shown in Figure 5. A SSSC has several advantages over a TCSC such as (a) elimination of bulky passive components – capacitors and reactors, (b) improved technical characteristics (c) symmetric capability in both inductive and capacitive operating modes (d) possibility of connecting an energy source on the DC side to exchange real power with the AC network.

The magnitude of VC can be controlled to regulate power flow. The negative resistance emulation can destabilize oscillations in power. The winding resistance and leakage reactance of the connecting transformer appear is series with the voltage source VC. If there is no energy source on the DC side, neglect losses in the converter and DC capacitor.

Figure 5. Circuit structure of SSSC device

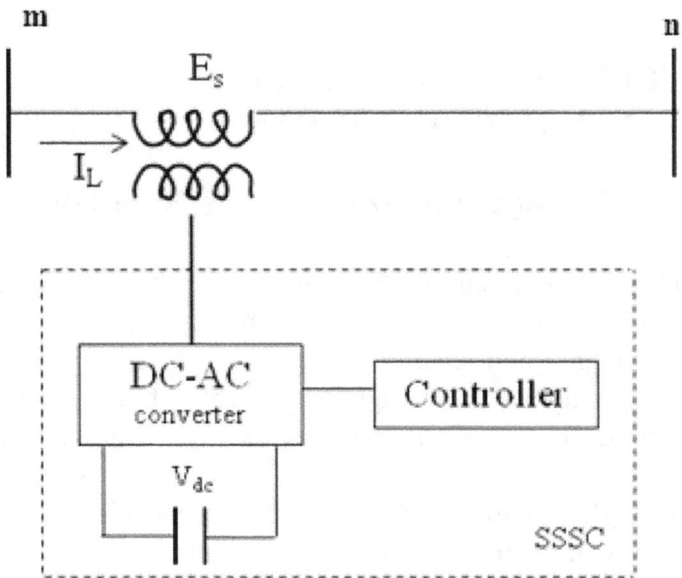

SSSC with an Energy Source- If an energy source such as battery or fuel cell is provided on the DC side of the VSC, it is possible to exchange real (active) power with the AC network. Here, the real (active) voltage component can be controlled in addition to the reactive voltage. This enables two degrees of freedom to control active and reactive power flow in the line. Alternately, it is possible to compensate for the line resistance to increase the power transfer capacity in the line. Instead of an energy source, it is possible to provide the real power from a shunt connected VSC which is connected to the series connected SSSC on the DC side.

The application of a SSSC would depend on the techno-economic evaluation and proven reliability based on operating experience. A major drawback with SSSC is the need for a coupling transformer (and an intermediate transformer if multi pulse converters are used). TCSCs don't require any magnetic devices or coupling transformers. However, the harmonics are better controlled with a SSSC. A SSSC requires protection against over currents. A high speed electronic Thyristor Bypass Switch (TBS) is installed in parallel with the converter terminals. When an overcurrent is detected, it operates quite fast. In case the system fault is not cleared by primary protection, then, the TBS is protected by a parallel connected low voltage breaker (LVB) which by-passes the TBS in about 6 cycles. If LVB fails to close when required, then its breaker failure protection closes the high side breaker (HSB) that by- passes the SSSC. After the fault is cleared, the SSSC is reinserted into the line by opening the LVB.

STATIC VOLTAGE AND PHASE ANGLE REGULATORS: TCVR AND TCPAR

Objective of Voltage and Phase Angle Regulators

The on-load tap changer and its electronic counterparts, play a significant role in transmission power flow control by providing an important functions of voltage regulation and reactive power management. However, whereas series reactive compensation is generally highly effective for power flow control, its application to certain transmission problems can be impractical, cumbersome, or economically not viable. These problems are related to the transmission angle. For example, the prevailing transmission angle may not be compatible with the transmission requirements of a given line or it may vary with daily or seasonal system loads over too large a range to maintain acceptable power flow in some affected lines. Other problems involve the control of real and reactive loop flows in a meshed network. Mechanical phase angle regulators (PARs) or phase shifting transformers (PSTs), using on-load tap changers with quadrature voltage injection, to solve power flow problems and increase the utilization of transmission lines. Whereas on-load tap changers with in-phase voltage injection control reactive power via voltage magnitude adjustment, those with quadrature voltage injection control real power via phase adjustment. Their combined use makes both the reactive and real power flow control possible. As a result, PARs have historically been used to redirect current flows and alleviate inherent loop flows in interconnected systems and thereby improve and balance the loading of interconnected transmission links. Apart from 'steady-state voltage and power flow control, the role of modern voltage phase angle regulators with fast electronic control can also be extended to handle dynamic system events. Potential areas of application include: transient stability improvement, power oscillation damping, and the minimization of post-disturbance overloads and the corresponding voltage dips. As compared to reactive compensators, voltage and phase angle regulators bring a new element to the control of dynamic events, the capability to exchange real power. At the same time, the voltage and angle regulators, based on the classical arrangement of tap-changing transformers, lack the ability of reactive compensators to supply or absorb reactive power and thus this burden is left to the power system to handle. In the following section, the basic techniques of voltage and angle regulation are reviewed to provide the necessary foundation for the treatment of power electronics-based approaches. The effect of phase angle control on the basic factors determine attainable maximal power transmission (steady-state power transmission limit, transient stability, and power oscillation damping) is also examined.

A phase-shifting transformer adjusted by Thyristor switches provides a rapidly variable phase angle. In general, phase shifting is obtained by adding a perpendicular voltage vector in series with a phase. This vector is derived from the other two phases via shunt connected transformers the perpendicular series voltage is made variable with a variety of power electronics topologies. A circuit concept that can handle voltage reversal can provide phase shift in either direction. This Controller is also referred to as Thyristor-Controlled Phase Angle Regulator (TCPAR).This is a broad based concept of series Controller, which can be designed to provide control of active and reactive power.

Thyristor-Controlled Voltage and Phase Angle Regulators (TCVRs and TCPARs)

There are two basic approaches to modern, power electronics-based reactive compensators: one that uses conventional thyristors to control current in reactive impedances, and the other that employs turn-off (GTO) thyristors in switching power converters to realize controllable synchronous voltage sources. This dual approach, in a different form, extends also to voltage and phase angle regulators, voltage and angle regulation are generally accomplished by in-phase and, respectively, quadrature voltage injection. Thus, both the conventional Thyristor-and GTO-controlled converter-based approaches insert a controlled voltage between the given bus and the controlled terminal or line. The major difference between the two is that whereas the Thyristor-based approach obtains the insertion voltage from appropriate taps of the regulating ("excitation") transformer, the GTO-based approach generates this voltage from a dc power supply. Consequently, the function of the Thyristor based controller in the first approach is that of an on-load tap-changer: selecting the proper tap of the regulating transformer and injecting the thus obtained voltage, usually by an insertion transformer, in series with the line. The function of the GTO-based voltage source is to generate the required voltage and inject it, also by an insertion transformer, in series with the line. At first sight, the difference between the two approaches appears to be minor. However, it will be seen that the two approaches result in important operational and performance differences, the most significant of which is the capability of the voltage-source-based approach to be self- sufficient in supplying or absorbing the reactive power the voltage or angle regulation demands. TCVR AND TCPAR are commonly known as TCPS device. Its schematic diagram is illustrated in Figure 6.

There are two main reasons for the application of Thyristor-Controlled Voltage and Phase Angle Regulators instead of mechanical on-load tap changers: One is the elimination of the expensive regular maintenance and the other is to provide the high speed of response necessary for dynamic system control. Voltage and angle regulators operate in the same manner, that is, they provide a voltage with a variable magnitude from a fixed voltage source. Of course, this is functionally the same as producing a voltage with a fixed magnitude from a source of variable magnitude.

Figure 6. Circuit structure of TCPS device

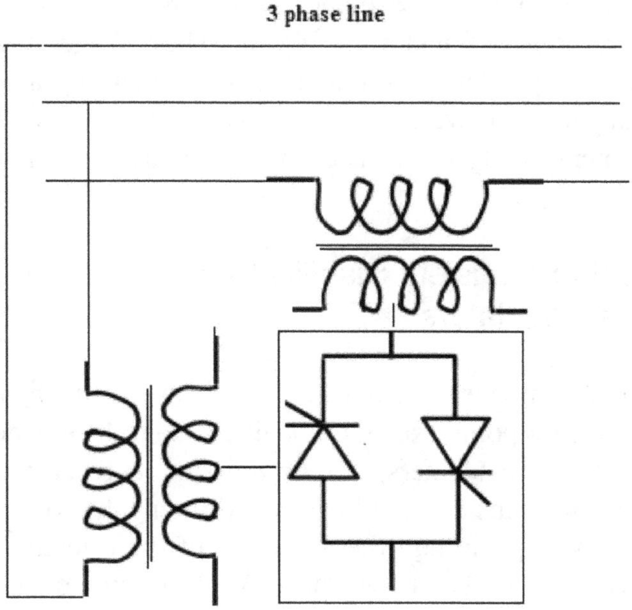

Basic Concept of Voltage and Phase Angle Regulation

The basic concept of voltage and phase angle regulation is the addition of an appropriate in-phase or a quadrature component to the prevailing terminal (bus) voltage in order to change (increase or decrease) its magnitude or angle to the value specified (or desired). Thus, voltage regulation could, theoretically, be achieved by a synchronous, in-phase voltage source with controllable amplitude, in series with the ac system and the regulated terminal.

An adjustable voltage is provided by means of a tap changer from a three-phase (auto) transformer (usually referred to as regulating or excitation transformer) for the primary of a series insertion transformer which injects it to achieve the required voltage regulation. In two winding transformers, on load tap changers are provided on the neutral side of the windings. In a similar manner, the arrangement can be used for phase- angle control simply by stipulating the injected voltage, dV, to have a phase of ±90° relative to the system voltage, V. With this stipulation, the injected voltage will change the prevailing phase-angle of the system voltage. For relatively small angular adjustments, the resultant angular change is approximately proportional to the injected voltage, while the voltage magnitude remains almost constant.

However, for large angular adjustments, the magnitude of the system voltage will appreciably increase and, for this reason, is often referred to as a quadrature booster transformer (QBT). The voltage magnitude could be maintained independent of the angular adjustment by a more complex winding arrangement. Nevertheless, because of its relative simplicity, the QBT arrangement has typically been used in conventional phase shifting applications.

Improvement of Transient Stability With Phase Angle Regulators

Controlled shunt compensation increases transient stability by increasing (or maintaining) the (midpoint) transmission line voltage during the accelerating swing of the disturbed machine(s) and controlled series reactive compensation is shown to improve transient stability by maximizing the power transmission during the first swing through the reduction of the effective transmission impedance. The capability of the Phase Angle Regulator to maintain the maximum effective transmission angle during the first swing can also be utilized effectively to increase the transient stability limit. The equal area criterion, is used here to assess the relative increase of the transient stability margin attainable by transmission angle control. Here, an ideal Phase Angle regulator is assumed for uniformity and clarity, a substantial increase in the transient stability margin which the Phase Angle Regulator can provide, It is worthwhile that in contrast to the shunt and series compensators, which provides the transient stability improvement via their inherent capability to increase the steady-state trans- mission limit of the uncompensated line, the angle regulator achieved by maintaining the power transmission at the steady-state limit of the uncompensated line during the first swing as if the sending-end and receiving-end system were asynchronously connected. The increase of transient stability margin is proportional to the angular range, and thereby ultimately the VA rating of the Phase Angle Regulator. It is pointed out again that under practical fault scenarios the pre-fault and post- fault systems are generally different and that the transient stability requirements would be established on the basis of the post-fault system.

Power Oscillation Damping With Phase Angle Regulators

Transmission angle control can also be applied to damp power oscillations. Power oscillation damping is achieved by varying the active power flow in the line(s) so as to counteract the accelerating and decelerating swings of the disturbed machine(s). That is, when the rotationally oscillating generator accelerates and angle δ increases the electric power transmitted must be increased to compensate for the excess mechanical input power. Conversely, when the generator decelerates and angle δ decreases. The electric power must be decreased to balance the insufficient mechanical input power.

COMBINED COMPENSATORS: UPFC AND IPFC

Objective of Combined Compensators

The objective of combined compensators is to investigate the application potential of multifunctional FACTS Controllers based on the back-to-back voltage-sourced converter

arrangement. This arrangement offers three basic possibilities. In the first Controller one converter of the back-to-back arrangement is in series and the other is in shunt with the transmission line. This arrangement in transmission applications is, generally referred to as the Unified Power Flow Controller (UPFC). UPFC identification is quite universal for this arrangement, although the UPFC, like the STATCOM and SSSC, could be implemented by a variety of different switching converters. In the second Controller, both converters of the back-to-back arrangement are connected in series with, usually, a different line. This arrangement is referred to as the Interline Power Flow Controller (IPFC). In the third Controller both converters are connected in shunt, each with a different power system. This arrangement functions as an asynchronous tie, sometimes referred to as a back-to-back STATCOM tie.

Unified Power Flow Controller (UPFC)

The Unified Power Flow Controller (UPFC) is the most versatile FACTS controller for the regulation of voltage and power flow in a transmission line. It consists of two voltage source converters (VSC) one shunt connected and the other series connected. The DC capacitors of the two converters are connected in parallel as shown in the Figure 7. If the switches 1 and 2 are open, the two converters work as STATCOM and SSSC controlling the reactive current and reactive voltage injected in shunt and series respectively in the line. The closing of the switches 1 and 2 enables the two converters to exchange real (active) power flow between the two converters. The active power can be either absorbed or supplied by the series connected converter. The shunt connected converter not only provides the necessary power required, but also the reactive current injected at the converter bus. Thus, a UPFC has 3 degrees of freedom unlike other FACTS controllers which have only one degree of freedom (control variable). The concept of combining two or more converters can be extended to provide flexibility and additional degrees of freedom. Even when only two converters are used, they may be designed to operate in the shunt or series mode. Depending on the requirement, there can be up to 11 configurations of the two converters resulting in (i) one or two STATCOMs (ii) one or two SSSCs (iii) combinations of a STATCOM and SSSC (iv) UPFC with series converter connected in either of the two lines and (v) IPFC. If two STATCOMs are provided in the middle of a long transmission line, the parallel connection of the two DC capacitors of the two shunt connected converters enable the power flow through the DC bus. The two STATCOMs now work as Back to Back (BTB) connected VSC based HVDC system. This is particularly useful when it becomes necessary to have asynchronous interconnection. The VSC converters are self-commutated and provide reactive power to the AC network. Thus, a HVDC link using VSC can be used to supply passive loads.

There is an upper limit on the current flowing through the GTOs which can be turned-off by applying a gate pulse. This limit is generally much higher than the nominal current rating of the converter. The latter is associated with continuous operation. When the line current increases rapidly above the critical level, under fault conditions, it is necessary to divert the current from the series converter. The bypassing of the converter may be achieved within microseconds by means of Thyristor bypass switch on the secondary side of the coupling transformer. The switch must be designed to carry the maximum fault current for short periods of time. If the fault is a transient fault, the Thyristor bypass switch opens when the current drops and the normal operation can be resumed. If the fault is of long duration, it is necessary to initiate the closing of a mechanical bypass switch (breaker) which is located across the primary winding of the coupling transformer, in order to relieve the loading on the thyristor bypass switch. Once the primary bypass switch is closed, the re-insertion of the series transformer requires the same procedure of start-up. The series converter, during the start-up has to drive a current through the primary winding of the series transformer, that matches the line current. When the match is achieved, the current in the mechanical bypass switch reduces to zero and it can be opened without creating any transients. The start-up can be achieved in a few cycles after the fault clearing. Although the current through the shunt connected converter is carefully controlled under normal operating conditions, it could exceed the critical level under fault condition. The protective action under such conditions is to interrupt the normal gating sequence. Generally, under transient conditions, it is possible to return to normal operation within a cycle (after the interruption of the normal gating sequence). In most cases, there is no need to trip the shunt breaker. However, a major internal failure may necessitate the opening of the shunt breaker. This is how UPFC is protected.

Interline Power Flow Controller

The combination of two or more Static Synchronous Series Compensators which are coupled via a common dc link to facilitate bi-directional flow of real power between the ac terminals of the SSSCs and are controlled to provide independent reactive compensation for the adjustment of real power flow in each line and maintain the desired distribution of reactive power flow among the lines. The IPFC (Gyugyi et al., 1999) structure may also include a STATCOM, coupled to the IFFC's common dc link, to provide shunt reactive compensation and supply or absorb the overall real power deficit of the combined SSSCs.

Figure 7. Circuit structure of UPFC device

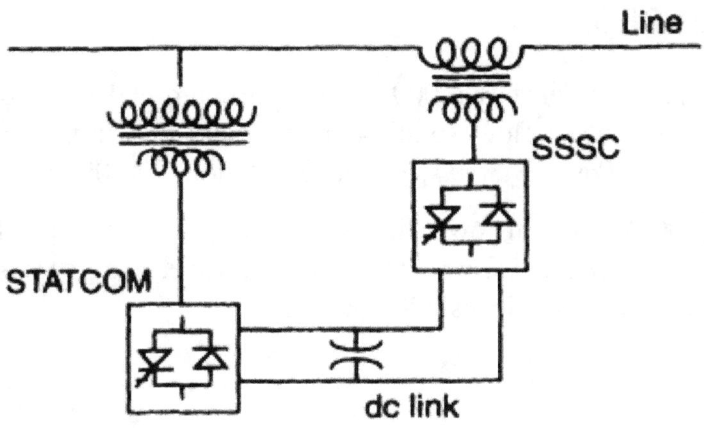

The objective of introducing this controller is to address the problem of compensating a number of transmission lines connected at a substation. While pure series reactive (controllable) compensation (in the form of a TCSC or SSSC) can be used to control or regulate the active power flow in a line, the control of reactive power is not feasible unless active (real) voltage in phase with the line current is not injected. The application of a TCSC (or SSSC with impedance emulation) results in the reduction of net series reactance of the line.

The Interline Power Flow Controller (IPFC) provides, in addition to the facility for independently controllable reactive (series) compensation of each individual line, a capability to directly transfer or exchange real power between the compensated lines. This is achieved by coupling the series connected VSCs in individual lines on the DC side, by connecting all the DC capacitors of individual converters in parallel. Since all the series converters are located inside the substation in close proximity, this is feasible.

An IPFC with two converters compensating two lines is similar to a UPFC in that the magnitude and phase angle of the injected voltage in the prime system (or line) can be controlled by exchanging real power with the support system (which is also a series converter in the second line). The basic difference with a UPFC is that the support system in the latter case is the shunt converter instead of a series converter. The series converter associated with the prime system (of one IPFC) is termed as the master converter while the series converter associated with the support system is termed as the slave converter. The master converter controls both active and reactive voltage (within limits) while the slave converter controls the DC voltage (across the capacitor) and the reactive voltage magnitude

REFERENCES

Acha, E., & Ambriz-Perez, H. (2000). A thyristor controlled series compensator model for the power flow solution of practical power networks. *IEEE Transactions on Power Systems*, *15*(1), 58–64. doi:10.1109/59.852101

Basu, M. (2008). Optimal power flow with FACTS devices using differential evolution. *International Journal of Electrical Power & Energy Systems*, *30*(2), 150–156. doi:10.1016/j.ijepes.2007.06.011

Ghadir, R., & Reshma, S. R. (2007). Power flow model/calculation for power systems with multiple FACTS controllers. *Electric Power Systems Research, 77*(12), 1521–1531. doi:10.1016/j.epsr.2006.10.008

Gyugyi, L., Sen, K. K., & Schauder, C. D. (1999). The interline power flow controller concept: A new approach to power flow management in transmission systems. *IEEE Transactions on Power Delivery, 14*(3), 1115–1123. doi:10.1109/61.772382

Hingorani, N. G. (1988). Power electronics in electrical utilities: Role of power electronics in future power systems. *Proceedings of the IEEE, 76*(4), 481–482. doi:10.1109/5.4432

Hingorani, N. G., & Gyugyi, L. (2000). *Understanding FACTS: concepts and technology of flexible AC transmission system*. New York: Wiley-IEEE Press.

Mathur, R. M., & Basati, R. S. (1981). A thyristor controlled static phase-shifter for AC power transmission. *IEEE Transactions on Power Apparatus and Systems, 100*(5), 2650–2655. doi:10.1109/TPAS.1981.316780

Mishra, S., Dash, P. K., Hota, P. K., & Tripathy, M. (2002). Genetically optimized Neuro-fuzzy IPFC for damping modal oscillations of power system. *IEEE Transactions on Power Systems, 17*(4), 1140–1147. doi:10.1109/TPWRS.2002.804958

Othman, A. M., & Gabbar, H. A. (2017). *FACTS-based high-performance AC/DC microgrids in Smart Energy Grid Engineering*. Cambridge, UK: Academic Press. doi:10.1016/B978-0-12-805343-0.00010-3

Peng, F. Z. (2017). Flexible AC Transmission Systems (FACTS) and Resilient AC Distribution Systems (RACDS) in Smart Grid. *Proceedings of the IEEE, 105*(11), 2099–2115. doi:10.1109/JPROC.2017.2714022

Wang, K., & Crow, M. L. (2013). Modern Flexible AC transmission system (FACTS) devices. In Electricity Transmission, Distribution and Storage systems (pp. 174-205). Philadelphia: Woodhead Publishing Limited.

Watanabe, E. H., Lima, F. K., Dias, R. F., Aredes, M., Barbosa, P. G., Barcelos, S. L. S. L., & Jr, G. S. (2018). 28 – flexible ac transmission systems. In Power Electronics Handbook (4th ed.; pp. 885–909). New York: Elsevier.

Chapter 6
Power Flow Modeling in Power System With Multiple FACTS Controller

ABSTRACT

Flexible AC transmission systems (FACTS) devices are integrated into power system networks to control power flow, increase transmission line capability to its thermal limit, and improve the security of transmission systems. Power flow is an important mathematical calculation for planning, operation, and control of power systems network. The focus of the chapter is to explore how to modify Newton-Raphson power flow method to include various FACTS devices such as static VAR compensator (SVC), static synchronous compensator (STATCOM), static synchronous series compensator (SSSC), thyristor-controlled series capacitor (TCSC), thyristor-controlled phase shifter (TCPS), unified power flow controller (UPFC) controllers. This chapter briefly describes the power flow equations of the aforesaid FACTS-based power system network, and how the conventional power flow calculation is systematically extended to include these controllers is also been discussed.

DOI: 10.4018/978-1-5225-6971-8.ch006

INTRODUCTION

Load flow (or Power flow) calculations are an important mathematical calculation in power systems network for planning, operational planning, and control (Scott, 1974; Tamimi et al., 2017) Power flow programs are the most frequently used computer routines for power systems calculations. Appropriate steady state model of the power system is needed for writing the computer programs. The model includes nonlinear algebraic equations, which must be solved iteratively. Power flow calculation is needed for both steady state power flow analysis and initializations for different dynamic analyses.

Flexible AC transmission system (FACTS) controllers are able to change the network parameters in a fast and effective way in order to achieve better system performance (Hingorani & Gyugyi, 2000; Peng, 2017) These controllers are used for enhancing dynamic performance of power systems in terms of voltage/angle stability while improving the power transfer capability and voltage profile in steady state. Different FACTS controllers are considered in this chapter which is connected either in series or shunt with the bus bar. These controllers are: Static synchronous Compensator (STATCOM), static synchronous series compensator (SSSC), and unified power flow controller (UPFC), Thyrister control series capacitor (TCSC), Static VAR Compensator (SVC) and Thyrister control phase shifter (TCPS).

MODELING OF POWER SYSTEM WITH MULTIPLE FACTS CONTROLLER

Power System With Static VAR Compensator (SVC)

The SVC is the most widely used employed FACTS controller. It is a shunt-connected static VAR generator or absorber whose output is adjusted to exchange capacitive or inductive current so as to maintain or control specific parameters of the electrical power system (typically bus voltage).

The SVC in general, may be a Thyristor Controlled Reactor (TCR) or Thyristor Switched Capacitor (TSC) or a combination of both. Some other configuration of SVC is Fixed Capacitor-TCR(FC-TCR) or TCR-Mechanically Switched Capacitor(TCR-MSC). The high voltage side system bus voltage is measured and filtered and compared with the reference voltage. The error voltage is processed through a gain-time constant regulator to provide the desired Susceptance requirement for the SVC. From the operational point of view, the SVC adjusts its value automatically in response to changes in the operating conditions of the network. By suitable control of its equivalent susceptance, it is possible to regulate the voltage magnitude at the SVC point of connection, thus enhancing significantly the performance of the power system.

The equivalent susceptance, which neglects harmonic current, can be expressed as

$$B_{eq} = B_L(\alpha) + B_C \tag{1}$$

where,

$$B_L(\alpha) = -\frac{1}{\omega L}(1 - \frac{2\alpha}{\pi}),\ B_C = \omega \times C \tag{2}$$

Let the complex power injected at s^{th} k^{th} $bus is$. s

Figure 1. Schematic diagram of SVC device

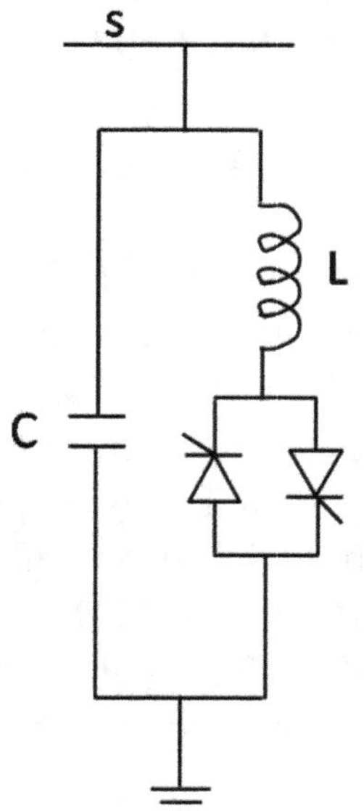

$$S_k = P_k + jQ_k = V_k \overset{*}{I_k}$$

Taking conjugate of S_k

$$\overset{*}{S_k} = P_k - jQ_k = V_k^* I_k$$

$$= V_k\angle - \delta_k[(V_k\angle\delta_k - 0)B(\alpha)]$$

$$= V_k\angle - \delta_k[(V_k\angle\delta_k - 0)(-jB_L)] \text{For inductor } Y = -jB_L$$

$$= -jV_k^2 B_L$$

$$Q_k = V_K^2 B_L \text{ (For inductor)}$$

The reactive power drawn by the SVC, which is also the reactive power injected at bus S, is given by

$$Q_{SVC} = Q_k = V_k^2 B_{SVC}$$

$$Q_{SVC} = -V_k^2 B_{SVC} \text{ (For capacitor} Y = jB_c)$$

The constraint on the reactive power at bus k is

$$B_{SVC}^{\min} \leq B_{SVC} \leq B_{SVC}^{\max}$$

Let us consider an IEEE-7 Bus system, where Bus 1 is swing bus (v, δ . δ own), bus 2, 5 are $\rightarrow$. gerator buses (P,V known), bus 3,4,6,7 are Load buses (P,Q known). Buses 3, 4 are Voltage control buses where the two SVC are connected.

The system matrix becomes

$$\begin{bmatrix} \Delta P_2 \\ \Delta P_3 \\ \Delta P_4 \\ \Delta P_5 \\ \Delta P_6 \\ \Delta P_7 \\ \Delta Q_3 \\ \Delta Q_4 \\ \Delta Q_6 \\ \Delta Q_7 \end{bmatrix} = \begin{bmatrix} \frac{\Delta P_2}{\Delta \delta_2} & \frac{\Delta P_2}{\Delta \delta_3} & \cdots & \frac{\Delta P_2}{\Delta B_{SVC3}} & \cdots & \frac{\Delta P_2}{\Delta V_7} \\ \frac{\Delta P_3}{\Delta \delta_2} & \frac{\Delta P_3}{\Delta \delta_3} & \cdots & \frac{\Delta P_3}{\Delta B_{SVC3}} & \cdots & \frac{\Delta P_3}{\Delta V_7} \\ \cdots & \cdots & \cdots & \cdots & \cdots & \cdots \\ \frac{\Delta Q_3}{\Delta \delta_2} & \frac{\Delta Q_3}{\Delta \delta_3} & \cdots & \frac{\Delta Q_3}{\Delta B_{SVC3}} & \cdots & \frac{\Delta Q_3}{\Delta V_7} \\ \cdots & \cdots & \cdots & \cdots & \cdots & \cdots \\ \frac{\Delta Q_7}{\Delta \delta_2} & \frac{\Delta Q_7}{\Delta \delta_3} & \cdots & \frac{\Delta Q_7}{\Delta B_{SVC3}} & \cdots & \frac{\Delta Q_7}{\Delta V_7} \end{bmatrix} \begin{bmatrix} \Delta \delta_2 \\ \Delta \delta_3 \\ \Delta \delta_4 \\ \Delta \delta_5 \\ \Delta \delta_6 \\ \Delta \delta_7 \\ \Delta B_{SVC3} \\ \Delta B_{SVC4} \\ \Delta V_6 \\ \Delta V_7 \end{bmatrix}$$

Power System With TCSC

Figure 2 presents the general circuit structure of TCSC (Halder et al., 2018) It consists of a series compensating capacitor shunted by a thyristor controlled reactor (TCR). TCR is a variable inductive reactor X_L controlled by firing angle α. The effective reactance of TCSC may be represented as follows:

$$X_{TCSC}(\alpha) = \frac{X_C X_L(\alpha)}{X_L(\alpha) - X_C} = -jX_C \,. \tag{3}$$

Figure 2. Circuit structure of TCSC device

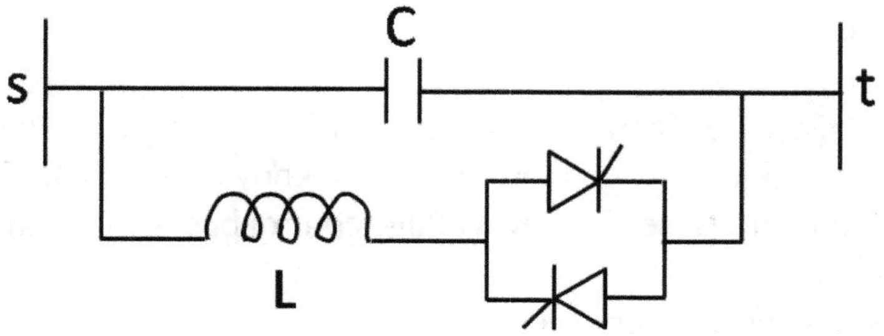

The effect of TCSC on the transmission network can be seen as a controllable reactance inserted in the related transmission line. The model of the network with TCSC is shown in Figure 3. TCSC is integrated in the OPF problem by modifying the line data. The maximum compensation by TCSC is limited to 70% of the reactance of the un-compensated line where TCSC is located. The modified line reactance (X_{new}) of the transmission line after incorporating TCSC is given as follows.

$$X_{t_{mn}} = X_{mn} - X_C \tag{4}$$

$$X_{t_{mn}} = (1-\tau)X_{mn} \tag{5}$$

where, $\tau = \dfrac{X_C}{X_{mn}}$ is the degree of series compensation and X_{mn} is the line reactance between the mth bus and the nth bus.

The power flow equations of the branch having TCSC are given by:

$$P_{mn} = V_m^{\,2} G_{mn} - V_m V_n \left(G_{mn} \cos\partial_{mn} + B_{mn} \sin\partial_{mn} \right) \tag{6}$$

$$Q_{mn} = -V_m^{\,2} B_{mn} - V_m V_n \left(G_{mn} \sin\partial_{mn} - B_{mn} \cos\partial_{mn} \right) \tag{7}$$

where

$$G_{mn} = \frac{R_{mn}}{R_{mn}^{\,2} + \left(X_{mn} - X_{t_{mn}} \right)^2}$$

Figure 3. Circuit model of TCSC connected between the mth bus and the nth bus

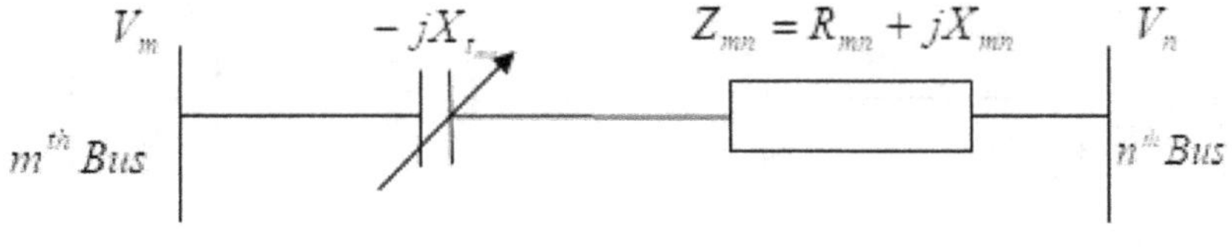

and

$$B_{mn} = \frac{\left(X_{mn} - X_{t_{mn}}\right)}{R_{mn}^{\ 2} + \left(X_{mn} - X_{t_{mn}}\right)^2}$$

P_{mn}, Q_{mn} are the active and reactive power flows, respectively, between the mth bus and the nth bus; V_m, V_n are the voltage magnitudes at the mth bus and the nth bus, respectively; ∂_{mn} is the angle differences between the mth bus and the nth bus voltages; R_{mn}, X_{mn} are the resistance and reactance, respectively, of transmission line that is connected between the mth bus and the nth bus; $X_{t_{mn}}$ is the reactance of TCSC placed in the transmission line that is connected between the mth bus and the nth bus.

Power System With TCPS

The static model of a TCPS connected between the mth bus and the nth bus, having a complex tapping ratio, $1 : a\angle\alpha$ and series admittance of $Y_{mn} = \left(G_{mn} - jB_{mn}\right)$ of the transformer is shown in Figure 4.

The real and reactive power flows from the mth bus to the nth bus may be derived similar to TCSC and can be expressed as:

$$P_{mn} = a^2 V_m^{\ 2} G_{mn} - a V_m V_n \left[G_{mn} \cos\left(\partial_{mn} + \alpha\right) + B_{mn} \sin\left(\partial_{mn} + \alpha\right)\right] \tag{8}$$

$$Q_{mn} = -a^2 V_m^{\ 2} B_{mn} - a V_m V_n \left[G_{mn} \sin\left(\partial_{mn} + \alpha\right) - B_{mn} \cos\left(\partial_{mn} + \alpha\right)\right] \tag{9}$$

Figure 4. Circuit model of TCPS connected between the mth bus and the nth bus

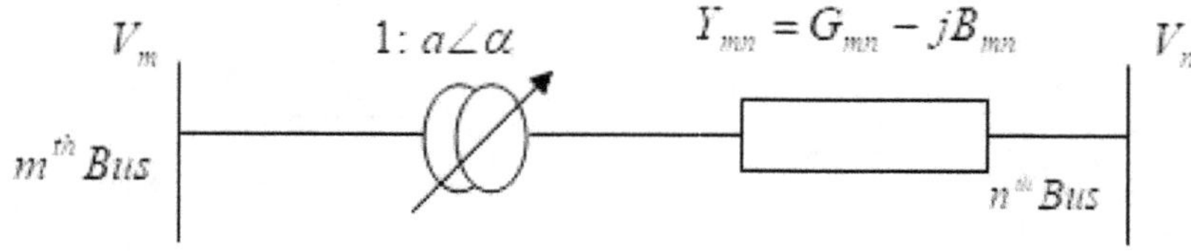

Similarly, the real and reactive power flows from the nth bus to the mth bus may be expressed as follows:

$$P_{nm} = V_n^2 G_{mn} - aV_m V_n \left[G_{mn}\cos\left(\partial_{mn} + \alpha\right) - B_{mn}\sin\left(\partial_{mn} + \alpha\right)\right] \tag{10}$$

$$Q_{nm} = -V_n^2 B_{mn} + aV_m V_n \left[G_{mn}\sin\left(\partial_{mn} + \alpha\right) + B_{mn}\cos\left(\partial_{mn} + \alpha\right)\right] \tag{11}$$

The injected power model of TCPS is shown in Figure 5. The injected real and reactive powers of TCPS at the mth bus and the nth bus are as follows:

$$P_{ms} = -a^2 V_m^2 G_{mn} - aV_m V_n \left[G_{mn}\sin\left(\partial_{mn}\right) - B_{mn}\cos\left(\partial_{mn}\right)\right] \tag{12}$$

$$Q_{ms} = a^2 V_m^2 B_{mn} + aV_m V_n \left[G_{mn}\cos\left(\partial_{mn}\right) + B_{mn}\sin\left(\partial_{mn}\right)\right] \tag{13}$$

$$P_{ns} = -aV_m V_n \left[G_{mn}\sin\left(\partial_{mn}\right) + B_{mn}\cos\left(\partial_{mn}\right)\right] \tag{14}$$

$$Q_{ns} = -aV_m V_n \left[G_{mn}\cos\left(\partial_{mn}\right) - B_{mn}\sin\left(\partial_{mn}\right)\right] \tag{15}$$

where G_{mn}, B_{mn} are the conductance and susceptance of transmission line, respectively, connected between the mth bus and the nth bus.

Power System With STATCOM

Although, there are several FACTS devices for controlling power flow and voltage profile in power system, for this study, initially, only STATCOM device has been considered to minimize the transmission loss, improve the voltage profile and voltage stability of the power system. The static model of this FACTS device is as described below.

Static synchronous compensator (STATCOM) is connected in parallel with the power system. The principal of STATCOM is to enhance the reactive

Figure 5. Injected power model of TCPS connected between the mth bus and the nth bus

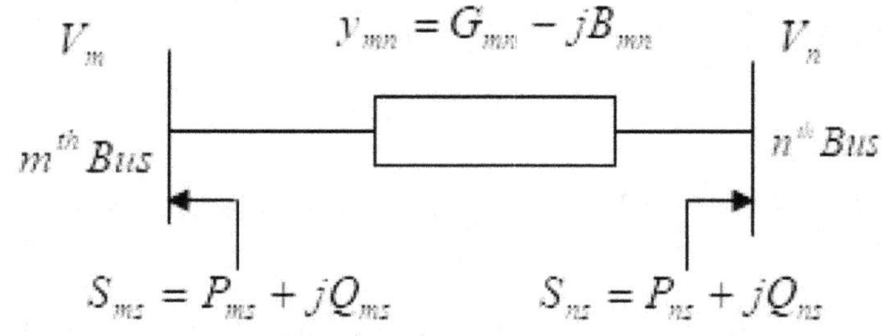

power compensation which adjusts the reactive power and voltage magnitude of power system network. It consists of three parts: transformer, voltage source converter (VSC), and capacitor. The STATCOM is modeled as a controllable voltage source (E_p) in series with impedance. The real part of this impedance represents the ohmic losses of the converter and coupling transformer, while the imaginary part of this impedance represents the leakage reactance of the coupling transformer. STATCOM absorbs or delivers the proper amount of reactive power from the grid to keep the voltage constant for all power system loading within the reasonable range. Figure 6 shows the circuit model of a STATCOM connected to the mth bus of a power system. The injected active and reactive power flow equation of the mth bus are given below:

$$P_m = G_p \left|V_m\right|^2 - \left|V_m\right|\left|E_p\right|\left|Y_p\right|\cos(\delta_m - \delta_p - \theta_p) + \sum_{j=1}^{N}\left|V_m\right|\left|V_j\right|\left|Y_{ij}\right|\cos(\delta_m - \delta_j - \theta_{ij}) \tag{16}$$

$$Q_m = -B_p \left|V_m\right|^2 - \left|V_m\right|\left|E_p\right|\left|Y_p\right|\sin(\delta_{m,} - \delta_p - \theta_p) + \sum_{j=1}^{N}\left|V_m\right|\left|V_j\right|\left|Y_{ij}\right|\sin(\delta_{mi} - \delta_j - \theta_{ij}) \tag{17}$$

The implementation of STATCOM in transmission system introduces two state variables ($\left|E_p\right|$ and δ_p); however, $\left|V_m\right|$ is known for STATCOM connected bus. Thus, one more equation is needed to solve the power flow problem. It is assume that the power consumed by the source E_p must be zero in steady state.

$$P_{E_p} = real[E_p I_p^*] \tag{18}$$

$$P_{E_p} = -G_p \left|E_p\right|^2 + \left|E_p\right|\left|V_m\right|\left|Y_p\right|\cos(\delta_m - \delta_p + \theta_p) = 0 \tag{19}$$

where V_m is the voltage magnitudes at the mth bus; Y_p is the admittance of the parallel component; G_p, B_p are the conductance and susceptance, respectively, of the parallel components of STATCOM; θ_{mn} is the admittance angle of transmission line connected between the mth bus and n^{th} bus, respectively; δ_p is the voltage source angle of STATCOM; E_p is the voltage sources of STATCOM converters.

Power System With SSSC

A SSSC is a solid-state voltage sourced converter (VSC), which generates a controllable AC voltage, and connected in series to power transmission lines in a power system. Figure 7 shows the circuit model of a SSSC connected between the *m^th^* and the *n^th^* buses of a power system.SSSC provides the virtual

Figure 6. Circuit structure of STATCOM device

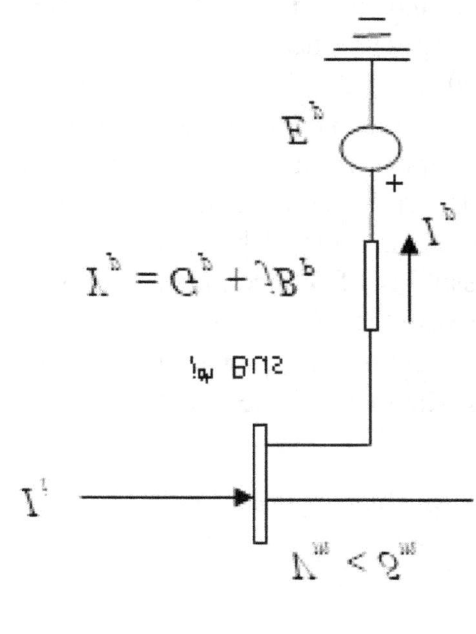

Figure 7. Schematic static model of SSSC

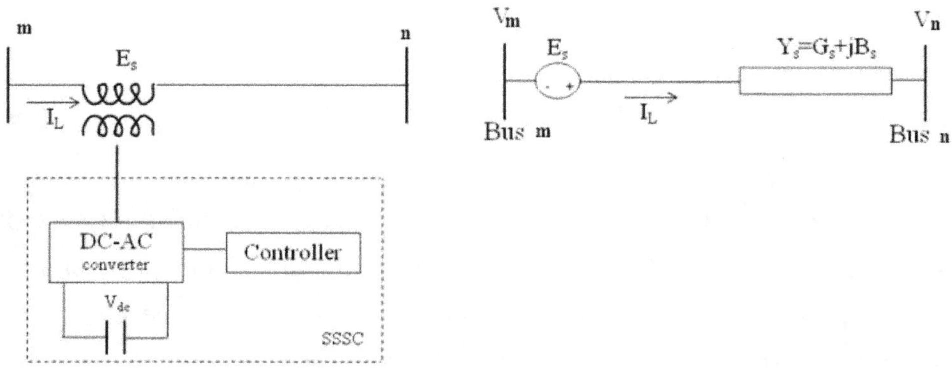

(a) Simplified diagram of SSSC

(b) Circuit model of SSSC connected between j^{th} bus and k^{th} bus

compensation of transmission line impedance by injecting the controllable voltage (E_s) in series with the transmission line. E_s is in quadrature with the line current, and emulates an inductive or a capacitive reactance so as to influence the power flow in the transmission lines. The virtual reactance inserted by E_s influences electric power flow in the transmission lines independent of the magnitude of the line current. The variation of E_s is performed by means of a VSC connected on the secondary side of a coupling transformer. The compensation level can be controlled dynamically by changing the magnitude and polarity of E_s and the device can be operated both in capacitive and inductive mode. The VSC uses forced-commutated power electronic devices to produce an AC voltage from a DC voltage source. A capacitor connected on the DC side of the VSC acts as a DC voltage source. To keep the capacitor charged and to provide transformer and VSC losses, a small active power is drawn from the line. A SSSC has several advantages over a TCSC such as (i) symmetric capability in both inductive and capacitive operating modes (ii) elimination of bulky passive components-capacitors and reactors (iii) possibility of connecting an energy source on the DC side exchange real power with the AC network.

The injected active and reactive power of the mth and the nth bus may be expressed as follows (Radman & Raje, 2007)

$$P_m = P_{mn} + \sum_{j=1}^{N} \left|V_m\right|\left|V_j\right|\left|Y_{mj}\right| \cos\left(\delta_m - \delta_j - \theta_{mj}\right) \tag{20}$$

$$Q_m = Q_{mn} + \sum_{j=1}^{N} |V_m||V_j||Y_{mj}| \sin(\delta_m - \delta_j - \theta_{mj}) \tag{21}$$

$$P_n = P_{nm} + \sum_{j=1}^{N} |V_n||V_j||Y_{nj}| \cos(\delta_n - \delta_j - \theta_{nj}) \tag{22}$$

$$Q_n = Q_{nm} + \sum_{j=1}^{N} |V_n||V_j||Y_{nj}| \sin(\delta_n - \delta_j - \theta_{nj}) \tag{23}$$

The active and the reactive power flow through the transmission line connected between the mth and n^{th} bus having SSSC may be derived as follows (Radman & Raje, 2007)

$$P_{mn} = |V_m|^2 (G_s) - |V_m||V_n||Y_s| \cos(\theta_s - \delta_m + \delta_n) + |V_m||E_s||Y_s| \cos(\theta_s - \delta_m + \delta_s) \tag{24}$$

$$Q_{mn} = -|V_m|^2 B_s - |V_m||E_s||Y_s| \sin(\theta_s - \delta_m + \delta_s) + |V_n||V_m||Y_s| \sin(\theta_s - \delta_n + \delta_m) \tag{25}$$

$$P_{nm} = |V_n|^2 (G_s) - |V_m||V_n||Y_s| \cos(\theta_s - \delta_m + \delta_n) + |V_n||E_s||Y_s| \cos(\theta_s - \delta_m + \delta_s) \tag{26}$$

$$Q_{nm} = -|V_n|^2 B_s + |V_n||E_s||Y_s| \sin(\theta_s - \delta_n + \delta_s) - |V_n||V_m||Y_s| \sin(\theta_s - \delta_n + \delta_m) \tag{27}$$

The power losses in SSSC may be neglected. Therefore, the net active power exchange in SSSC is zero which may be expressed as follows (Radman & Raje 2007)

$$\left|E_s\right|^2\left(G_s\right)+\left|V_m\right|\left|E_s\right|\left|Y_s\right|\cos\left(\theta_s-\delta_m+\delta_s\right)+\left|V_n\right|\left|E_s\right|\left|Y_s\right|\cos\left(\theta_s-\delta_m+\delta_s\right)=0 \quad (28)$$

where, V_m, V_n are the voltage magnitudes at the mth and the nth bus, respectively; Y_s is the summation of the admittance of the transmission line connected between the *m-n* bus and the admittance of the series component of the SSSC; G_s, B_s are the conductance and susceptance, respectively, of the series components of SSSC; θ_{mj} is the admittance angle between the mth and the jth bus of the transmission line; θ_s is the admittance angle of the admittance that includes the admittance of the line *m-n* and the admittance of the series component of the SSSC; δ_s are the voltage source angle of the series components of the SSSC; E_s are the voltage sources of series converters of the SSSC devices.

Power System With UPFC

The static model of the UPFC and its mathematical expressions are described below. The UPFC, the most versatile FACTS device presently available for transmission system control, is capable of providing active and reactive load flow control between its terminals. It may also provide reactive power compensation to the node to which it is connected. The device consists of two voltage source converters connected by a common DC link. The parallel component is similar to STATCOM while the series component is similar to the SSSC. The circuit representation of UPFC is shown in Figure 8 where UPFC is connected between the mth and the nth bus. The injected active and reactive power of the mth and the nth bus may be expressed as follows (Radman & Raje 2007; Bahamani et al., 2017)

$$\begin{aligned}P_m &= (G_P+G_S)\left|V_m\right|^2-\left|V_m\right|\left|E_p\right|\left|Y_p\right|\cos(\theta_p-\delta_m+\delta_p)+\left|V_m\right|\left|E_s\right|\left|Y_s\right|\cos(\theta_s+\delta_s-\delta_m)\\ &-\left|V_m\right|\left|V_n\right|\left|Y_s\right|\cos(\theta_s-\delta_m+\delta_n)+\sum_{k=1}^{NB}\left|V_m\right|\left|V_k\right|\left|Y_{mk}\right|\cos(\theta_{mk}-\delta_m+\delta_k)\end{aligned} \quad (29)$$

Figure 8. Circuit model for UPFC

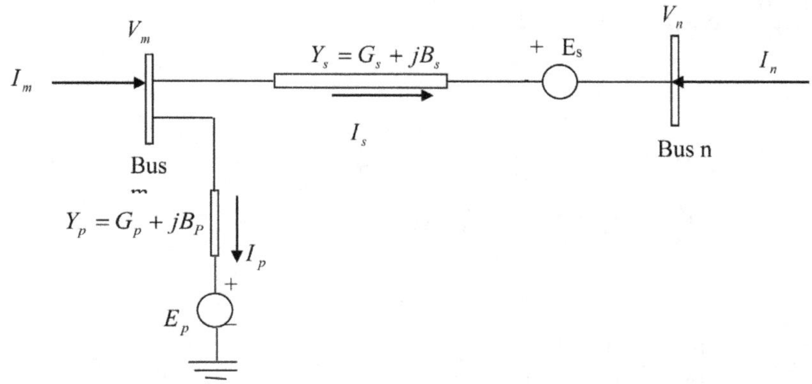

$$Q_m = -(B_P + B_S)|V_m|^2 + |V_m||E_p||Y_p|\sin(\theta_p - \delta_m + \delta_p)$$
$$-|V_m||E_s||Y_s|\sin(\theta_s + \delta_s - \delta_m) + |V_m||V_n||Y_s|\sin(\theta_s - \delta_m + \delta_n) + \sum_{k=1}^{NB}|V_m||V_k||Y_{mk}|\sin(\theta_{mk} - \delta_m + \delta_k) \quad (30)$$

$$P_n = +G_S|V_n|^2 - |V_n||V_m||Y_s|\cos(\theta_s - \delta_n + \delta_m)$$
$$-|V_n||E_s||Y_s|\cos(\theta_s + \delta_s - \delta_m) + \sum_{k=1}^{NB}|V_n||V_k||Y_{nk}|\cos(\theta_{nk} - \delta_n + \delta_k) \quad (31)$$

$$Q_n = -B_S|V_n|^2 + |V_n||V_m||Y_s|\sin(\theta_s - \delta_n + \delta_p)$$
$$-|V_n||E_s||Y_s|\sin(\theta_s + \delta_s - \delta_m) + \sum_{k=1}^{NB}|V_n||V_k||Y_{nk}|\sin(\theta_{nk} - \delta_n + \delta_k) \quad (32)$$

The active and the reactive power flow through the transmission line connected between the mth and the nth bus having UPFC may be derived as follows (Radman & Raje 2007).

$$P_{mn} = (G_P + G_S)\left|V_m\right|^2 - \left|V_m\right|\left|E_p\right|\left|Y_p\right|\cos(\theta_p - \delta_m + \delta_p) + \left|V_m\right|\left|E_s\right|\left|Y_s\right|\cos(\theta_s + \delta_s - \delta_m) - \left|V_m\right|\left|V_n\right|\left|Y_s\right|\cos(\theta_s - \delta_m + \delta_n) \quad (33)$$

$$Q_{mn} = -(B_P + B_S)\left|V_m\right|^2 + \left|V_m\right|\left|E_p\right|\left|Y_p\right|\sin(\theta_p - \delta_m + \delta_p) - \left|V_m\right|\left|E_s\right|\left|Y_s\right|\sin(\theta_s + \delta_s - \delta_m) + \left|V_m\right|\left|V_n\right|\left|Y_s\right|\sin(\theta_s - \delta_m + \delta_n) \quad (34)$$

$$P_{nm} = +G_S\left|V_n\right|^2 - \left|V_n\right|\left|V_m\right|\left|Y_s\right|\cos(\theta_s - \delta_n + \delta_m) - \left|V_n\right|\left|E_s\right|\left|Y_s\right|\cos(\theta_s + \delta_s - \delta_m) \quad (35)$$

$$Q_{nm} = -B_S\left|V_n\right|^2 + \left|V_n\right|\left|V_m\right|\left|Y_s\right|\sin(\theta_s - \delta_n + \delta_p) - \left|V_n\right|\left|E_s\right|\left|Y_s\right|\sin(\theta_s + \delta_s - \delta_n) \quad (36)$$

The power losses in UPFC may be neglected. Therefore, the net active power exchange in UPFC is zero which may be expressed as follows (Radman & Raje 2007)

$$G_P\left|E_p\right|^2 + G_S\left|E_S\right|^2 - \left|E_p\right|\left|V_m\right|\left|Y_p\right|\cos(\theta_p + \delta_m - \delta_p) + \left|V_m\right|\left|E_s\right|\left|Y_s\right|\cos(\theta_s + \delta_m - \delta_s) - \left|E_S\right|\left|V_n\right|\left|Y_s\right|\cos(\theta_s - \delta_S + \delta_n) = 0 \quad (37)$$

CONCLUSION

This chapter demonstrated how the conventional power flow solution could systematically be modified to include multiple FACTS controllers: TCSC, SVC, TCPS, STATCOM, SSSC, and UPFC. It was shown that the impacts of FACTS controllers on power flow can be accommodated for by adding new entries and modifying some existing entries in the linearized Jacobian equation of the underlying system with no FACTS controllers.

REFERENCES

Bahamani, A. K., Reddy, G. M. S., & Ganesh, V. (2017). Comparative of performance for UPFC with DPFC. *Proceedings of IEEE International Conference on Electrical, Instrumentation and Communication Engineering (ICEICE)*, 1 – 4. 10.1109/ICEICE.2017.8191843

Halder, A., Pal, N., & Mondal, D. (2018). Transient Stability Analysis of a Multimachine Power System with TCSC Controller – A Zero Dynamic Design Approach. *International Journal of Electrical Power & Energy Systems*, *97*, 51–71. doi:10.1016/j.ijepes.2017.10.030

Hingorani, N. G., & Gyugyi, L. (2000). Understanding FACTS: concepts and technology of flexible AC transmission systems. New York: Wiley–IEEE Press.

Peng, F. Z. (2017). Flexible AC Transmission Systems (FACTS) and Resilient AC Distribution Systems (RACDS) in Smart Grid. *Proceedings of the IEEE*, *105*(11), 2099–2115. doi:10.1109/JPROC.2017.2714022

Radman, G., & Raje, R. S. (2007). Power flow model/calculation for power systems with multiple FACTS controllers. *Electric Power Systems Research*, *77*(12), 1521–1531. doi:10.1016/j.epsr.2006.10.008

Scott, B. (1974). Review of load flow calculation methods. *Proceedings of the IEEE*, *62*(7), 916–929. doi:10.1109/PROC.1974.9544

Tamimi, B., Cañizares, C., & Battistelli, C. (2017). Hybrid Power Flow Controller Steady-State Modeling, Control, and Practical Application. *IEEE Transactions on Power Systems*, *32*(2), 1483–1492.

Chapter 7
Optimal Power Flow and Optimal Reactive Power Dispatch Incorporating TCSC-TCPS Devices Using Different Evolutionary Optimization Techniques

ABSTRACT

The introduction of flexible AC transmission system (FACTS) has added a new dimension in power system operation and planning. Various types of FACTS controllers such as static compensator (STATCOM), static synchronous series compensator (SSSC), thyristor control series compensator (TCSC), thyristor control phase shifter (TCPS), unified power flow controller (UPFC), etc. are successfully used by various researchers in order to get optimal performance of power system. In this chapter, the various population-based nature-inspired techniques such as grey wolf optimization (GWO), teaching-learning-based optimization (TLBO), biogeography-based optimization (BBO), krill herd algorithm (KHA), chemical reaction optimization (CRO), and hybrid CRO (HCRO) are used to find the optimal size of TCSC and TCPS devices in order to find the optimum performance of IEEE 30-bus power system. The simulation results of various cases demonstrate the effectiveness and robustness of the proposed techniques to solve TCSC-TCPS-based OPF and ORPD problems.

DOI: 10.4018/978-1-5225-6971-8.ch007

INTRODUCTION

The developing countries like India, commercial as well as industrial power demand grows unpredictably; thereby supplying reliable and secure power to the consumer is a challenging task. The McKinsey & Company's 'Electric Power and Natural Gas Practice' made a survey on India of 'Powering India: The Road to 2017'. Above analyses suggested that if India grows at an average rate of 8% for the next 10 years, then India's demand for power is likely to soar from around 120 GW at present to 315–335 GW by 2017, 100 GW higher than most current estimates. Four main factors behind the analysis are: (i) Rise of Industrial Power demand, (ii) Commercial demand rising at 14% over the next 10 years, (iii) 'Power for all by 2012' and (iv) the realization of demand suppressed due to load shedding. To achieve the demand by 2017, India needs a generation capacity of 415–440 GW (McKinsey & Company, 2017).

In recent years Environment, deregulation of power utilities has delayed the construction of both generation facilities and new transmission lines. However, due to the increasing load demand, more and more power is required to be pushed over the existing transmission lines but the increase in the power flow above a certain operating level increases power loss, reduces the voltage profile and decreases the overall stability of the system. To achieve better power flow control over the existing transmission lines without violating the stability margin of the system, application of flexible ac transmission system (FACTS) may be adopted.

The optimal power flow (OPF) problem has been well known since 1960s. The first person to introduce the concept of OPF was Carpentier in the year 1962 (Carpentier, 1962). Nowadays, modern methods like differential evolution (DE) (Panda, 2009), evolutionary algorithm (EA) (Panda, 2009), genetic algorithm (GA) (Rashed et al., 2007; Narayana et al., 2012), particle swarm optimization (PSO) (Singh et al., 2015; Shayeghi et al., 2010), bacctoria foraging algorithm (BFA) (Edward et al., 2013), symbiotic organisms search algorithm (SOA), (Prasad & Mukherjee, 2016), Brainstorm optimisation algorithm (BSOA) (Jordehi, 2015), Cat Swarm Optimization (CSO) (Nireekshana et al., 2016), whale optimization algorithm (WOA) (Raj, 2017) to solve OPF must be adjusted in order to satisfy the new challenges of the power industry coming from market and technological transformations. In fact, the development of competitive electricity markets has had an important impact on classical analysis tools.

In this chapter, thyristor based FACTS controllers (TCSC and TCPS) have been introduced for the OPF and ORPD formulations. The basic objective of OPF is to find proper adjustments of the control variables that would minimize the fuel cost. Reactive power optimization is a sub problem of the optimal power flow (OPF) calculation, which determines all kinds of controllable variables, such as reactive power outputs of generators and static VAr compensators, tap ratios of transformers, outputs of shunt capacitors/reactors, etc., and minimizes transmission losses or other appropriate objective functions, while satisfying a given set of physical and operating constraints. Since transformer tap ratios and outputs of shunt capacitors/reactors are discrete variables, the reactive power optimization problem can be exactly formulated as a mixed integer nonlinear programming one.

However, the task of searching and finding the optimal values of control variables for the FACTS controllers is a complex optimization problem which cannot be easily solved.

LITERATURE SURVEY

A method based on mathematical decomposition and network compensation to solve OPF with a phase shifter and series compensator was proposed by Taranto *et al.* (1992). The proposed methodology was applied to a reduced network of the Brazilian Southern-Southeastern system with 124 buses. Ambriz-Perez *et al.* (2006) presented Newton's method for solving OPF with the thyristor controlled series compensator (TCSC). The method was tested on 5-bus and 14-bus systems and the results clearly showed the technique's flexibility and reliability towards the convergence. Mahdad *et al.* (2010) reported a parallel GA technique for solving OPF incorporating static VAr compensator (SVC). This method was developed for fuel cost minimization and voltage profile improvement of the system. This method was tested on IEEE 30-bus, IEEE 118-bus systems and 15 units with prohibited operating zones. The superiority of the method in comparison to other competitive global approaches was reported. A DE technique for solving OPF with TCSC and thyristor controlled phase shifter (TCPS) was introduced by Basu (2008). The method was validated on IEEE 30-bus system and it was observed that the proposed technique was capable of producing better results compared to EP and GA. A hybrid TS and SA approach for solving OPF with TCSC and TCPS devices were reported by Ongsakul and Bhasaputra (2002). The proposed approach was applied to IEEE 30-bus system and computational results obtained by the proposed approach were compared to those obtained by GA, SA and TS; and it was shown that the proposed method gave a better optimal solution. A sequential quadratic programming (SQP) used to solve OPF incorporating SVC and UPFC was developed by Palma-Behnke *et al.* (2004). Recently, Ghahremani and Kamwa (2013) developed a graphical user interface (GUI) based on GA to maximize the system static loadability using five different FACTS devices namely, SVC, TCSC, thyristor controlled voltage regulators (TCVR), thyristor control phase shifting transformer (TCPST) and UPFC. Newton Raphson method power flow analysis (NRPFA) was introduced by Meikandasivam *et al.* (2013) to fine micro tuning of line reactance by implementing split TCSC in place of single TCSC to compensate the small change in power demand. Basu (2011) presented multi-objective differential evolution to optimize the cost of generation, emission and active power transmission loss of TCSC and TCPS based OPF problem. The proposed approach was tested on the modified IEEE 30-bus and IEEE 57-bus test systems. Singh *et al.* (2015) developed PSO with an aging leader and challengers (ALC-PSO) to solve the OPF problem of power system equipped with TCSC and TCPS. This method was tested on modified IEEE 30-bus and IEEE 57-bus power systems with four different objectives. Shaoyun and Chung (1998) decomposed the OPF problem into two stages, the first stage was a power flow control problem and the second stage was a normal OPF problem using controllable FACTS devices like series compensators and phase shifters.

Roy *et al.* (2011) proposed BBO to solve TCSC and TCPS based ORPD problem for minimizing voltage deviation and transmission loss of IEEE 30-bus test system. Padhy *et al.* (2004) used TCSC, to minimize an objective function that was composed of complex power lost in transmission lines (i.e. real and reactive line losses) and the generated reactive power. Benabid *et al.* (2009) reported non-dominated sorting PSO to identify optimal location and settings of SVC and TCSC devices for multi-objective optimization problem. This method was used to improve voltage stability margin, to reduce real power losses, and to reduce load voltage deviation. The proposed method was tested on IEEE 30-bus system and Algerian 114-bus power system and showed excellent results. Bacteria foraging optimization (BFO) technique was developed by Tripathy & Mishra (2007) to solve the multi-objective OPF problem that optimized the real power losses and voltage stability index simultaneously. The control variables considered were the UPFC optimal placement and setting in addition to the existing transformer settings. Marouani *et al.* (2009) presented a multi-objective evolutionary algorithm to solve ORPD problem with FACTS devices.

Motivation of the Present Work

In recent years, the fast progress in the field of power electronics and microelectronics has resulted into a new opportunity for more flexible operation of power systems. These new devices have made the present transmission and distribution of electricity more reliable, more controllable and more efficient. OPF is non-linear problems and can be non-convex in some cases. Moreover, incorporation of FACTS devices complicates the problem further. Such complicated problem needs a well-efficient optimization technique for the optimal solution.

Traditional optimization techniques such as Newton's method (Ambriz-Perez et al., 2006), gradient method (Yu et al., 1986) and LP (Urdaneta et al., 1999) approach are used to solve problems of OPF with FACTS assuming continuous, differentiable and monotonically increasing cost function. However, these methods have failed in handling non-convex and non-linear engineering optimization problems and tended to get stuck at local optimum solutions. Since OPF incorporating FACTS devices with valve-point discontinuities are highly non-linear optimization problems with non-differentiable features, stochastic search techniques such as GA (Ramirez & Onate, 2003), PSO (Singh et al., 2015), DE (Ming et al., 2011), TS (Wennan et al., 2008) and SA (Keyan et al., 2006), EP (Lai & Ma, 1995) etc were used to solve these problems.

Despite to the fact that most of the aforesaid evolutionary techniques have good convergence characteristic for solving nonlinear constraint optimization problem, their degree of optimization depends on the initializations of input parameters and initial populations which means the solutions may trap into local minima/maxima instead of global solutions. This chapter presents BBO, KHA, TLBO, GWO, CRO and HCRO techniques to solve FACTS based OPF problem in order to minimize the generation cost and ORPD problem in order to minimize the transmission loss, voltage deviation. The motivation behind the use of these techniques is their inherent simplicity, robustness, stable convergence characteristics and excellent records in locating global or near global optimal solutions of non-linear, non-differentiable problems. Modified IEEE 30-bus system with multiple TCSC and TCPS devices in different locations are presented to show the proposed techniques' flexibility and reliability towards the convergence.

This chapter is organized as follows: the section titled "Static Model of FACTS devices" presents the modeling of FACTS devices for load flow studies followed by the formulation of OPF problems with FACTS in the section titled "Mathematical Problem Formulation," the section titled "Evolutionary optimization techniques applied to OPF and ORPD incorporating TCSC and TCPS devices" is devoted to the solution methodology where the optimization technique is explained. IEEE 30-bus test system is used for the case studies in the "Simulation results and discussions" section, and then the chapter is concluded in the "Conclusion" section.

STATIC MODEL OF FACTS DEVICES

Static models of the TCSC and TCPS devices have already been discussed in chapter 6.

MATHEMATICAL PROBLEM FORMULATION

Objective Function

Objective Function of OPF With FACTS

The conventional formulation of OPF problem with TCSC and TCPS determines the optimal settings of control variables such as real power generations, generators' terminal voltages, transformers' tap settings, reactive power injections of shunt regulators, reactance value of TCSC devices and phase shifting angle of TCPS devices while minimizing the total fuel cost while satisfying all equality and inequality constraints. Total fuel cost of generating units having quadratic cost functions is given by:

$$FC = \sum_{i=1}^{NG} \left(a_i P_{Gi}^{\ 2} + b_i P_{Gi} + c_i \right) \tag{1}$$

For more practical and accurate model of the cost function, multiple valve steam turbines are incorporated for flexible operational facilities. Total generation cost of generating units with valve point discontinuities is given by:

$$FC = \sum_{i=1}^{NG} a_i P_{Gi}^{\ 2} + b_i P_{Gi} + c_i + \left| d_i \times \sin\left(e_i \times \left(P_{Gi}^{\min} - P_{Gi} \right) \right) \right| \tag{2}$$

where a_i, b_i, c_i, d_i, e_i: cost coefficients of the ith generator; P_{Gi} is the active power generation of the ith generator; $P_{Gi}^{\min}$ is the minimum active power generation limit of the ith generator.

Objective Function of ORPD With FACTS

The conventional formulation of ORPD problem determines the optimal setting of control variables such as generator terminal voltages, transformers tap setting, reactive power of shunt compensators, reactance values of TCSCs and phase shifting angles of TCPSs while minimizing two objective functions.

Minimize the Voltage Deviation

The first objective function is used to minimize the voltage deviation of all load buses and it may be expressed as follows:

$$f_1(x, y) = \sum_{i=1}^{NL} \left| \left(V_{l_i} - V_{ref} \right) \right| \tag{3}$$

where, $f_1(x, y)$ is the voltage deviation minimization objective function; x is the vector of dependent variable consisting of load voltages ($V_{l_1}, \cdots, V_{l_{NL}}$), generators' reactive powers ($Q_{g_1}, \cdots, Q_{g_{NG}}$), transmission lines' loadings ($S_{l_1}, \cdots, S_{l_{NTL}}$), reactance of TCSC devices ($x_{k_1}, \ldots, x_{k_{NTCSC}}$) and phase shifting angle of TCPS devices ($\alpha_1, \ldots, \alpha_{NTCPS}$); y is the vector of independent variables consisting of generators' voltages ($V_{g_1}, \cdots, V_{g_{NG}}$), transformers' tap settings ($T_1, \cdots, T_{NT}$), reactive power injections ($Q_{C_1}, \cdots, Q_{C_{NC}}$), active power flow through transmission line having TCSC devices ($PF_{TCSC_1}, \cdots, PF_{TCSC_{NTCSC}}$)

and active power flow through transmission line having TCPS devices ($PF_{TCPS_1},\cdots,PF_{TCPS_{NTCPS}}$).

Therefore x and y can be expressed as:

$$x=[V_{l_1},\cdots,V_{l_{NL}},Q_{g_1},\cdots,Q_{g_{NG}},S_{l_1},\cdots,S_{l_{NTL}},x_{k_1},\ldots,x_{k_{NTCSC}},\alpha_1,\ldots,\alpha_{NTCPS}] \tag{4}$$

$$y=[V_{g_1},\cdots,V_{g_{NG}},T_1,\cdots,T_{NT},Q_{C_1},\cdots,Q_{C_{NC}},PF_{TCSC_1},\cdots,PF_{TCSC_{NTCSC}},PF_{TCPS_1},\cdots,PF_{TCPS_{NTCPS}}] \tag{5}$$

where NG, NL are the number of generator and load buses; NTL, NT, NC are the number of transmission lines, regulating transformers and shunt compensators; V_{ref}, V_{L_i} are the reference voltage of the load buses and load voltage of ith bus.

Minimize the Transmission Loss

The second objective function is used to minimize the transmission loss and is given by

$$f_2(x,y)=P_{loss}=\sum_{k=1}^{NTL}G_k\left[\;V_m^{\,2}+V_n^{\,2}-2|V_m||V_n|\cos\left(\partial_m-\partial_n\right)\;\right] \tag{6}$$

where, $f_2(x,y)$ is the transmission loss minimization objective function; P_{loss} is the total power loss; G_k is the conductance of kth line connected between m^{th} and n^{th} buses; V_m, V_n are the voltage of mth and n^{th} buses respectively; ∂_m, ∂_n are the phase angle of mth and n^{th} bus voltages.

Constraints

The OPF with TCSC and TCPS are subjected to the following constraints.

Equality Constraints

These constraints represent the load flow equations as given below:

$$\begin{cases} \sum_{i=1}^{NB}\left(P_{Gi} - P_{Li}\right) + \sum_{i=1}^{NTCPS} P_{is} = \sum_{i=1}^{NB} \sum_{j=1}^{NB} \left|V_i\right|\left|V_j\right|\left|Y_{ij}\right| \cos\left(\theta_{ij} - \delta_{ij}\right) \\ \sum_{i=1}^{NB} Q_{Gi} - Q_{Li} + \sum_{i=1}^{NTCPS} Q_{is} = -\sum_{i=1}^{NB} \sum_{j=1}^{NB} \left|V_i\right|\left|V_j\right|\left|Y_{ij}\right| \sin\left(\theta_{ij} - \delta_{ij}\right) \end{cases} \quad (7)$$

where P_{Li} is the active power demand of the ith bus; Q_{Gi}, Q_{Li} are the reactive power generation and demand of the ith bus, respectively; P_{is}, Q_{is} are the injected active and reactive powers of TCPS at the ith bus, respectively; Y_{ij} is the admittance of transmission line connected between the ith bus and the jth bus; θ_{ij} is the admittance angle of transmission line connected between the ith bus and the jth bus; NB is the number of buses; $NTCPS$ is the number of TCPSs.

Inequality Constraints

Generator constraints:

$$\begin{cases} V_i^{\min} \leq V_i \leq V_i^{\max} \\ PG_i^{\min} \leq PG_i \leq PG_i^{\max} \qquad i \in NG \\ Q_{Gi}^{\min} \leq Q_{Gi} \leq Q_{Gi}^{\max} \end{cases} \quad (8)$$

Load bus constraints:

$$V_{Li}^{\min} \leq V_{Li} \leq V_{Li}^{\max} \qquad i \in NL \quad (9)$$

Transmission line constraints:

$$S_{Li} \leq S_{Li}^{\max} \qquad i \in NTL \quad (10)$$

Transformers tap constraints:

$$TC_i^{\min} \le TC_i \le TC_i^{\max} \qquad i \in NT \tag{11}$$

Shunt compensator constraints:

$$QC_i^{\min} \le QC_i \le QC_i^{\max} \qquad i \in NC \tag{12}$$

TCSC reactance constraints:

$$X_{ti}^{\min} \le X_{ti} \le X_{ti}^{\max} \qquad i \in NTCSC \tag{13}$$

TCPS phase shift constraints:

$$\alpha_i^{\min} \le \alpha_i \le \alpha_i^{\max} \qquad i \in NTCPS \tag{14}$$

where $V_{Gi}^{\min}, V_{Gi}^{\max}$ are the minimum and maximum voltage limits, respectively, of the ith generator bus; $P_{Gi}^{\min}, P_{Gi}^{\max}$ are the minimum and maximum active power generation limits, respectively, of the ith bus; $Q_{Gi}^{\min}, Q_{Gi}^{\max}$ are the minimum and maximum reactive power generation limits, respectively, of the ith bus; $V_{Li}^{\min}, V_{Li}^{\max}$ are the minimum and maximum voltage limits, respectively, of the ith load bus; $S_{Li}, S_{Li}^{\max}$ are the apparent power flow and maximum apparent power flow limit, respectively, of the ith branch; $T_i^{\min}, T_i^{\max}$ are the minimum and maximum tap setting limits, respectively, of the ith regulating transformer; $Q_{Ci}^{\min}, Q_{Ci}^{\max}$ are the minimum and maximum reactive power injection limits, respectively of the ith shunt compensator; $X_{ti}^{\min}, X_{ti}^{\max}$ are the minimum and maximum reactance limits, respectively, of the ith TCSC; $\alpha_i^{\min}, \alpha_i^{\max}$ the minimum and maximum phase shift angle limits, respectively, of the ith TCPS; NG is the number of generator buses; NL is the number of load buses; NTL is the number of transmission line; NT is the number of regulating transformers; NC is the number of shunt compensators; $NTCSC$, $NTCPS$ are the number of TCSC and TCPS devices, respectively.

EVOLUTIONARY OPTIMIZATION TECHNIQUES APPLIED TO OPF AND ORPD INCORPORATING TCSC AND TCPS DEVICES

BBO Algorithm Applied to OPF and ORPD With TCSC and TCPS Devices

Firstly, HDE-BBO algorithm is employed to solve constrained OPF problem incorporating multiple FACTS devices like TCSC and TCPS and to find the optimal solutions satisfying both equality and inequality constraints. The algorithm of the proposed method is as enumerated below.

Step 1: Initialize the population randomly in the search space, i.e. the initial positions of *SIVs* (independent variables such as active powers of all generators except slack bus, generators' voltages, tap settings of regulating transformers, reactive power injections of shunt capacitors, power flow through the transmission lines where TCSC and TCPS devices are placed and power injections of TCPS devices for OPF and generators' voltages, tap settings of regulating transformers, reactive power injections for ORPD) of each habitat should be randomly selected while satisfying different operational constraints. Several numbers of habitats depending upon the population size are being generated. Each habitat represents a potential solution.

Step 2: Perform TCSC and TCPS based load flow using these *SIVs* and determine all dependent variables such as load voltages, active power of slack bus, generators' reactive powers, reactance of TCSC and phase shifting angle of TCPS etc.

Step 3: Evaluate the fuel cost (i.e. *HSI*) for each habitat of the population set.

Step 4: Sort the habitats from the best to the worst using *HSI* values.

Step 5: Identify few elite habitats based on the *HSI* values. These best habitat sets are kept as it is without any modification.

Step 6: Assign number of species to each habitat based on *HSI*

Step 7: Modify the independent variables (*SIVs*) of the non-elite habitats using hybrid migration operation as explained in Step 6 of the section "BBO Algorithm" in Chapter 1.

Step 8: Update species count probability of each habitat. Based on habitat modification probability, modify the independent variables (*SIVs*) of non-elite habitats by probabilistic mutation operation.

Step 9: Go to Step 2 for the next iteration.
Step 10: Stop iterations after a predefined number of iterations.

TLBO Algorithm Applied to OPF and ORPD With TCSC and TCPS Devices

The main steps of the proposed TLBO approach applied to FACTS based OPF and ORPD problem are listed as follows.

Step 1: Depending upon the population size, several numbers of initial solutions are initialized randomly within their specified operating limits.
Step 2: The independent variables of OPF problem (active power of all generator buses except slack bus, voltage of generator buses, reactive power injection, tap settings of transformers, reactance of TCSC, power injections of TCPS devices) of each initial solution string are updated using the following expression to
Step 3: The dependent variables of the problem are evaluated using load flow analysis.
Step 4: The fitness values of the current population *(P)* and oppositional population *(OP)* sets are evaluated.
Step 5: n_p number of fittest individuals are selected from $\{P \cup OP\}$.
Step 6: Sort the requisite number of fittest vectors are ed from best to worst.
Step 7: Few elite solutions are selected based on the fitness values.
Step 8: The independent variables (active power of all generator buses except slack bus, voltage of generator buses, reactive power injection, tap settings of transformers, reactance of TCSC) of each non-elite solution are modified by performing teaching and learning phase of TLBO algorithm.
Step 9: The dependent variables as mentioned above of TCSC and TCPS based OPF are evaluated. Afterward, the fitness value of the population set is calculated.
Step 10: Based on a jumping rate j_r (i.e. jumping probability), the opposite population of the current population is evaluated and fitness values of the opposite population are calculated.
Step 11: n_p fittest individuals are selected from the union of the current population and the opposite population.
Step 12: Go to Step 8 for the next iteration until termination criterion is made.

HCRO Algorithm Applied to OPF and ORPD With TCSC and TCPS Devices

The HCRO algorithm is applied to seek out the optimal or near-optimal solutions for the OPF problem with TCSC and TCPS devices. Its computational procedure can be illustrated using the following steps:

Step 1: Confine the search space, i.e., specify the lower and upper limits of each decision variable such as active powers of all generators except slack bus, generators' voltages, tap settings of regulating transformers, reactive power injections of shunt capacitors, power flow through the transmission lines where TCSC and TCPS devices are placed and power injections of TCPS devices.

Step 2: Initialize the input parameters of the proposed HCRO approach.

Step 3: Run the load flow to find the slack bus power, load voltages, generators' reactive powers, and reactance of TCSC and phase shifting angle of TCPS etc. In this study load flow is performed by Newton-Raphson method. Each feasible solution represents the molecular structure of each molecule.

Step 4: Evaluate the fitness value. Based on the objective value, sort the molecules and few are kept as elite molecules.

Step 5: Assign potential energy same as fitness value for each molecule of the population set.

Step 6: Apply on-wall ineffective collision operations, decomposition operation, inter-molecular ineffective collision and synthesis collision operation on non elite molecules as described in Steps 4, 5, and 6 of the section "HCRO Algorithm Applied to ELD" in Chapter 3 to modify the independent variables i.e. generators' active power, voltage of generator buses, reactive power injection, tap settings of transformers, power flow through the transmission lines where TCSC and TCPS devices are placed and power injections of TCPS devices.

Step 7: To check the feasibility, run load flow and find the dependent variables. If any dependent variable violates the operating limit, replace that solution set by new feasible solution.

Step 8: Compute fitness value for each newly generated feasible solution set.

Step 9: If the stopping criteria is satisfied, go to Step 10. Otherwise, go to Step 4.

Step 10: Output optimal settings of the control variables and the corresponding cost.

GWO Algorithm Applied to OPF and ORPD With TCSC and TCPS Devices

Step 1: Read line data, bus data, fuel cost co-efficient, load flow parameters, etc. of the power system and specify the upper and lower limits of each control variables of OPF like active power generation, generators' bus voltage, load bus voltages, reactive power generations, tap settings of the tap changing transformers, VAr injection of the shunt compensating devices and line flow through each transmission lines. Afterward, randomly select the independent control variables within their operating limits. Thereafter, Newton Raphson based load flow method is run to evaluate the slack bus power and the other dependent variables. Check whether the control variables are within the limits or not. If any of them violates the limits, corresponding set is discarded and initialise again. In this fashion, a feasible initial solution is generated. Now several initial solution set depending upon the population size are generated. A feasible solution set (control variables) represents the position of different GWO individuals.

Step 2: The input parameters such as a, A and C of GWO (Singh et al., 2017) algorithm is initialized.

Step 3: Evaluate the fitness function of the initial solution set and elite solutions are identified.

Step 4: Find the updated best position of alpha, beta and delta wolves and independent variables of OPF are modified. Thereafter, Newton Raphson based load flow is run to obtain the dependent variables. The feasibility of the controlled variables is checked. If any one of them does not satisfy the inequality constraints, the particular set is replaced by the best feasible solution set.

Step 5: Evaluate the fitness value of the feasible population set.

Step 6: Check if the number of iteration is less than the maximum specified iteration or not.

Step 7: When the maximum iteration is reached display the fitness value and the updated final position of the wolves denoted by X(t+1) from the prey which is generally a circle around the prey.

Step 8: If the maximum iteration is not reached, iteration cycle is incremented.

Step 9: Go to Step 4 until a stopping criterion is satisfied.

KHA Algorithm Applied to OPF and ORPD With TCSC and TCPS Devices

The procedure for implementing the KHA (Mukherjee & Mukherjee 2016; Mehdinejad & Shayanfar, 2017) algorithm in solving optimal power flow problem to get minimum cost may be summarized by the following steps:

Step 1: All independent variables such as all generators' active power except slack bus, generators' voltages, tap settings of regulating transformers, should be randomly selected between their operating limits

Step 2: Run the Newton-Raphson load to determine the slack bus power, load voltages, power flow through the transmission line, and check whether they satisfy the operating limits or not. If any of these parameters violate the operating limits; discard that population set and re-initialize the corresponding set. Depending upon the population size, several solutions are generated. Each feasible solution set represents the initial position of each krill individuals.

Step 3: Evaluate the fitness value and sort the solutions from best to worst.

Step 4: Select few best solutions for elitism.

Step 5: Evaluate the three motion index, namely, motion induced by other individual; foraging motion and random diffusion.

Step 6: The non-elite krill individuals' position are modified .

Step 7: Applying crossover and mutation to update the position of each non-elite krill individuals.

Step 8: Check whether the independent variables violate the operating limits or not. If any independent variable is less than the minimum level it is made equal to minimum value and if it is greater than the maximum level it is made equal to maximum level.

Step 9: Run Newton-Raphson based load flow to determine the dependent variables and check whether they satisfy the system operating constraints or not. The infeasible solutions are replaced by the best feasible solutions.

Step 10: Go to Step 4 until the current iteration number reaches the pre specified maximum iteration number.

SIMULATION RESULTS AND DISCUSSIONS

IEEE 30 Bus System

The applicability and validity of the different techniques is tested on IEEE 30-bus system for OPF problems having multiple numbers of TCSC and TCPS devices. IEEE 30-bus system contains 6 generating units connected at bus number. 1, 2, 5, 8, 11 and 13; 30 buses, 4 regulating transformers connected between the line numbers 6-9, 6-10, 4-12 and 27-28; 2 shunt compensators connected at bus number 10 and 24; 41 transmission lines. The single line diagram of IEEE 30-bus system including two TCSC and two TCPS devices is illustrated in Figure 1. The system bus data and line data, the generators' capacities and cost coefficient data including valve point discontinuities are taken form (Basu, 2008). The generators' active powers (except slack bus-1) and generators' voltages, tap settings of the regulating transformers, VAr injections of the shunt capacitors, power flow through the transmission lines where TCSC and TCPS devices are placed and power injections of TCPS devices are considered as controlled variables. The voltage magnitudes of the generator and load buses are considered within the range of [0.95 *p.u.*, 1.1 *p.u.*] and [0.95 *p.u.*, 1.05 *p.u.*], respectively (Basu, 2008). Tap settings of regulating transformers are within the range of $[0.9p.u., 1.1p.u.]$ (Roy et al., 2011). The VAr injection of the shunt capacitors are taken within the interval of [0 MVAr, 30 MVAr] (Roy et al., 2011). The reactance limits of TCSC in *p.u.* is $0 \le x_t \le 0.15$. The limits of phase shift angle (radian) of TCPS are taken as $-0.075 \le \alpha \le 0.075$. The limits of transmission line loading are adopted from. The reactive power generation limits of different generator buses are taken from . The base values of apparent power and voltage used for this test system are taken as 100 MVA and 11 KV, respectively. All the programs are developed in MATLAB language, using MATLAB 7.1 and the system configuration is on core 2 duo processor, 2.0 GHz with 1GB RAM. The feasibility of the proposed BBO, KHA, TLBO, GWO, CRO, and HCRO method are tested on modified IEEE 30-bus system in five different cases (incorporating single and multiple TCSC and TCPS devices) and its performance is compared to other population-based optimization techniques like RGA and PSOIWA . For the sake of a fair comparison, same system with the same number of FACTS devices and same location of FACTS devices are used in this simulation study. The simulation results shown in different Tables are the best solutions obtained in fifty runs. Seven cases have been considered and analyzed, with FACTS devices located at different transmission lines and buses; Case 1: OPF with one TCSC, Case 2: OPF with two TCSCs, Case 3: OPF with one TCSC and one TCPS, Case 4: OPF with two TCSCs two TCPSs, Case 5 (ORPD with three TCSCs), Case 6 (ORPD with one TCSC and one TCPS) and Case 7 (ORPD with two TCSCs and three TCPSs). The results obtained with all the above cases are tabulated and presented in Table 1, 2, 3, 4, 5, 6 and 7.

Case 1 (OPF With One TCSC)

In this case, one TCSC device is placed in the transmission line that is connected between the buses 4-12. The optimal settings of control variables (i.e. active powers and voltages of generator buses, tap settings of the regulating transformers, VAr injections of the shunt capacitors and reactance of TCSC) and the total fuel costs which are obtained by the different methods are shown in the Table 1. The results clearly show that the total costs obtained by all methods have significantly improved in this case, compared to Case 1. The simulation results also show that HCRO method gives the lowest fuel

Figure 1. Single line diagram of IEEE 30 system with two TCSC and two TCPS devices

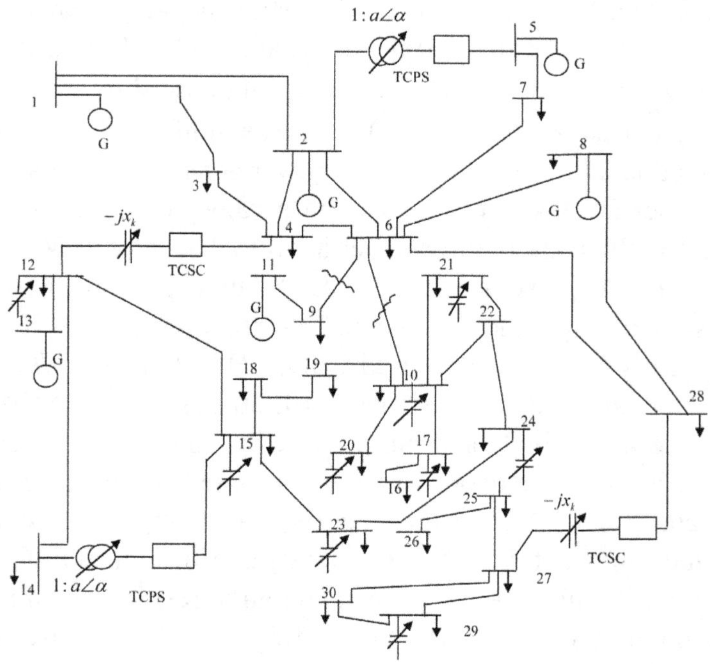

cost (820.0346$/hr.) as compared to other population-based optimization techniques. The information presented in the result, provides the additional indicators to evaluate the accuracy in finding the optimal solution and the convergence rate (number of time that the algorithm is able to converge to a feasible solution).

Case 2 (OPF With Two TCSCs)

Two TCSC devices, one in the transmission line connected between the buses 4-12 and the other in the transmission line connected between the buses 27-28 are placed in this case. The simulation results which are obtained by the

Table 1. Comparative simulation results obtained by various algorithms (Case 1)

Algorithms	With TCSC at Line 4-12					
	HCRO	CRO	GWO	TLBO	KHA	BBO
PG_1 (MW)	218.9528	215.6516	218.9757	218.2293	219.7853	219.8112
PG_2 (MW)	25.2056	29.0082	26.7143	26.4941	25.6947	26.3050
PG_5 (MW)	17.1536	15.9043	15.2664	15.7660	16.2711	15.7990
PG_8 (MW)	10.4728	10.9436	10.3417	10.5071	10.0000	10.9100
PG_{11} (MW)	10.0677	10.1375	10.9083	10.8315	10.9202	10.0010
PG_{13} (MW)	12.3551	12.3298	12.2203	12.4758	12.0000	12.0010
V_1 (*p.u.*)	1.0898	1.0842	1.0951	1.0818	1.0997	1.0956
V_2 (*p.u.*)	1.0666	1.0585	1.0691	1.0686	1.0836	1.0824
V_5 (*p.u.*)	1.0323	1.0315	1.0254	1.0360	1.0559	1.0591
V_8 (*p.u.*)	1.0414	1.0359	1.0312	1.0426	1.0632	1.0618
V_{11} (*p.u.*)	1.0830	1.0694	0.9982	1.0759	1.0998	1.0997
V_{13} (*p.u.*)	1.0756	1.0542	1.0999	1.0038	1.0993	1.0965
QC_{10} (MVAr)	0.1792	0.1570	0.1748	0.0655	29.6934	29.4600
QC_{24} (MVAr)	0.1504	0.1708	0.1523	0.1480	10.2316	9.6200
TC_{6-9}	1.0170	1.0583	0.9682	1.0185	1.0296	1.0021
TC_{6-10}	0.9936	1.0528	0.9450	0.9715	0.9787	1.0253
TC_{4-12}	1.0185	1.0543	0.9569	1.0100	0.9704	0.9771
TC_{27-28}	0.9940	0.9718	0.9612	0.9634	0.9566	0.9582
$TCSC_{4-12}$ (*p.u.*)	0.014521	0.009837	0.017845	0.024313	0.024346	0.01395
Cost ($/hr)	820.0346	820.3164	820.5807	820.8124	820.9801	821.2299

different methods are shown in the Table 2. The results clearly show that the total costs obtained by all methods have significantly improved in this case, compared to Cases 1. It can be inferred from the results that the solution obtained from the HCRO (819.8345$/hr.) method is better than those obtained by other techniques.

Table 2. Comparative simulation results obtained by various algorithms (Case 2)

Algorithms	With TCSC at Lines 4-12 and 27-28					
	HCRO	CRO	GWO	TLBO	KHA	BBO
PG_1 (MW)	218.4932	218.4643	215.9875	218.2143	219.7096	219.8161
PG_2 (MW)	25.8622	27.4943	28.1688	26.3407	26.5917	25.6980
PG_5 (MW)	16.3968	15.7552	16.5561	15.2462	15.8797	16.4490
PG_8 (MW)	10.1253	10.0086	10.9348	10.4639	10.0000	10.3040
PG_{11} (MW)	10.2447	10.3790	10.2552	11.7250	10.5283	10.1210
PG_{13} (MW)	12.9419	12.2537	12.0805	12.1747	12.0000	12.3270
V_1 (*p.u.*)	1.0689	1.0967	1.0974	1.0731	1.0998	1.1000
V_2 (*p.u.*)	1.0459	1.0820	1.0737	1.0492	1.0827	1.0840
V_5 (*p.u.*)	1.0113	1.0538	1.0389	1.0301	1.0523	1.0543
V_8 (*p.u.*)	1.0163	1.0570	1.0286	1.0462	1.0582	1.0595
V_{11} (*p.u.*)	1.0711	1.0686	1.0299	1.0469	1.0949	1.0823
V_{13} (*p.u.*)	1.0764	1.0831	1.0985	1.0926	1.0977	1.0998
QC_{10} (MVAr)	0.1608	0.2968	0.2845	0.2806	29.0426	28.6513
QC_{24} (MVAr)	0.1073	0.0840	0.0995	0.1572	10.6541	10.7032
TC_{6-9}	1.0200	1.0605	1.0782	1.0176	1.0144	1.0082
TC_{6-10}	0.9507	0.9763	0.9874	0.9607	0.9944	1.0400
TC_{4-12}	0.9672	0.9822	1.0825	1.0688	0.9753	1.0113
TC_{27-28}	0.9600	0.9760	1.0048	0.9566	0.9565	0.9808
$TCSC_{4-12}$ (*p.u.*)	0.023406	0.010325	0.018293	0.009843	0.015628	0.006373
$TCSC_{27-28}$ (*p.u.*)	0.025906	0.032156	0.020045	0.021382	0.032954	0.016782
Cost ($/hr)	819.8345	820.0046	820.2543	820.6937	820.6805	821.0336

Case 3 (OPF With One TCSC and One TCPS)

In order to further verify the effectiveness of the proposed techniques, both TCSC and TCPS are placed in IEEE 30-bus system in this case study. One TCSC device is placed in the line connected between the buses 4-12 and one TCPS device is placed in the line connected between the buses 2-5. The comparison results are given in the Table 3. The simulation results show that the HCRO technique gives more reduction in cost (819.5943 \$/hr), compared to BBO based cost (820.5558 \$/hr), KHA based cost (820.2682 \$/hr), TLBO based cost (820.1344 \$/hr), GWO based cost (819.9873\$/hr) and CRO based cost (819.8034 \$/hr) without violating any operational constraints. It is also clear that by inserting TCPS based FACTS device with TCSC device, the fuel cost is further reduced as compared to cases 1-2.

Case 4 (OPF With Two TCSCs and Two TCPSs)

To investigate the effectiveness of the proposed techniques for more complicated systems, two TCSC devices and two TCPS devices are incorporated in IEEE 30-bus system. Two TCSC devices, one in the transmission line connected between the buses 4-12 and the other in the transmission line connected between the buses 27-28 are placed and two TCPS devices, one in the transmission line connected between the buses 2-5 and the other in the transmission line connected between the buses 14-15 are placed. The optimal control variable settings of active power and voltage of generator buses, tap settings of the regulating transformers, VAr injections of the shunt capacitors, reactances of TCSCs and phase angles of TCPSs and the fuel costs obtained by different methods are given in the Table 4. It is found from the results that the fuel costs are reduced further for all the techniques in this case compared to the other cases. It is also noticed that HCRO technique outperforms the other techniques.

Case 5 (ORPD With Three TCSCs)

Three TCSC devices, one in the transmission line connected between the buses 4-12 and the others in the transmission line connected between buses 16-17 and 27-28 are placed in IEEE 30-bus system. Optimization capability of BBO, KHA, TLBO, GWO, CRO and HCRO technique are compared.

Table 3. Comparison of simulation results obtained by various algorithms (Case 3)

Algorithms	With TCSC at Line 4-12 and TCPS at Line 2-5					
	HCRO	CRO	GWO	TLBO	KHA	BBO
PG_1 (MW)	218.4596	219.1190	218.0653	217.9431	219.7052	219.6723
PG_2 (MW)	27.7235	25.8863	28.0649	28.6189	26.2723	26.8680
PG_5 (MW)	15.3716	16.5786	15.2008	15.2558	16.6557	16.2031
PG_8 (MW)	10.1350	10.1543	10.5511	10.1533	10.0000	10.0000
PG_{11} (MW)	10.4317	10.3662	10.4074	10.2606	10.0000	10.0180
PG_{13} (MW)	12.1282	12.1708	12.0008	12.1572	12.0000	12.0000
V_1 (*p.u.*)	1.0936	1.0975	1.0996	1.0996	1.1000	1.0991
V_2 (*p.u.*)	1.0835	1.0748	1.0770	1.0770	1.0841	1.0739
V_5 (*p.u.*)	1.0437	1.0404	1.0219	1.0219	1.0566	1.0451
V_8 (*p.u.*)	1.0541	1.0489	1.0551	1.0551	1.0633	1.0459
V_{11} (*p.u.*)	1.0934	1.0562	1.0892	1.0892	1.0995	1.0999
V_{13} (*p.u.*)	1.0673	1.0929	1.0909	1.0909	1.0999	1.1000
QC_{10} (MVAr)	0.2844	0.2885	0.1811	0.2127	27.7983	26.5812
QC_{24} (MVAr)	0.1741	0.1068	0.0745	0.1451	10.1931	10.6091
$TC_{6\text{-}9}$	1.0476	1.0185	0.9867	1.0185	1.0810	1.0337
$TC_{6\text{-}10}$	1.0281	1.0274	1.0445	1.0734	0.9142	0.9256
$TC_{4\text{-}12}$	1.0005	1.0008	1.0287	0.9934	0.9703	0.9517
$TC_{27\text{-}28}$	1.0132	0.9872	0.9735	0.9498	0.9564	0.9424
$TCSC_{4\text{-}12}$ (*p.u.*)	0.025643	0.008314	0.02183	0.032413	0.0066	0.003052
$TCPS_{2\text{-}5}$ (radian)	0.014622	-0.013421	0.04562	-0.034621	0.0345	0.033186
Cost ($/hr)	819.5943	819.8034	819.9873	820.1344	820.2682	820.5558

Minimization of Voltage Deviation

In this case, the different algorithms are applied for improvement of voltage profile. The optimal settings of control variables (i.e. voltages of generator buses, tap settings of the regulating transformers, VAr injections of the shunt capacitors, reactances of TCSC devices), transmission loss and SVD that are obtained by the different methods are shown in the Table 5. The results

Table 4. Comparison of simulation results obtained by various algorithms (Case 4)

Algorithms	With TCSC at Lines 4-12 and 27-28 and TCPS at Lines 2-5 and 14-15					
	HCRO	**CRO**	**GWO**	**TLBO**	**KHA**	**BBO**
PG_1 (MW)	215.6987	214.2469	213.4518	213.4917	219.6314	219.8166
PG_2 (MW)	25.6473	26.8562	26.6255	26.6747	28.0625	27.7740
PG_5 (MW)	16.3728	16.5357	16.5702	16.0500	15.0000	15.0530
PG_8 (MW)	11.2040	11.7066	11.9003	11.7065	10.0000	10.0890
PG_{11} (MW)	11.3024	11.2023	11.5034	11.3911	10.0000	10.0000
PG_{13} (MW)	12.9285	12.5416	12.8247	13.5347	12.0000	12.0000
V_1 (*p.u.*)	1.0854	1.0948	1.0865	1.0980	1.0997	1.1000
V_2 (*p.u.*)	1.0714	1.0588	1.0536	1.0634	1.0840	1.0841
V_5 (*p.u.*)	1.0413	1.0358	1.0229	1.0193	1.0556	1.0565
V_8 (*p.u.*)	1.0457	1.0445	1.0351	1.0321	1.0630	1.0624
V_{11} (*p.u.*)	1.0044	0.9500	1.0109	1.0290	1.0990	1.0849
V_{13} (*p.u.*)	1.0588	1.0729	1.0095	1.0792	1.1000	1.1000
QC_{10} (MVAr)	0.1002	0.1004	0.2765	0.1616	28.6015	29.9153
QC_{24} (MVAr)	0.1127	0.1219	0.1194	0.1292	10.1727	10.7924
$TC_{6\text{-}9}$	1.0559	0.9534	1.0203	0.9108	1.0646	1.0531
$TC_{6\text{-}10}$	0.9546	0.9518	1.0278	0.9900	0.9327	0.9407
$TC_{4\text{-}12}$	1.0309	1.0508	0.9784	1.0265	0.9700	0.9736
$TC_{27\text{-}28}$	1.0179	0.9672	0.9620	0.9206	0.9561	0.9564
$TCSC_{4\text{-}12}$ (*p.u.*)	0.021923	0.021684	0.041362	0.031462	0.010241	0.008396
$TCSC_{27\text{-}28}$ (*p.u.*)	0.014621	0.017804	0.024672	0.028845	0.041469	0.027792
$TCPS_{2\text{-}5}$ (radian)	0.021672	-0.009887	-0.033561	0.014793	0.010429	0.032757
$TCPS_{14\text{-}15}$ (radian)	-0.013945	0.034712	0.025091	0.009346	-0.010811	-.004436
Cost ($/hr)	819.3657	819.6011	819.8347	819.9997	820.0751	820.2042

clearly show that the voltage deviation has improved after incorporating TCSC as compared to without FACTS. It may also be noticed in Table 5 that the objective function value (SVD) obtained by HCRO is better than those obtained by BBO, KHA, TLBO, GWO and CRO. In this case, the voltage deviations are reduced from 0.1648 in the initial state to 0.109784 with a reduction of 5.5% for HCRO algorithm. The comparison with the results reported in chapter 4 is given in Table 10. From the comparison, the proposed algorithm gives the best results for voltage deviations which demonstrate the effectiveness of the proposed algorithm.

Minimization of Transmission Loss

The results obtained for transmission loss minimization objective function by BBO, KHA, TLBO, GWO, CRO and HCRO with three TCSC are reported in the Table 6. It may be seen that BBO, KHA, TLBO, GWO, CRO and HCRO give significantly better transmission losses as compared to without FACTS. It is also evident that HCRO approach outperforms then BBO, KHA, TLBO, GWO and CRO techniques. The convergence characteristic of the HCRO algorithm is better than the other algorithms. The HCRO algorithm reaches a minimum loss of 6.3240MW. The optimal values of the control variables are given in the different column for the different algorithms in Table 6.

Case 6 (ORPD With One TCSC and One TCPS)

This case study constitutes 14 control variables which need to be optimized. In this case, one TCSC device (at 4-12) and one TCPS (at 2-5) have been used to test the effectiveness of ABC, BBO, HDE-BBO, TLBO, OTLBO, CRO and HCRO methods over PSOIWA and RGA.

Minimization of Voltage Deviation

The simulation results that are obtained by the different methods are shown in the Table 7. In this case, the proposed HCRO-based approach is applied for improvement of voltage profile. The convergence characteristic of the algorithm for this case is better. The optimal values of the control variable settings obtained in this case are given in the second column of Table 7. In this case, the voltages deviations are reduced to 0.110849 with respect other algorithms. From the comparison, the proposed algorithm gives the best results for voltage deviations which demonstrate the effectiveness of the proposed algorithm. It is also evident that HCRO outperforms BBO, KHA, TLBO, GWO and CRO.

Minimization of Transmission Loss

The results obtained for transmission loss minimization objective function by BBO, KHA, TLBO, GWO, CRO and HCRO with three TCSC are reported in the Table 8. It may be seen that PSOIWA, RGA, ABC, BBO, HDE-BBO, TLBO, OTLBO, CRO and HCRO give significantly better transmission losses as compared to Case 1 and case 2. It is also evident that HCRO approach outperforms then BBO, KHA, TLBO, GWO and CRO techniques incorporating one TCSC and one TCPS.

Table 5. Comparative results obtained by various algorithms for voltage deviation minimization objective (Case 2)

Algorithms	Minimization of voltage deviation					
	HCRO	CRO	GWO	TLBO	KHA	BBO
V_1 (*p.u.*)	1.0337	0.9785	1.0270	1.0185	1.0203	1.0155
V_2 (*p.u.*)	0.9831	1.0579	1.0434	1.0048	0.9931	0.9970
V_5 (*p.u.*)	1.0182	1.0167	1.0051	1.0166	1.0151	1.0204
V_8 (*p.u.*)	1.0171	0.9865	1.0053	1.0194	1.0215	1.0093
V_{11} (*p.u.*)	1.0054	1.0505	0.9952	0.9642	0.9935	0.9943
V_{13} (*p.u.*)	1.0005	1.0353	0.9945	1.0197	1.0150	1.0234
QC_{10} (MVAr)	0.1531	0.1809	0.2353	0.0495	0.0822	0.2109
QC_{24} (MVAr)	0.1983	0.1469	0.1071	0.1862	0.1479	0.1640
$TC_{6\text{-}9}$	1.0169	1.0717	1.0047	0.9765	1.0058	1.0070
$TC_{6\text{-}10}$	0.9574	0.9428	0.9693	0.9003	0.9114	0.9672
$TC_{4\text{-}12}$	0.9288	0.9972	0.9200	0.9656	0.9499	0.9729
$TC_{27\text{-}28}$	0.9593	0.9495	0.9458	0.9587	0.9561	0.9518
$TCSC_{4\text{-}12}$ reactance (*p.u.*)	0.010213	0.013962	0.034562	0.012847	0.021645	0.031462
$TCSC_{16\text{-}17}$ reactance (*p.u.*)	0.073823	0.134257	0.074955	0.174562	0.114006	0.139621
$TCSC_{27\text{-}28}$ reactance (*p.u.*)	0.063678	0.010069	0.008961	0.072349	0.010541	0.009763
Voltage deviation (*p.u.*)	0.109784	0.110370	0.111628	0.115143	0.116931	0.118214
Transmission Loss (MW)	9.6543	10.6824	8.2567	8.2361	8.6007	8,2846

Case 7 (ORPD With Two TCSCs and Three TCPSs)

To test the potential of the BBO, KHA, TLBO, GWO, CRO and HCRO techniques in solving complicated systems, the standard IEEE 30-bus test system with two TCSCs (one in the transmission line connected between buses 4-12, the second one in the transmission line connected between the buses 27-28) and three TCPSs (one in the transmission line connected between the buses 2-5, the second one in the transmission line connected between the buses 6-10 and the third one in the transmission line connected between the buses 14-15) is considered.

Table 6. Comparative results obtained by various algorithms for transmission loss minimization (Case 5)

Algorithms	Minimization of transmission loss					
	HCRO	CRO	GWO	TLBO	KHA	BBO
V_1 (*p.u.*)	1.0996	1.1000	1.1000	1.0879	1.0714	1.0990
V_2 (*p.u.*)	1.0894	1.0901	1.0903	1.0705	1.0612	1.0825
V_5 (*p.u.*)	1.0646	1.0652	1.0656	1.0432	1.0318	1.0476
V_8 (*p.u.*)	1.0751	1.0754	1.0758	1.0635	1.0441	1.0665
V_{11} (*p.u.*)	1.0996	1.0998	1.1000	1.0981	1.0862	1.0860
V_{13} (*p.u.*)	1.1000	1.1000	1.1000	1.0900	1.0757	1.0964
QC_{10} (MVAr)	0.2954	0.2954	0.2718	0.1989	0.2694	0.2342
QC_{24} (MVAr)	0.0979	0.0979	0.1000	0.0899	0.1069	0.1259
$TC_{6\text{-}9}$	1.0103	1.0492	1.0894	1.0472	0.9813	1.0407
$TC_{6\text{-}10}$	0.9943	0.9425	0.9000	0.9213	1.0484	0.9609
$TC_{4\text{-}12}$	0.9630	0.9614	0.9619	0.9563	0.9712	0.9893
$TC_{27\text{-}28}$	0.9557	0.9574	0.9578	0.9660	0.9530	0.9868
$TCSC_{4\text{-}12}$ reactance (*p.u.*)	0.021462	0.019470	0.013471	0.008437	0.035421	0.013421
$TCSC_{16\text{-}17}$ reactance (*p.u.*)	0.018399	0.021856	0.008996	0.018440	0.018295	0.009643
$TCSC_{27\text{-}28}$ reactance (*p.u.*)	0.010006	0.009753	0.062834	0.010064	0.009362	0.034217
Voltage deviation (*p.u.*)	1.954691	1.971736	1.934732	1.873428	1.932654	1.915437
Transmission Loss (MW)	6.3240	6.3281	6,3306	6.3357	6.3397	6.3592

Minimization of Voltage Deviation

The comparison results are given in Table 9. From the results, it is noticed that the voltage deviation has been improved significantly (BBO, KHA, TLBO, GWO, CRO and HCRO) for all the techniques, compared to Case 1, case 2 and case 3. Table 9 also shows the objective function value obtained by HCRO method is the best (0.106732*p.u.*) amongst BBO (0.117396*p.u*), KHA (0.115288*p.u*), TLBO (0.113500*p.u*), GWO (0.110317*p.u*), CRO (0.109381*p.u*) and HCRO (0.106732*p.u*).

Table 7. Comparative results obtained by various algorithms for voltage deviation minimization objective (Case 6)

Algorithms	Minimization of voltage deviation					
	HCRO	CRO	GWO	TLBO	KHA	BBO
V_1 (*p.u.*)	1.0180	1.0071	1.0076	1.0063	0.9864	1.0278
V_2 (*p.u.*)	1.0094	1.0305	0.9756	1.0655	1.0039	0.9660
V_5 (*p.u.*)	1.0085	1.0197	1.0128	1.0186	1.0388	1.0197
V_8 (*p.u.*)	1.0101	0.9802	1.0150	0.9955	0.9966	1.0157
V_{11} (*p.u.*)	1.0608	1.0783	1.0252	1.0171	1.0565	1.0073
V_{13} (*p.u.*)	0.9921	1.0433	1.0555	0.9825	1.0649	1.0502
QC_{10} (MVAr)	0.2923	0.2385	0.0366	0.1389	0.1223	0.1549
QC_{24} (MVAr)	0.2114	0.1032	0.2218	0.1868	0.1698	0.1347
TC_{6-9}	1.0882	1.0748	1.0376	1.0397	1.0792	1.0234
TC_{6-10}	1.0413	1.0253	0.9092	0.9325	0.9296	0.9328
TC_{4-12}	0.9206	0.9786	1.0327	0.9096	1.0441	1.0084
TC_{27-28}	0.9646	0.9350	0.9634	0.9545	0.9480	0.9570
$TCSC_{4-12}$ reactance (*p.u.*)	0.028361	0.034165	0.017634	0.023157	0.032781	0.013628
$TCPS_{2-5}$ angle (radian)	0.036247	-0.013030	0.014327	0.009362	0.011905	0.009983
Voltage deviation (*p.u.*)	0.110849	0.114612	0.11596	0.11715	0.119803	0.120399
Transmission Loss (MW)	8.9413	8.9245	9.2068	10.9341	9.1336	8.6743

Minimization of Transmission Loss

The simulation results of the control variable, transmission loss and voltage deviation which are obtained BBO, KHA, TLBO, GWO, CRO and HCRO are shown in the Table 10. The simulation results show that the HCRO technique gives more reduction in loss (6.2343MW), compared to BBO based loss (6.2933 MW), KHA based loss (6.2816 MW), TLBO based loss (6.2705MW), GWO based loss (6.2748 MW), CRO based loss (6.2748 MW). So, HCRO technique gives the best result among other techniques.

Table 8. Comparative results obtained by various algorithms for transmission loss minimization (Case 6)

Algorithms	Minimization of transmission loss					
	HCRO	CRO	GWO	TLBO	KHA	BBO
V_1 (*p.u.*)	1.0884	1.0795	1.0931	1.0788	1.0974	1.0991
V_2 (*p.u.*)	1.0739	1.0633	1.0802	1.0646	1.0839	1.0880
V_5 (*p.u.*)	1.0489	1.0329	1.0523	1.0382	1.0555	1.0626
V_8 (*p.u.*)	1.0630	1.0444	1.0621	1.0425	1.0655	1.0701
V_{11} (*p.u.*)	1.0848	1.0959	1.0927	1.0671	1.0777	1.0977
V_{13} (*p.u.*)	1.0966	1.0983	1.0983	1.0890	1.0985	1.0978
QC_{10} (MVAr)	0.2795	0.2784	0.2814	0.2287	0.2950	0.2877
QC_{24} (MVAr)	0.0935	0.1103	0.1049	0.0917	0.1082	0.1026
$TC_{6\text{-}9}$	1.0016	0.9839	1.0618	1.0339	1.0729	1.0133
$TC_{6\text{-}10}$	1.0022	0.9958	0.9091	0.9521	0.9056	0.9920
$TC_{4\text{-}12}$	0.9736	0.9507	0.9569	1.0187	0.9662	0.9653
$TC_{27\text{-}28}$	0.9582	0.9329	0.9503	0.9627	0.9525	0.9562
$TCSC_{4\text{-}12}$ reactance (*p.u.*)	0.031723	0.01682	0.026205	0.013651	0.065336	0.012134
$TCPS_{2\text{-}5}$ angle (radian)	-0.013864	0.009116	-0.01043	0.007638	-0.009611	0.009328
Voltage deviation (*p.u.*)	1.887504	1.734129	1.812622	1.663245	1.783103	1.841135
Transmission Loss (MW)	6.2417	6.2659	6.2803	6.2837	6.2962	6.3096

CONCLUSION

In this chapter BBO, KHA, TLBO, GWO, CRO and HCRO techniques have been proposed, developed and successfully implemented to solve TCSC and TCPS based OPF problems considering the valve point discontinuities of the generating units and to solve the FACTS based optimal reactive power dispatch (ORPD) problem. The proposed methods have been tested on the modified IEEE 30-bus test system for different cases. The simulation results demonstrate the effectiveness and robustness of the proposed techniques to solve TCSC-TCPS based OPF problem. It is also observed that the FACTS

Table 9. Comparative results obtained by various algorithms for voltage deviation minimization objective (Case 7)

Algorithms	Minimization of voltage deviation					
	HCRO	CRO	GWO	TLBO	KHA	BBO
V_1 (*p.u.*)	1.0075	1.0089	1.0012	1.0106	0.9944	1.0064
V_2 (*p.u.*)	1.0359	1.0101	1.0146	1.0060	1.0026	0.9801
V_5 (*p.u.*)	1.0164	1.0212	1.0211	1.0126	1.0209	1.0213
V_8 (*p.u.*)	1.0061	0.9952	1.0025	1.0124	1.0013	1.0058
V_{11} (*p.u.*)	0.9915	1.0422	0.9947	1.0178	1.0310	1.0293
V_{13} (*p.u.*)	1.0011	1.0438	1.0473	1.0157	1.0608	1.0511
QC_{10} (MVAr)	0.1533	0.1176	0.2047	0.2328	0.1636	0.2276
QC_{24} (MVAr)	0.2005	0.1905	0.1007	0.1162	0.1194	0.1746
TC_{6-9}	1.0044	1.0451	1.0099	1.0390	1.0520	1.0450
TC_{6-10}	0.9586	0.9610	0.9416	0.9586	0.9190	0.9873
TC_{4-12}	0.9356	0.9943	0.9973	0.9556	1.0369	1.0215
TC_{27-28}	0.9582	0.9534	0.9466	0.9531	0.9495	0.9451
$TCSC_{4-12}$ reactance (*p.u.*)	0.013628	0.025744	0.034824	0.014562	0.021639	0.014521
$TCSC_{27-28}$ reactance (*p.u.*)	0.014321	0.007893	0.016935	0.022054	0.014782	0.009832
$TCPS_{2-5}$ angle (radian)	0.024779	-0.009380	0.018934	-0.011363	0.010070	-0.015237
$TCPS_{6-10}$ angle (radian)	-0.011632	0.014689	-0.008347	0.006328	-0.009217	0.019421
$TCPS_{14-15}$ angle (radian)	0.065487	0.010604	-0.013962	0.013482	-0.012833	0.008507
Voltage deviation (*p.u.*)	0.106732	0.109381	0.110317	0.113500	0.115288	0.117396
Transmission Loss (MW)	8.9147	8.0642	8.4109	7.9635	8.1347	8.7962

devices (TCSC and TCPS) can reduce the fuel cost compared to the normal system without FACTS. Hence, the results presented explicitly show the superiority of new techniques in order to achieve better performances of finding optimal solutions. All those algorithms are tested on the modified 6 generator IEEE 30-bus test system for different cases for voltage deviation minimization and transmission loss minimization objective functions. It is observed that the FACTS devices (TCSC and TCPS) can reduce the transmission loss and can improve the voltage profile compared to the normal system without FACTS devices. It is also clear from the simulation results obtained by different algorithms that the HCRO method has good convergence property.

Table 10. Comparative results obtained by various algorithms for transmission loss minimization (Case 7)

Algorithms	Minimization of transmission loss					
	HCRO	CRO	GWO	TLBO	KHA	BBO
V_1 (*p.u.*)	1.0755	1.0907	1.0995	1.1000	1.0928	1.0994
V_2 (*p.u.*)	1.0601	1.0733	1.0871	1.0882	1.0837	1.0884
V_5 (*p.u.*)	1.0309	1.0492	1.0605	1.0529	1.0644	1.0632
V_8 (*p.u.*)	1.0444	1.0608	1.0690	1.0680	1.0694	1.0733
V_{11} (*p.u.*)	1.0998	1.0984	1.0914	1.0907	1.0953	1.0863
V_{13} (*p.u.*)	1.0972	1.0994	1.0992	1.0990	1.0977	1.0966
QC_{10} (MVAr)	0.2003	0.2695	0.2929	0.2943	0.2900	0.2895
QC_{24} (MVAr)	0.1034	0.1032	0.0993	0.1012	0.1023	0.1101
TC_{6-9}	1.0406	1.0151	0.9887	1.0426	0.9780	1.0277
TC_{6-10}	0.9100	0.9463	1.0258	0.9399	1.0450	0.9812
TC_{4-12}	0.9468	0.9553	0.9743	0.9613	0.9667	0.9774
TC_{27-28}	0.9437	0.9493	0.9600	0.9545	0.9553	0.9644
$TCSC_{4-12}$ reactance (*p.u.*)	0.014112	0.019735	0.023412	0.013427	0.017438	0.017633
$TCSC_{27-28}$ reactance (*p.u.*)	0.003613	0.008037	0.017136	0.019234	0.008954	0.005482
$TCPS_{2-5}$ angle (radian)	0.013769	0.012390	-0.008427	0.009104	0.031208	-0.01324
$TCPS_{6-10}$ angle (radian)	-0.008351	0.007123	0.023628	0.014521	0.006710	0.009932
$TCPS_{14-15}$ angle (radian)	0.014006	0.006134	-0.018632	0.008213	-0.018321	0.021659
Voltage deviation (*p.u.*)	1.873412	1.831152	1.790361	1.793216	1.81307	1.757314
Transmission Loss (MW)	6.2343	6.2748	6.2748	6.2705	6.2816	6.2933

REFERENCES

Ambriz-Perez, H., Acha, E., & Fuerte-Esquivel, C. R. (2006). Incorporation of a TCSC firing model for optimal power flow solutions using Newton's method. *International Journal of Electrical Power & Energy Systems*, *28*(2), 77–85. doi:10.1016/j.ijepes.2005.10.003

Basu, M. (2008). Optimal power flow with FACTS devices using differential evolution. *International Journal of Electrical Power & Energy Systems*, *30*(2), 150–156. doi:10.1016/j.ijepes.2007.06.011

Basu, M. (2011). Multi-objective optimal power flow with FACTS devices. *Energy Conversion and Management*, *52*(2), 903–910. doi:10.1016/j.enconman.2010.08.017

Benabid, R., Boudour, M., & Abido, M. A. (2009). Optimal location and setting of SVC and TCSC devices using non-dominated sorting particle swarm optimization. *Electric Power Systems Research*, *79*(12), 1668–1677. doi:10.1016/j.epsr.2009.07.004

Bouktir, T., Belkacemi, M., & Zehar, K. (2000). Optimal power flow using modified gradient method. *Proceedings of ICEL*, 2, 436-442.

Carpentier, J. (1962). Contributional'étude dudispatchingéconomique,series 8: Bulletin delaSociétéFrançaisedesÉlectriciens,Août, Vol.3, pp. 431–47.

Dutta, S., & Roy, P. K. (2017). Optimal power flow with FACTS devices using a novel grey wolf algorithm. *Third International Conference on Science Technology Engineering & Management (ICONSTEM), IEEE Conferences*, 501 – 506.

Edward, J. B., Rajasekar, N., & Sathiyasekar, K. N. (2013). An enhanced bacterial foraging algorithm approach for optimal power flow problem including FACTS devices considering system loadability. *ISA Transactions*, *52*(5), 622–628. doi:10.1016/j.isatra.2013.04.002 PMID:23759251

Ghahremani, E., & Kamwa, I. (2013). Optimal placement of multiple-type FACTS devices to maximize power system loadability using a generic graphical user interface. *IEEE Transactions on Power Systems*, *28*(2), 764–778. doi:10.1109/TPWRS.2012.2210253

Jordehi, A. R. (2015). Brainstorm optimisation algorithm (BSOA): An efficient algorithm for finding optimal location and setting of FACTS devices in electric power systems. *International Journal of Electrical Power & Energy Systems*, *69*, 48–57. doi:10.1016/j.ijepes.2014.12.083

Keyan, L., Wanxing, S., & Yunhua, L. (2006). Research on reactive power optimization based on adaptive genetic simulated annealing algorithm. *International conference on power system technology, PowerCon*, 1–6.

Lai, L. L., & Ma, J. T. (1995). Power flow control in FACTS using evolutionary programming. *IEEE International Conference Evolutionary Computation*, 109–113. 10.1109/ICEC.1995.489126

Mahdad, B., Srairi, K., & Bouktir, T. (2010). Optimal power flow for large-scale power system with shunt FACTS using efficient parallel GA. *International Journal of Electrical Power & Energy Systems*, *32*(5), 507–517. doi:10.1016/j.ijepes.2009.09.013

Marouani, I., Guesmi, T., Abdallah, H. H., & Ouali, A. (2009). *Application of a multi-objective evolutionary algorithm for optimal location and parameters of FACTS devices considering the real power loss in transmission lines and voltage deviation buses.* 6th International Multi-conference on Systems, Signals and Devices, Djerba.

McKinsey & Company's Electric Power and Natural Gas Practice. (n.d.). *Powering India: The Road to 2017*. Retrieved from https://www.scribd.com/document/258507171/Powering-India-the-Road-to-2017

Mehdinejad, M., & Shayanfar, H. (2017). Commentary of Chaotic Krill Herd algorithm for optimal reactive power dispatch considering FACTS devices. *Applied Soft Computing*, *44*, 163–190.

Meikandasivam, S., Nema, R. K., & Jain, S. K. (2013). Fine power flow control by split TCSC. *International Journal of Electrical Power & Energy Systems*, *45*(1), 519–529. doi:10.1016/j.ijepes.2012.09.012

Ming, H. C., Ju, C. S., Chang, H. Y., & Pei, Y. S. (2011). Optimal active-reactive power dispatch using an enhanced differential evolution algorithm. *6th IEEE Conference on Industrial Electronics Applications (ICIEA)*, 1869–74.

Mukherjee, A., & Mukherjee, V. (2016). Chaotic krill herd algorithm for optimal reactive power dispatch considering FACTS devices. *Applied Soft Computing*, *44*, 163–190. doi:10.1016/j.asoc.2016.03.008

Narayana, P. P., Abdel Moamen, M. A., & Praveen, B. J. (2012). Optimal location and initial parameter settings of multiple TCSCs for reactive power planning using genetic algorithm. *IEEE Power Engineering Society General Meeting*, *33*, 1864 –1878.

Nireekshana, T., Rao, G., Kesava, S., & Sivanaga, R. (2016). Available transfer capability enhancement with FACTS using Cat Swarm Optimization. *Ain Shams. Engineering Journal (New York)*, *7*(1), 159–167.

Ongsakul, W., & Bhasaputra, P. (2002). Optimal power flow with FACTS devices by hybrid TS/SA approach. *International Journal of Electrical Power & Energy Systems*, *24*(10), 851–857. doi:10.1016/S0142-0615(02)00006-6

Padhy, N. P., Abdel-Moamen, M. A., & Kumar, B. J. (2004). Optimal location and initial parameter settings of multiple TCSCs for reactive power planning using genetic algorithms. *IEEE Power Engine Society General Meeting*, 1110–1114.

Palma-Behnke, R., Vargas, L. S., Perez, J. R., Nunez, J. D., & Torres, R. A. (2004). OPF with SVC and UPFC modeling for longitudinal systems. *IEEE Transactions on Power Systems*, *19*(4), 1742–1753. doi:10.1109/TPWRS.2004.836260

Panda, S. (2009). Multi-objective evolutionary algorithm for SSSC-based controller design. *Electric Power Systems Research*, *79*(6), 937–944. doi:10.1016/j.epsr.2008.12.004

Panda, S. (2009). Differential evolutionary algorithm for TCSC-based controller design. *Simulation Modelling Practice and Theory*, *17*(10), 1618–1634. doi:10.1016/j.simpat.2009.07.002

Prasad, D, and Mukherjee, V, (2016). A novel symbiotic organisms search algorithm for optimal power flow of power system with FACTS devices. *Engineering Science and Technology, an International Journal*, *19*(1), 79-89.

Raj, S., & Bhattacharyya, B. (2017). Optimal placement of TCSC and SVC for reactive power planning using Whale optimization algorithm. *Swarm and Evolutionary Computation*.

Ramirez, J. M., & Onate, P. E. (2003). Short term hydrothermal generation scheduling model using a genetic algorithm. *IEEE Transactions on Power Systems*, *18*(4), 1256–1264. doi:10.1109/TPWRS.2003.819877

Rashed, G. I., Shaheen, H. I., & Cheng, S. J. (2007). Optimal Location and Parameter Setting of TCSC by Both Genetic Algorithm and Particle Swarm Optimization. *2nd IEEE Conference on Industrial Electronics and Applications (ICIEA)*, Harbin, China.

Roy, P. K., Ghoshal, S. P., & Thakur, S. S. (2011). Optimal reactive power dispatch considering flexible AC transmission system devices using biogeography based optimization. *Electric Power Components and Systems*, *39*(8), 733–750. doi:10.1080/15325008.2010.541410

Shaoyun, G., & Chung, T. S. (1998). Optimal active power flow incorporating FACTS devices with power flow control constraints. *International Journal of Electrical Power & Energy Systems*, *20*(5), 321–326. doi:10.1016/S0142-0615(97)00081-1

Shayeghi, H., Shayanfar, H. A., Jalilzadeh, S., & Safari, A. (2010). TCSC robust damping controller design based on particle swarm optimization for a multi-machine power system. *Energy Conversion and Management*, *51*(10), 1873–1882. doi:10.1016/j.enconman.2010.02.017

Singh, R. P., Mukherjee, V., & Ghoshal, S. P. (2015). Particle swarm optimization with an aging leader and challengers algorithm for optimal power flow problem with FACTS devices. *International Journal of Electrical Power & Energy Systems*, *64*, 1185–1196. doi:10.1016/j.ijepes.2014.09.005

Taranto, G. N., Pinto, L. M. V. G., & Pereira, M. V. F. (1992). Representation of FACTS devices in power system economic dispatch. *IEEE Transactions on Power Systems*, *7*(2), 572–576. doi:10.1109/59.141761

Tripathy, M., & Mishra, S. (2007). Bacteria foraging-based solution to optimize both real power loss and voltage stability limit. *IEEE Transactions on Power Systems*, *22*(1), 240–248. doi:10.1109/TPWRS.2006.887968

Urdaneta, A. J., Gomez, J. F., Sorrentino, E., Flores, L., & Diaz, R. (1999). A hybrid genetic algorithm for optimal reactive power planning based upon successive linear programming. *IEEE Transactions on Power Systems*, *14*(4), 1292–1298. doi:10.1109/59.801887

Wennan, L., Yihua, L., Xingtao, X., & Maojun, I. (2008). Reactive power optimization in area power grid based on improved Tabu search algorithm. *Third International Conference on Electric Utility Deregulation Restructuring And Power Technologies*, 1472–1477. 10.1109/DRPT.2008.4523637

Yu, D. C., Fagan, J. E., Foote, B., & Aly, A. A. (1986). An optimal load flow study by the generalized reduced gradient approach. *Electric Power Systems Research*, *10*(1), 47–53. doi:10.1016/0378-7796(86)90048-9

Chapter 8
Optimal Location of TCSC, TCPS, and SVC Devices for Solving OPF and ORPD Problem Using Different Evolutionary Optimization Techniques

ABSTRACT

The secure operation of power system has become a topmost issue in today's largely complicated interconnected power systems. This chapter presents the implementation of grey wolf optimization (GWO), teaching-learning-based optimization (TLBO), biogeography-based optimization (BBO), krill herd algorithm (KHA), chemical reaction optimization (CRO), and hybrid CRO (HCRO) approaches to find the optimal location of various FACTS devices such as thyristor control series compensator (TCSC), thyristor control phase shifter (TCPS), and static VAR compensator (SVC) to solve optimal power flow (OPF) and optimal reactive power dispatch (ORPD) in power system. In this chapter, a standard IEEE 30-bus test system with multiple TCSC and TCPS and SVC devices are used for different single and multi-objective functions to validate the performance of the proposed methods. The simulation results validate the ability of the HCRO to produce better optimal solutions compared to GWO, TLBO, BBO, KHA, and CRO algorithms.

DOI: 10.4018/978-1-5225-6971-8.ch008

INTRODUCTION

In the last few years, voltage collapse problems in power systems have been of permanent concern for electric utilities: several major blackouts throughout the world have been directly associated to this phenomenon. The voltage instability can occur when a power system is heavily loaded in transmission lines and/or lacks in local reactive power sources (Shin et al., 2007). The capacity of transmission lines is becoming the main bottleneck of electricity transmission in the deregulated power industry. The competition of electricity may aggravate load ability of some transmission lines. To meet the load demands in a power system and to satisfy the stability and reliability criteria, the existing transmission lines must be utilized more efficiently. The purpose of the transmission network is to pool power plants and load centres in order to supply the load at a required reliability and maximum efficiency at a lower cost. A technically attractive solution to above problems is to use some efficient controls with the help of FACTS (Flexible AC Transmission Systems) devices.

The location of FACTS devices can be based on static or dynamic performances of the system. The sensitivity factor methods are generally used to find the best location to enhance the static performance of the system. FACTS offer a viable solution to these problems, and are now finding wide usage in power systems worldwide. In the deregulated power industry, private power producers are increasing rapidly to meet the increase demand. The purpose of the transmission network is to pool power plants and load centers in order to supply the load at a required reliability, maximum efficiency and at a lower cost. As power transfer increases, the power system becomes increasingly more difficult to operate and insecure with unscheduled power flows and higher losses. FACTS (Dutta et al., 2016; Kavitha & Neela 2017) devices such as Static Var Compensator (SVC), Thyristor Controlled Series Compensator (TCSC) and Thyrister controlled phase shifter (TCPS) can be very effective for power system security. Proper location of SVC, TCSC and TCPS plays a key role for enhancement of system performance without violating the security of the system.

In this chapter, the optimal location of thyristor based FACTS controllers (SVC, TCSC and TCPS) have been introduced for the OPF formulations. The basic objective of OPF is to find proper adjustments of the control variables that would minimize the fuel cost.

LITERATURE SURVEY

A linear programming (LP) method was introduced by Sundar and Ravikumar (2012) to locate the optimal position of TCSC devices to solve OPF under normal and network contingency conditions. To test the effectiveness, the proposed approach was implemented on standard 6-bus, IEEE 14-bus and IEEE 118-bus test systems. Lima et al. (2003) proposed mixed integer linear programming (MILP) to find optimal setting and location of the thyristor control phase shifter (TCPST) for determining maximum loadability of the transmission system. Sharma et al. (2006) presented the MINLP approach for optimal location and sizing of TCSC in the transmission system. Duong et al. (2013) introduced min cut algorithm to select the optimal position and sizing of TCSC devices in standard 6-bus, IEEE 14-bus, IEEE 30-bus and IEEE 118-bus test systems to solve the OPF under normal and contingencies operating condition. Cai et al. (2003) applied GA technique to find optimal choice and allocation of FACTS devices. This method identified the optimal locations of the FACTS devices, their types and ratings. The feasibility of the proposed method was demonstrated on IEEE 14-bus system with different FACTS devices (UPFC, TCSC, TCPST and SVC). A PSO approach to find the optimal locations of FACTS devices (UPFC, TCSC and SVC) with a minimum cost of installation of FACTS devices and to improve system loadability (SL) was reported by Saravanan et al. (2007). The proposed technique was implemented on standard 6-bus and 69-bus, IEEE 30-bus and IEEE 118-bus systems. Sebaa et al. (2014) proposed the cross-entropy (CE) approach for optimal location and tuning of TCPSTs and SVCs for optimal power flow. Lie and Deng (1997) presented decomposition-coordination method and the network compensation technique to find the optimal locations of variable series capacitors and static phase shifters. Singh & David (2001) presented a simple and efficient model for optimizing the location of FACTS devices used for congestion management by controlling the device parameters. Proposed methods were demonstrated using a 5-bus system. Balamurugan et al. (2015) proposed new methodology for proper placement of FACTS devices in the restructured electricity market using reliable evolutionary programming (EP) and differential evolution (DE) techniques based economic dispatch for pool electricity market is used to improve the social welfare. Wirmond et al. (2011) developed GA to find allocation TCPST at congested transmission systems. This methodology was tested at a 291 bus system that is equivalent, in heavy load, to the electric network of the State of Paraná in Brazil. Hooshmand et al. (2015) proposed a hybrid method that is the combination of the bacterial foraging (BF) algorithm with Nelder–Mead (NM) method (BF-NM) to solve the OPF problems for minimizing fuel and emission penalty cost of generators and also determine the optimal location of TCSC.

Motivation of the Present Work

OPF is a non-linear problem and can be non-convex in some cases. Moreover, incorporating FACTS devices complicates the problem further. Such complicated problem needs a well-efficient optimization technique for solving. BBO, GWO, KHA, TLBO, CRO and HCRO are such efficient technique employed for this task in this chapter.

This chapter presents new schemes of OPF and ORPD with multiple types optimally allocate FACTS devices, using evolutionary techniques, considering minimization of the generator fuel cost and transmission loss as well as voltage deviation as the objective functions. The motivation behind the use of these techniques is their inherent simplicity, robustness, stable convergence characteristics and excellent records in locating global or near global optimal solutions of non-linear, non-differentiable problems. Modified IEEE 30-bus system with SVC, TCSC and TCPS devices in optimal locations are presented to show the technique's flexibility and reliability towards the convergence. The simulation results show that BBO, GWO, KHA, TLBO, CRO and HCRO are superior in terms of solution quality compared to PSOIWA and RGA.

This chapter is organized as follows: the section titled "Static Model of TCSC, TCPS, and SVC presents the modeling of FACTS devices for load flow studies followed by the formulation of OPF problems with FACTS in the section titled "Mathematical Problem Formulation." The section titled "Evolutionary optimization techniques applied to OPF and ORPD incorporating optimal location of TCSC, TCPS and SVC" is devoted to the solution methodology where the optimization technique is explained. IEEE 30-bus test system is used for the case studies in section titled "Simulation Results and Discussions" and then the chapter is concluded in the "Conclusion" section.

STATIC MODEL OF TCSC, TCPS AND SVC

Static models of the SVC, TCSC and TCPS devices have already been discussed in chapter 6.

Mathematical Problem Formulation

Objective Function of OPF With FACTS

The conventional formulation of OPF problem with TCSC and TCPS determines the optimal settings of control variables such as real power generations, generators' terminal voltages, transformers' tap settings, reactive power injections of shunt regulators, reactance value of TCSC devices and phase shifting angle of TCPS devices while minimizing the total fuel cost while satisfying all equality and inequality constraints. Total fuel cost of generating units having quadratic cost functions is given by:

$$FC = \sum_{i=1}^{NG} \left(a_i P_{Gi}^{\ 2} + b_i P_{Gi} + c_i \right) \tag{1}$$

For more practical and accurate model of the cost function, multiple valve steam turbines are incorporated for flexible operational facilities. Total generation cost of generating units with valve point discontinuities is given by:

$$FC = \sum_{i=1}^{NG} a_i P_{Gi}^{\ 2} + b_i P_{Gi} + c_i + \left| d_i \times \sin\left(e_i \times \left(P_{Gi}^{\min} - P_{Gi} \right) \right) \right| \tag{2}$$

where a_i, b_i, c_i, d_i, e_i: cost coefficients of the ith generator; P_{Gi} is the active power generation of the ith generator; $P_{Gi}^{\min}$ is the minimum active power generation limit of the ith generator.

Objective Function of ORPD With FACTS

The main goal of ORPD is to minimize active power losses, improve the voltage profile and voltage stability by proper setting of generator bus voltages, VAR compensators, and transformer taps. The mathematical problem formulations of different single and multi-objective functions used in this chapter are as follows:

Single Objective Optimization

Minimization of Total Real Power Loss

The primary objective of ORPD (Mahdad & Srairi 2016) is minimization of network active power loss, while satisfying the operating constraints. This objective function may be expressed as:

$$F_1 = \min(P_{loss}) = \min\left[\sum_{k=1}^{N_{TL}} G_p\left(V_s^2 + V_t^2 - 2V_sV_t\cos\theta_{st}\right)\right] \tag{3}$$

where P_{loss} is the total active power loss; G_p is the conductance of the pth branch connected between the sth and t^{th} bus; θ_{st} is the admittance angle of the transmission line connected between the sth and the tth bus; N_{TL} is the number of transmission lines; V_s, V_t are the voltage magnitudes of the sth and the tth bus, respectively.

Minimization of Voltage Deviation

Minimization of deviations of voltages from desired values is done since bus voltage is one of the most important security and service quality indexes. The objective function of voltage profile improvement, i.e. voltage deviation minimization at load buses, may be expressed as:

$$F_2 = \min\left(\sum_{f=1}^{N_L}\left|V_{L_f} - V_{L_f}^{sp}\right|\right) \tag{4}$$

where, V_{L_f} is the voltage at the Sth load bus; $V_{L_f}^{sp}$ is the desired voltage at the Sth load bus, usually set to 1.0 p.u.

Minimization of L-Index

It is very important to maintain constantly acceptable bus voltage at each bus under normal operating conditions, or when the system is being subjected to a disturbance. The non-optimized control variables may lead to a progressive and uncontrollable drop in voltage resulting in an eventual widespread voltage collapse. In this work, voltage stability enhancement is achieved through minimizing the voltage stability indicator L-*index*. The indicator values vary in the range between 0 (the no-load case) and 1 which corresponds to voltage collapse (Zhang & Liu, 2008).

For the multi-node system,

$$I_{bus} = Y_{bus}.V_{bus} \tag{5}$$

Segregating the nodes into two categories, that is, the load bus set ψ_L and generator bus set ψ_G can be written as

$$\begin{bmatrix} I^L \\ I^G \end{bmatrix} = \begin{bmatrix} Y_1 & Y_2 \\ Y_3 & Y_4 \end{bmatrix} \cdot \begin{bmatrix} V^L \\ V^G \end{bmatrix} \tag{6}$$

The transmission system itself is linear and allows a representation in terms of a hybrid matrix H:

$$\begin{bmatrix} V^L \\ I^G \end{bmatrix} = H. \begin{bmatrix} I^L \\ V^G \end{bmatrix} = \begin{bmatrix} Z^{LL} & F^{LG} \\ K^{GL} & Y^{GG} \end{bmatrix} \cdot \begin{bmatrix} I^L \\ V^G \end{bmatrix} \tag{7}$$

where V^L and I^L are vectors of voltages and currents at load buses; V^G and I^G are vectors of voltages and currents at generator buses; Z^{LL}, F^{LG}, K^{GL}, Y^{GG} are sub-matrices the H-matrix generated from Y_{bus} partial inversion, respectively. A voltage collapse can take place in systems or subsystems and can appear quite abruptly (Kesselm & Glavitsch, 1986).

L-index L_j of the jth bus can be expressed through the following equation:

$$\begin{cases} L_j = \left| 1 - \sum_{i=1}^{N_{PV}} F_{ji} \dfrac{V_i}{V_j} \right|, & J = 1, 2, \ldots, N_{PQ} \\ F_{ji} = -\left[Y_1\right]^{-1} \left[Y_2\right] & \end{cases} \tag{8}$$

where, N_{PV} and N_{PQ} are number of PV and PQ bus, respectively. L-*index* is calculated for all the load buses. Thus, a global power system representative L describing the stability of the complete system is represented as follows:

$$L = \max(L_j),\ j = 1,2,...,N_{PQ} \tag{9}$$

In the optimal RPD problem, incorrect tuning of continuous and discrete control variable settings may boost the value of L-*index,* that is may reduce the voltage stability outskirt of the system (Bhattacharya & Chattopadhyay, 2011). Let the maximum value of L-*index*, be denoted as $L_{\max}$. Therefore, to enhance the voltage stability and to move the system far from the voltage collapse margin, the following objective function can be calculated as:

$$F_3 = \min\quad L_{\max} \tag{10}$$

Multi-Objective Optimization

Many real-world optimization problems involve simultaneous optimization of several conflicting objectives. Multi-objective optimization problems with such conflicting objectives give rise to a set of the optimal solution, rather than a single optimal solution. Because no solution can be considered to be better than other solutions without an information. In this chapter, the multi-objective ORPD problem is converted to a single objective function by introducing penalty factors approach as follows:

$$Minimize \lambda = F_1 + pf_1 \times F_2 + pf_2 \times F_3 \tag{11}$$

where, pf_1 and pf_2 are the penalty factor of load-bus voltage violation and voltage stability index respectively. With this penalty, any violation of bus voltage and L-index will be reflected on the fitness function. The choice of pf_1, pf_2 determines the accuracy and speed of convergence.

Constraints

The OPF with TCSC and TCPS are subjected to the following constraints.

Equality Constraints

These constraints represent the load flow equations as given below:

$$\begin{cases} \sum_{i=1}^{NB}\left(P_{Gi}-P_{Li}\right)+\sum_{i=1}^{NTCPS}P_{is}=\sum_{i=1}^{NB}\sum_{j=1}^{NB}\left|V_i\right|\left|V_j\right|\left|Y_{ij}\right|\cos\left(\theta_{ij}-\delta_{ij}\right) \\ \sum_{i=1}^{NB}Q_{Gi}-Q_{Li}+\sum_{i=1}^{NTCPS}Q_{is}=-\sum_{i=1}^{NB}\sum_{j=1}^{NB}\left|V_i\right|\left|V_j\right|\left|Y_{ij}\right|\sin\left(\theta_{ij}-\delta_{ij}\right) \end{cases} \tag{12}$$

where P_{Li} is the active power demand of the ith bus; Q_{Gi}, Q_{Li} are the reactive power generation and demand of the ith bus, respectively; P_{is}, Q_{is} are the injected active and reactive powers of TCPS at the ith bus, respectively; Y_{ij} is the admittance of transmission line connected between the ith bus and the jth bus; θ_{ij} is the admittance angle of transmission line connected between the ith bus and the jth bus; NB is the number of buses; $NTCPS$ is the number of TCPSs.

Inequality Constraints

Some inequality constraints such as generator constraints, load bus constraints, transmission line constraints, transformer tap constraints, and shunt compensator constraints, TCSC reactance constraints and TCPS phase shift constraints have already been discussed in chapter 7.

EVOLUTIONARY OPTIMIZATION TECHNIQUES APPLIED TO OPF AND ORPD INCORPORATING OPTIMAL LOCATION OF TCSC, TCPS AND SVC

BBO Algorithm Applied to OPF and ORPD With SVC, TCSC and TCPS

In this chapter, BBO algorithm is employed to solve constrained OPF and ORPD problem incorporating multiple FACTS devices like SVC, TCSC and TCPS and to find the optimal solutions satisfying both equality and inequality constraints.

The algorithm of the proposed method is as enumerated below.

Step 1: Initialize the population randomly in the search space, i.e. the initial positions of *SIVs* (independent variables such as active powers of all generators except slack bus, generators' voltages, tap settings of regulating transformers, reactive power injections of shunt capacitors, power flow through the transmission lines where TCSC & TCPS devices are placed and power injections of TCPS devices, reactance of TCSC, phase shift angle of TCPS, location of TCSC devices and TCPS devices for OPF and generators' voltages, tap settings of regulating transformers, reactive power injections for ORPD) of each habitat should be randomly selected while satisfying different operational constraints. Several numbers of habitats depending upon the population size are being generated. Each habitat represents a potential solution.

Step 2: Perform TCSC and TCPS based load flow using these *SIVs* and determine all dependent variables such as load voltages, active power of slack bus, generators' reactive powers, reactance of TCSC and phase shifting angle of TCPS etc.

Step 3: Evaluate the fitness (i.e. *HSI*) for each habitat of the population using (8.1) – (8.11)

Step 4: Sort the habitats from the best to the worst using *HSI* values.

Step 5: Identify few elite habitats based on the *HSI* values. These best habitat sets are kept as it is without any modification.

Step 6: Assign number of species to each habitat based on *HSI*

Step 7: Modify the independent variables (*SIVs*) of the non-elite habitats using hybrid migration operation as explained in an earlier chapter.

Step 8: Update species count probability of each habitat. Based on habitat modification probability, modify the independent variables (*SIVs*) of non-elite habitats by probabilistic mutation operation.

Step 9: Go to Step 2 for the next iteration.

Step 10: Stop iterations after a predefined number of iterations.

TLBO Algorithm Applied to OPF and ORPD With SVC, TCSC, and TCPS Devices

The main steps of the proposed OTLBO approach applied to OPF and ORPD problem are listed as follows.

Step 1: Depending upon the population size, several numbers of initial solutions are initialized randomly within their specified operating limits.

Step 2: The independent variables of OPF problem (active power of all generator buses except slack bus, voltage of generator buses, reactive power injection, tap settings of transformers, reactance of TCSC, phase shift angle of TCPS, location of TCSC devices and TCPS devices) of each initial solution string are updated using the following expression to generate oppositional population.

$$OP_{r,s} = a_s + b_s - P_{rs}$$

where $r = 1, 2, ..., n_p; s = 1, 2, ..., n_d; P_{r,s}$ is the sth independent variables of the rth vector of the population; $OP_{r,s}$ is the sth independent variables of the rth vector of the opposite population; n_p is the population size and n_d is the number of independent variables.

Step 3: The dependent variables of OPF problem (slack bus power, load bus voltages, generator reactive power outputs, and transmission line loadings) are evaluated using load flow analysis.

Step 4: The fitness values of the current population *(P)* and oppositional population *(OP)* sets are evaluated.

Step 5: n_p number of fittest individuals are selected from $\{P \cup OP\}$.

Step 6: n_p number of fittest vectors are sorted from best to worst.

Step 7: Few elite solutions are selected based on the fitness values.

Step 8: The independent variables (active power of all generator buses except slack bus, voltage of generator buses, reactive power injection, tap settings of transformers, reactance of TCSC and location of TCSC devices) of each non-elite solution are modified by using teaching and learning phase of TLBO.

Step 9: The dependent variables of OPF (slack bus power, load bus voltages, generator reactive power outputs, transmission line loadings) are evaluated. Afterward, the fitness value of the population set is calculated.

Step 10: Based on a jumping rate j_r (i.e. jumping probability), the opposite population of the current population is evaluated and its fitness values are calculated.

Step 11: n_p fittest individuals are selected from the union of the current population and the opposite population.

Step 12: Go to Step 8 for the next iteration until termination criterion is made.

HCRO Algorithm Applied to OPF and ORPD With SVC, TCSC and TCPS

The HCRO algorithm is applied to seek out the optimal or near-optimal solutions for the OPF and ORPD problem with TCSC and TCPS devices. Its computational procedure can be illustrated using the following steps:

Step 1: Confine the search space, i.e., specify the lower and upper limits of each decision variable such as active powers of all generators except slack bus, generators' voltages, tap settings of regulating transformers, reactive power injections of shunt capacitors, power flow through the transmission lines where TCSC & TCPS devices are placed and power injections of TCPS devices, reactance of TCSC, phase shift angle of TCPS, location of TCSC devices and TCPS devices.

Step 2: Initialize the individuals of the population. The molecular structure of each molecule should be randomly initialized such that each candidate solution lies within the feasible operating region.

Step 2.1: Depending upon the number of TCSC or TCPS, their positions are randomly generated as follows (see Box 1)

Step 3: Run the Newton-Raphsonload flow to find the slack bus power, load voltages, generators' reactive powers, and reactance of TCSC and phase shifting angle of TCPS etc.

Step 4: Evaluate the fitness value using $FC = \sum_{i=1}^{NG} \left(a_i P_{Gi}^{\ 2} + b_i P_{Gi} + c_i \right)$. Based on the objective value, sort the molecules and few are kept as elite molecules.

Step 5: Assign potential energy same as fitness value for each molecule of the population set.

Step 6: Apply on-wall ineffective collision operations, decomposition operation, inter-molecular ineffective collision and synthesis collision operation on non elite molecules (as described earlier) to modify the independent variables i.e generators' active power

Step 7: To check the feasibility, run load flow and find the dependent variables. If any dependent variable violates the operating limit, replace that solution set by new feasible solution.

Step 8: Compute fitness value for each newly generated feasible solution set.

Step 9: If the stopping criteria are satisfied, then go to Step 10. Otherwise, go to Step 4.

Step 10: Output optimal settings of the control variables and the corresponding cost.

GWO Algorithm Applied to OPF and ORPD With SVC, TCSC, and TCPS Devices

Step 1: Read line data, bus data, fuel cost co-efficient, load flow parameters, etc. of the power system and specify the upper and lower limits of each control variables of OPF like active power generation, generators' bus voltage, load bus voltages, reactive power generations, tap settings of the tap changing transformers, VAr injection of the shunt compensating devices and line flow through each transmission lines. Afterward, randomly select the independent control variables within their operating limits. Thereafter, Newton Raphson based load flow method is run to evaluate the slack bus power and the other dependent variables. Check whether the control variables are within the limits or not. If any of them violates the limits, corresponding set is discarded and initialise again. In this fashion, a feasible initial solution is generated. Now several initial solution set depending upon the population size are generated. A feasible solution set (control variables) represents the position of different GWO individuals.

Step 2: The input parameters such as a, A and C of GWO algorithm is initialized.

Step 3: Evaluate the fitness function of the initial solution set and elite solutions are identified.

Step 4: Find the updated best position of alpha, beta and delta wolves and independent variables of OPF are modified. Thereafter, Newton Raphson based load flow is run to obtain the dependent variables. The feasibility

Box 1.

$$
\begin{array}{l}
\textit{for } i = 1 : N_P \\
\quad a = randperm(NTL) \\
\qquad \textit{for } j = 1 : NUPFC \\
\qquad\quad loc(j) = a(j) \\
\qquad \textit{end} \\
\textit{end}
\end{array}
$$

of the controlled variables is checked. If any one of them does not satisfy the inequality constraints, the particular set is replaced by the best feasible solution set.

Step 5: Evaluate the fitness value of the feasible population set.

Step 6: Check if the number of iteration is less than the maximum specified iteration or not.

Step 7: When the maximum iteration is reached display the fitness value and the updated final position of the wolves denoted by X(t+1) from the prey which is generally a circle around the prey.

Step 8: If the maximum iteration is not reached, iteration cycle is incremented.

Step 9: Go to Step 4 until a stopping criterion is satisfied.

KHA Algorithm Applied to OPF and ORPD With SVC, TCSC, and TCPS Devices

The procedure for implementing the KHA algorithm in solving optimal power flow problem to get minimum cost may be summarized by the following steps:

Step 1: All independent variables such as all generators' active power except slack bus, generators' voltages, tap settings of regulating transformers, should be randomly selected between their operating limits

Step 2: Run the Newton-Raphson load to determine the slack bus power, load voltages, power flow through the transmission line, and check whether they satisfy the operating limits or not. If any of these parameters violate the operating limits; discard that population set and re-initialize the corresponding set. Depending upon the population size, several solutions are generated. Each feasible solution set represents the initial position of each krill individuals.

Step 3: Evaluate the fitness value and sort the solutions from best to worst.

Step 4: Select few best solutions for elitism.

Step 5: Evaluate the three motion index, namely, motion induced by other individual; foraging motion and random diffusion.

Step 6: The non-elite krill individuals' position are modified .

Step 7: Apply crossover and mutation to update the position of each non-elite krill.

Step 8: Check whether the independent variables violate the operating limits or not. If any variable is less than the minimum level it is made equal to minimum value and if it is greater than the maximum level it is made equal to maximum level.

Step 9: Run Newton-Raphson based load flow to determine the dependent variables and check whether they satisfy the system operating constraints or not. The infeasible solutions are replaced by the best feasible solutions.

Step 10: Go to Step 4 until the current iteration number reaches the pre specified maximum iteration number.

SIMULATION RESULTS AND DISCUSSIONS

IEEE 30 Bus System

The applicability and validity of the different techniques have been tested on IEEE 30-bus system for OPF problems having multiple numbers of TCSC and TCPS devices. IEEE 30-bus system contains 6 generating units connected at bus numbers 1, 2, 5, 8, 11 and 13; 30 buses, 4 regulating transformers connected between the line numbers 6-9, 6-10, 4-12 and 27-28; 9 shunt compensators connected at bus numbers 10, 12, 15, 17, 20, 21, 23, 24 and 29; 41 transmission lines. The full system data are taken from (Basu, 2008) and fuel cost data, transmission line data. The generators' active powers except for slack (for OPF with FACTS) and generators' voltages, tap settings of the regulating transformers, VAr injections of the shunt capacitors, power flows through the transmission lines where TCSC and TCPS devices are placed and power injections of TCPS devices are considered as control variables. The voltage magnitudes of all buses are considered within the range of $[\,0.95p.u., 1.05p.u.$. Tap settings of regulating transformers are within the range of $[\,0.9p.u., 1.1p.u.\,]$. The VAr injections of the shunt capacitors are taken within the interval of $[0 \text{ MVAr}, 30 \text{ MVAr}]$. The reactance limits of TCSC in *p.u.* are $0 \leq x_t \leq 0.20$. The limits of phase shift angle (radian) of TCPS are taken as $-0.1 \leq \alpha \leq 0.1$. The susceptance limits of SVC in p.u. is assumed to be $0 \leq B_{SVC} \leq 0.15$.

The feasibility of the proposed BBO, TLBO, GWO, KHA, CRO, and HCRO methods is tested on modified IEEE 30-bus system in three different cases (incorporating TCSC and TCPS devices) and its performance is compared to other population-based optimization techniques like PSOIWA and RGA. The simulation results shown in different Tables are the best solutions obtained in fifty runs.

Four cases have been considered and analyzed, with FACTS devices located at different transmission lines and buses; Case 1: OPF without FACTS devices, Case 2: OPF with TCSC, Case 3: OPF with TCPS and Case 4: ORPD with FACTS devices, The results obtained with all the above cases are tabulated and presented in Table 1, 2, 3 and 4.

Case 1 (OPF Without FACTS Devices)

In order to show the effectiveness of HCRO, the comparison with other technique which is mentions in the table are performed to determine the optimal values of control variables along with active transmission loss (TL), and active generation cost. These simulation results are listed in Table 1. From these results, it is observed that the fuel cost obtained using HCRO is the lowest (1625.7069 $/hr) as compared to GWO, TLBO, KHA, BBO, and CRO. The results clearly show that HCRO and CRO outperform other methods in terms of solution quality. The convergence characteristics of the using HCRO are better than BBO, TLBO, CRO, KHA and GWO techniques.

Case 2 (OPF With TCSC)

In order to check the feasibility of the proposed method for a complicated network, OPF is solved for the same test system with TCSC device included in it. The optimal values of active power generation cost, optimal location of TCSC, and power loss obtained by the proposed algorithm and those other methods found from the references are listed in Table 2. From the results, it is seen that an improved secured OPF solution is achieved using KHA, GWO, TLBO, BBO,,CRO and HCRO with TCSC placed at line 18-19, 10-21, 23-24, 21-22, 18-19 and 10-21 respectively. The fuel cost obtained using HCRO is lowest (1611.4410$/hr) as compared to with the other algorithms. Therefore, it can finally be concluded that the algorithm presented in this chapter works satisfactorily for TCSC based IEEE 30-bus test system with 160% of rated load.

Case 3 (OPF With TCPS)

In order to check the feasibility of the proposed method for a complicated network, it is applied to solve OPF with TCPS of the same test system. The optimal values of active power generation cost, optimal location of TCPS,

Table 1. Comparative simulation results obtained by various algorithms (without FACTS)

Algorithms	HCRO	CRO	GWO	TLBO	KHA	BBO
PG_1(MW)	199.5722	199.6236	202.5445	211.0333	1.9959	1.9960
PG_2 (MW)	116.7613	115.1427	84.5429	82.4621	1.1565	1.1779
PG_5 (MW)	26.4929	31.8596	28.4920	28.9477	0.2508	0.2584
PG_8 (MW)	67.7843	73.2215	102.9954	77.0484	0.6984	0.6044
PG_{11} (MW)	27.6706	32.3238	27.9735	33.8591	0.3280	0.3601
PG_{13} (MW)	32.9667	19.3683	25.2506	37.3553	0.3200	0.3425
V_1 (*p.u.*)	1.0785	1.0535	1.0699	1.0944	1.0604	1.0678
V_2 (*p.u.*)	1.0664	1.0342	1.0693	1.0775	1.0403	1.0481
V_5 (*p.u.*)	0.9960	0.9931	1.0823	1.0261	1.0935	1.0108
V_8 (*p.u.*)	1.0214	1.0003	1.0640	1.0568	1.0314	0.9746
V_{11} (*p.u.*)	1.0507	1.0711	0.9605	1.0409	1.0848	0.9554
V_{13} (*p.u.*)	1.0975	1.0544	1.0209	1.0403	0.9600	0.9902
$TC_{6\text{-}9}$	1.0340	1.0628	1.0985	1.0194	1.0405	1.0478
$TC_{6\text{-}10}$	0.9083	0.9602	1.0828	1.0349	0.9740	0.9384
$TC_{4\text{-}12}$	1.0154	0.9857	1.0924	0.9736	1.0221	1.0977
$TC_{27\text{-}28}$	0.9378	0.9001	1.0332	1.0015	1.0252	1.0422
QC_{10} (MVAr)	4.7642	4.9118	3.1092	4.2825	0.0457	0.0109
QC_{12} (MVAr)	4.9131	2.5056	0.1631	2.6260	0.0043	0.0417
QC_{15} (MVAr)	4.8528	4.4091	4.1471	2.4008	0.0012	0.0455
QC_{17} (MVAr)	4.3373	3.7836	3.0943	1.0737	0.0301	0.0343
QC_{20} (MVAr)	3.6842	3.5713	1.777	3.7732	0.0052	0.0019
QC_{21} (MVAr)	3.2906	4.6485	3.2309	0.8925	0.0333	0.0162
QC_{23} (MVAr)	4.6413	4.8107	0.4271	0.0767	0.0022	0.0274
QC_{24} (MVAr)	4.4028	3.4428	1.4046	3.2724	0.0026	0.0181
QC_{29} (MVAr)	4.3435	3.3766	0.0059	3.9121	0.0036	0.0239
Fuel cost ($/hr)	1625.7069	1631.1002	1637.0716	1640.3350	1641.8090	1644.4939
TL (MW)	17.8080	18.0995	18.3590	17.2660	21.5279	20.4930

and power loss obtained by the proposed algorithm and those other methods found from the references are listed in Table 3 From the results it seen that an improved secured OPF solution is achieved using BBO, TLBO, GWO, KHA, CRO, and HCRO with TCPS placed at line 12-16, 25-27, 27-30, 10-17, 15-23 and 25-27 respectively. The fuel cost obtained using HCRO is lowest (1611.8577$/hr) as compared to with the other algorithms. Therefore, it

Table 2. Comparative simulation results obtained by various algorithms (with TCSC)

Algorithms	HCRO	CRO	GWO	TLBO	KHA	BBO
PG_1(MW)	190.7236	195.2289	197.7463	198.7765	1.9946	199.90
PG_2 (MW)	114.8342	115.9537	84.2668	83.5482	0.8412	116.15
PG_5 (MW)	30.6428	32.9788	232.2727	35.1262	0.3790	32.19
PG_8 (MW)	63.2334	72.4132	72.480	94.1109	0.7858	66.26
PG_{11} (MW)	40.5605	24.9846	46.2977	26.0408	0.3611	39.01
PG_{13} (MW)	22.8361	25.5683	28.3171	30.9283	0.3321	0.1642
V_1 (*p.u.*)	1.0478	1.0690	1.0997	1.0950	1.0568	1.1000
V_2 (*p.u.*)	1.0217	1.0448	1.0857	1.0686	1.0327	1.0925
V_5 (*p.u.*)	0.9770	1.0276	1.0461	0.9820	0.9920	1.0409
V_8 (*p.u.*)	1.0022	1.0087	1.0696	1.0368	1.0286	1.0667
V_{11} (*p.u.*)	1.0743	1.0785	1.0977	1.0910	1.0964	1.0929
V_{13} (*p.u.*)	1.0895	1.0912	1.0944	1.0529	1.0978	1.0954
$TC_{6\text{-}9}$	1.0308	0.9281	1.0371	1.0505	1.0969	1.0276
$TC_{6\text{-}10}$	0.9029	0.9514	0.9000	1.0506	0.9352	0.9226
$TC_{4\text{-}12}$	1.0363	0.9843	1.0311	1.0874	0.9740	1.0250
$TC_{27\text{-}28}$	0.9242	0.9693	0.9627	0.9569	0.9686	0.9299
QC_{10} (MVAr)	1.1341	1.7908	4.2437	2.4763	0.0499	0.0497
QC_{12} (MVAr)	4.2217	0.1754	4.8751	2.6627	0.0496	0.0450
QC_{15} (MVAr)	1.9660	1.5689	4.8481	0.2614	0.0498	0.0498
QC_{17} (MVAr)	3.0928	1.8527	4.8930	4.4927	0.0496	0.0484
QC_{20} (MVAr)	3.7546	2.4258	4.8127	0.5067	0.0494	0.0482
QC_{21} (MVAr)	4.2417	4.6411	4.7800	3.5357	0.0498	0.0485
QC_{23} (MVAr)	3.4765	1.6249	4.9568	2.4867	0.0492	0.0423
QC_{24} (MVAr)	2.6392	4.4952	4.8668	1.9262	0.0500	0.0496
QC_{29} (MVAr)	2.1824	4.2236	4.9065	2.1812	0.0499	0.0052
Location of TCSC	10-21	18-19	10-21	23-24	18-19	21-22
TCSC reactance (*p.u.*)	0.03759	0.03183	0.04632	0.03872	0.04632	0.03872
Fuel Cost ($/hr.)	1611.4410	1615.3489	1619.4029	1622.1166	1621.3510	1624.1507
TL (MW)	9.3906	13.6875	20.7938	15.0908	15.9423	16.5143

can finally be concluded that the algorithm presented in this chapter works satisfactorily for TCPS based IEEE 30-bus test system with 160% of rated load.

Case 4 (ORPD With FACTS Devices)

Single Objective Optimization

- **Case 1: Loss Minimization**

The results obtained for transmission loss minimization objective function with SVC, with TCSC and with both SVC & TCSC devices by BBO, CRO and HCRO are reported in 2nd -4th columns of Tables 4, 5 and 6 respectively. It may be seen from simulation results that HCRO gives significantly better transmission loss compared to BBO and CRO approaches. Moreover, it is noticed that the proposed HCRO method with SVC and TCSC gives best results among the results of all the discussed cases of different algorithms.

- **Case 2: Voltage Deviation Minimization**

In the second case, the proposed HCRO based approach is applied for improvement of voltage profile. The optimal values of the control variables, TL, VD, L-index obtained by BBO, CRO and HCRO for SVC based ORPD, TCSC based ORPD and both SVC and TCSC based ORPD are listed in the 5nd -7th columns of Tables 4, 5 and 6 respectively. For all these three cases, it is observed from the simulation results that the voltage deviation obtained by HCRO formulation is the smallest among the three formulations, while that of BBO is the largest. Furthermore, the results show that incorporating SVC and TCSC in the optimal location with the optimal parameter setting can significantly improve the voltage deviation of the system. It is also observed from the results that the computing time calculated by HCRO is superior to that computed by other approaches, such as CRO and BBO in the mentioned test system for all the three cases.

- **Case III: Minimization of L-index voltage stability**

To test the superiority of BBO, CRO and HCRO techniques, they are applied for L-index minimization on IEEE 30-bus system. The optimal control variables, TL, VD, L-index and CT values are obtained using BBO, CRO and HCRO approaches in the IEEE 30-bus power system for L-index minimization

Table 3. Comparative simulation results obtained by various algorithms (with TCPS)

Algorithms	HCRO	CRO	GWO	TLBO	KHA	BBO
PG_1(MW)	195.2298	195.4013	202.2346	203.0989	195.8681	198.0227
PG_2 (MW)	84.1363	80.8543	83.2203	83.2743	82.7393	82.1322
PG_5 (MW)	31.8192	29.0499	31.1868	33.0138	27.7426	33.9517
PG_8 (MW)	80.8157	85.3641	75.4007	89.8425	98.7783	79.9053
PG_{11} (MW)	39.9643	38.0144	42.5013	36.7313	23.6657	39.9796
PG_{13} (MW)	32.8318	35.4826	33.3439	22.8367	37.8461	33.9328
V_1 (*p.u.*)	1.0930	1.0984	1.0941	1.0978	1.0748	1.0801
V_2 (*p.u.*)	1.0715	1.0887	1.0787	1.0829	1.0418	1.0571
V_5 (*p.u.*)	1.0188	1.0521	1.0424	1.0408	1.0325	1.0414
V_8 (*p.u.*)	0.9753	1.0806	1.0617	1.0668	1.0577	1.0647
V_{11} (*p.u.*)	1.0938	1.1000	1.0776	1.0985	1.0700	1.0960
V_{13} (*p.u.*)	1.0983	1.0984	1.0978	1.0971	1.0961	1.0703
TC_{6-9}	1.0559	1.0266	1.0098	1.0179	0.9585	1.0843
TC_{6-10}	0.9860	0.9279	0.9000	0.9333	0.9277	0.9922
TC_{4-12}	1.0829	1.0412	1.0381	1.0317	0.9858	1.0310
TC_{27-28}	0.9671	0.9732	0.9440	0.9513	0.9083	1.0213
QC_{10} (MVAr)	4.0047	4.8332	4.6046	4.8828	0.2013	2.4827
QC_{12} (MVAr)	5.0000	4.6647	4.7151	4.8397	1.5046	0.6407
QC_{15} (MVAr)	4.9523	4.8952	4.8423	4.5634	2.5958	3.5942
QC_{17} (MVAr)	4.9188	4.9238	4.6894	4.9652	2.8835	1.4393
QC_{20} (MVAr)	4.8837	4.8906	4.8261	5.0000	3.6404	0.5736
QC_{21} (MVAr)	4.7449	4.9827	4.8538	4.8982	2.7157	2.9042
QC_{23} (MVAr)	4.8820	4.7964	4.8442	4.9036	1.4365	1.8058
QC_{24} (MVAr)	4.5837	4.8253	4.8728	4.8879	4.8636	4.4637
QC_{29} (MVAr)	4.7921	4.7508	4.6257	4.7611	0.3843	1.6781
Location of TCPS	25-27	15-23	27-30	25-27	10-17	12-16
TCPS angle (radian)	-0.02047	0.009346	-0.01565	0.02193	0.01865	-0.01342
Fuel Cost ($/hr.)	1611.8577	1615.2133	1621.7361	1623.8676	1621.5009	1623.8019
TL (MW)	11.3571	10.7266	14.4476	15.3575	13.2001	14.4843

objective of SVC/TCSC based ORPD are elaborated in the columns 8th -10th of Tables 4, 5 and 6 respectively. The results clearly demonstrate that the L-index reduction accomplished using the HCRO approach is better than that obtained by the other approaches. Hence, the conclusion can be drawn that HCRO is better than all the other listed algorithms in terms of global search capacity and local search precision. Furthermore, it can be seen that all the control variables optimized by the various discussed methods are acceptably kept within the limits. It shows that, compared with BBO and CRO, HCRO has faster convergence speed and needs lesser iteration cycles to achieve the optimal L-index level. From simulation results, it can be seen that the average computing time of HCRO is less than that of other algorithms.

Multi-Objective Optimization

Finally, to verify the feasibility, effectiveness and performance of the proposed HCRO method for solving the multi-objective problem, simultaneous minimization of power loss, voltage deviation and L-index is considered for FACTS based ORPD problems. The comparative results obtained by BBO, CRO and HCRO are given in 11th-13th columns of Table 4, 5 and 6. It is quite evident from the results that the incorporating FACTS devices in the optimal location with optimal settings using various discussed algorithms gives better transmission loss, voltage deviation and L-index than the without FACTS device. Transmission loss, voltage deviation and L-index are 5.0238 MW, 0.1116p.u, 0.1187, respectively using HCRO algorithm with SVC. The results presented in Tables 4-7 show that HCRO is superior to CRO, BBO, TLBO and HCRO in finding the global best solution. It is also observed that the HCRO is the fastest algorithm since it less computational time as compared to that requires for CRO and BBO algorithms. It is also observed that all parameter settings of SVC and TCSC device are within the specified limits. Transmission loss, voltage deviation and L-index are 5.0685 MW, 0.1149p.u, 0.1179 using HCRO algorithm with TCSC, respectively. After incorporating both SVC and TCSC the transmission loss, voltage deviation and L-index are found to be 4.9937 MW, 0.1039 p.u, 0.1173, respectively using HCRO algorithm. From the above results, it is clear that if we incorporate more FACTS devices, above objective function will be minimized more significantly.

CONCLUSION

In this chapter, GWO, TLBO, BBO, KHA, CRO and HCRO techniques have been proposed, developed and successfully implemented to solve OPF and ORPD problems incorporating FACTS devices and valve point discontinuities. A SVC is a shunt FACTS device which can be controlled the bus voltage for reducing the transmission loss, voltage deviation and voltage stability. A

Table 4. Simulation results obtained by different techniques for both single and multi-objective ORPD withSVC device (IEEE 30-bus system)

Techniques	Single objective									Multi-objective		
	Transmission loss minimization			Voltage deviation minimization			L-index minimization					
	BBO	CRO	HCRO	BBO	CRO	HCRO	BBO	CRO	HCRO	BBO	CRO	HCRO
V_{g1} (p.u.)	1.0906	1.0982	1.0934	1.0186	1.0075	1.0102	1.0997	1.0981	1.1000	1.0691	1.0596	1.0375
V_{g2} (p.u.)	1.0991	1.0977	1.0992	1.0105	1.0026	1.0113	1.0827	1.0857	1.1000	1.0595	1.0501	1.0280
V_{g5} (p.u.)	1.0833	1.0799	1.0837	1.0016	0.9947	1.0124	1.0790	1.0839	1.1000	1.0340	1.0261	1.0088
V_{g8}(p.u.)	1.0792	1.0712	1.0698	1.0013	1.0045	0.9996	1.0545	1.0337	1.0761	1.0342	1.0227	1.0056
V_{g11}(p.u.)	1.0823	1.0946	1.0847	1.0190	1.0104	1.0148	1.0992	1.0916	1.0837	1.0355	1.0540	0.9861
V_{g13}(p.u.)	1.0947	1.0997	1.1000	1.0090	1.0308	1.0139	1.0843	1.0921	1.0845	1.0043	1.0091	1.0328
T_{6-9}	1.0421	1.0492	1.0552	1.0315	1.0279	1.0953	0.9425	0.9154	0.9594	1.0889	1.0925	1.0009
T_{6-10}	0.9892	0.9638	0.9737	0.9028	0.9022	0.9015	0.9029	0.9196	0.9001	0.9242	0.9237	0.9194
T_{4-12}	0.9947	1.0147	1.0346	0.9793	1.0119	0.9863	0.9076	0.9089	0.9376	1.0073	1.0027	1.0245
T_{6-9}	0.9329	0.9867	0.9917	0.9612	0.9657	0.9635	0.9202	0.9043	0.9365	0.9841	0.9809	0.9502
Q_{c10}	0.0487	0.0487	0.0476	0.0499	0.0338	0.0500	0.0449	0.0010	0.0314	0.0163	0.0087	0.0495
Q_{c12}	0.0500	0.0482	0.0437	0.0200	0.0012	0.0167	0.0311	0.0384	0.0482	0.0286	0.0292	0.0000
Q_{c15}	0.0437	0.0500	0.0500	0.0500	0.0451	0.0500	0.0113	0.0016	0.0411	0.0429	0.0473	0.0440
Q_{c17}	0.0493	0.0500	0.0500	0.0001	0.0009	0.0000	0.0025	0.0181	0.0306	0.0263	0.0344	0.0134
Q_{c20}	0.0488	0.0462	0.0488	0.0500	0.0500	0.0500	0.0016	0.0001	0.0000	0.0494	0.0462	0.0499
Q_{c21}	0.0500	0.0500	0.0374	0.0499	0.0500	0.0500	0.0437	0.0500	0.0482	0.0493	0.0475	0.0498
Q_{c23}	0.0489	0.0396	0.0462	0.0500	0.0496	0.0500	0.0000	0.0029	0.0018	0.0447	0.0358	0.0444

continued on following page

Table 4. Continued

Techniques	Single objective									Multi-objective		
	Transmission loss minimization			Voltage deviation minimization			L-index minimization					
	BBO	CRO	HCRO	BBO	CRO	HCRO	BBO	CRO	HCRO	BBO	CRO	HCRO
Q_{c24}	0.0500	0.0500	0.0500	0.0500	0.0500	0.0500	0.0209	0.0500	0.0138	0.0498	0.0471	0.0500
Q_{c29}	0.0453	0.0465	0.0437	0.0211	0.0498	0.0255	0.0028	0.0113	0.0047	0.0115	0.0215	0.0136
Optimal location of SVC	22	22	23	27	26	27	21	21	21	22	23	23
Optimal setting of SVC (p.u.)	0.1036	0.1283	0.0994	0.0847	0.1004	0.0821	0.1368	0.1445	0.1189	0.1082	0.0649	0.0837
TL (MW)	4.4921	4.4815	4.4437	5.5318	5.5461	5.5392	5.5496	5.1347	4.9904	5.1121	5.0836	5.0238
VD (p.u.)	2.0012	1.9967	2.0128	0.0892	0.0882	0.0876	2.0608	2.1143	2.4215	0.1267	0.1183	0.1116
L-index	0.1249	0.1247	0.1253	0.1458	0.1413	0.1488	0.1113	0.1096	0.1083	0.1211	0.1198	0.1187
CT (Sec.)	16.3427	7.9923	7.8137	16.5010	7.9815	7.6937	16.5506	8.0641	7.9870	16.4862	7.9995	7.9361

Table 5. Comparative results for both single and multi-objective ORPD with TCSC device (IEEE 30-bus system)

Techniques	Single objective									Multi-objective		
	Transmission loss minimization			Voltage deviation minimization			L-index minimization					
	BBO	CRO	HCRO	BBO	CRO	HCRO	BBO	CRO	HCRO	BBO	CRO	HCRO
V_{g1} (p.u.)	1.0926	1.0851	1.0948	1.0113	0.9994	1.0022	1.0846	1.0857	1.1000	1.0481	1.0587	1.0392
V_{g2} (p.u.)	1.0915	1.0998	1.0911	1.0016	1.0101	1.0007	1.0667	1.0768	1.0999	1.0382	1.0510	1.0305
V_{g5} (p.u.)	1.0865	1.0782	1.0854	1.0147	1.0017	1.0114	1.0609	1.0480	1.0942	1.0126	1.0190	1.0076
V_{g8}(p.u.)	1.0728	1.0827	1.0849	1.0182	0.9913	0.9988	1.0256	1.0631	1.0629	1.0142	1.0265	1.0042
V_{g11}(p.u.)	1.0983	1.0968	1.0912	1.0138	1.0288	1.0043	1.0673	1.0650	1.0688	1.0432	1.0296	1.0324
V_{g13}(p.u.)	1.0997	1.0991	1.1000	1.0283	1.0131	1.0011	1.0954	1.0674	1.0999	1.0249	0.9946	1.0242
$T_{6\text{-}9}$	1.0516	1.0456	1.0461	1.0137	1.0812	1.0792	0.9233	1.0954	0.9502	1.0043	1.0736	1.0481
$T_{6\text{-}10}$	1.0356	1.0136	1.0374	0.9895	0.9178	0.9073	0.9271	0.9855	0.9001	0.9936	0.9256	0.9314
$T_{4\text{-}12}$	0.9577	0.9105	0.9276	1.0312	0.9863	0.9658	0.9017	0.9262	0.9335	1.0196	0.9848	1.0028
$T_{6\text{-}9}$	0.9912	0.9889	1.0213	0.9839	0.9619	0.9652	0.9087	0.9224	0.9283	0.9664	0.9765	0.9513
Q_{c10}	0.0500	0.0498	0.0500	0.0218	0.0500	0.0498	0.0117	0.0490	0.0418	0.0032	0.0479	0.0487
Q_{c12}	0.0491	0.0472	0.0500	0.0325	0.0220	0.0007	0.0497	0.0048	0.0500	0.0014	0.0477	0.0006
Q_{c15}	0.0500	0.0500	0.0481	0.0500	0.0500	0.0499	0.0029	0.0353	0.0085	0.0478	0.0468	0.0326
Q_{c17}	0.0499	0.0500	0.0495	0.0017	0.0060	0.0500	0.0153	0.0414	0.0119	0.0428	0.0343	0.0360
Q_{c20}	0.0485	0.0479	0.0470	0.0500	0.0500	0.0500	0.0044	0.0028	0.0000	0.0500	0.0492	0.0499
Q_{c21}	0.0500	0.0500	0.0500	0.0463	0.0500	0.0500	0.0043	0.0100	0.0348	0.0496	0.0399	0.0498
Q_{c23}	0.0497	0.0468	0.0372	0.0500	0.0500	0.0500	0.0198	0.0407	0.0500	0.0439	0.0433	0.0413
Q_{c24}	0.0473	0.0492	0.0500	0.0483	0.0500	0.0500	0.0004	0.0341	0.0470	0.0488	0.0494	0.0493

continued on following page

Table 5. Continued

Techniques	Single objective									Multi-objective		
	Transmission loss minimization			Voltage deviation minimization			L-index minimization					
	BBO	CRO	HCRO	BBO	CRO	HCRO	BBO	CRO	HCRO	BBO	CRO	HCRO
Q_{c29}	0.0438	0.0408	0.0364	0.0500	0.0276	0.0302	0.0428	0.0363	0.0252	0.0095	0.0122	0.0099
Optimal location of TCSC	12-15	12-16	10-20	15-23	22-24	15-23	15-18	15-18	18-19	23-24	23-24	23-24
Optimal setting of TCSC (p.u.)	0.0982	0.1782	0.1369	0.2147	0.1836	0.1347	0.1144	0.1807	0.1078	0.1243	0.0935	0.1057
TL (MW)	4.5031	4.4942	4.4727	5.5543	5.5317	5.5280	5.5782	5.4631	5.1673	5.1198	5.1037	5.0685
VD (p.u.)	2.1362	2.0935	2.2040	0.0907	0.0898	0.0891	2.1144	2.1603	2.2192	0.1311	0.1216	0.1149
L-index	0.1246	0.1248	0.1252	0.1433	0.1427	0.1446	0.1098	0.1086	0.1075	0.1204	0.1184	0.1179
CT (Sec.)	16.4121	7.9772	7.8246	16.4413	7.9204	7.7132	16.7112	7.9864	7.9213	16.5523	7.9756	7.9012

Table 6. Comparative results for both single and multi-objective ORPD with SVC and TCSC device (IEEE 30-bus system)

Techniques	Single objective									Multi-objective		
	Transmission loss minimization			Voltage deviation minimization			L-index minimization					
	BBO	CRO	HCRO	BBO	CRO	HCRO	BBO	CRO	HCRO	BBO	CRO	HCRO
V_{g1} (p.u.)	1.0886	1.0903	1.0872	1.0078	1.0046	1.0138	1.0887	1.0979	1.0835	1.0456	1.0596	1.0581
V_{g2} (p.u.)	1.0938	1.0849	1.0836	1.0237	1.0098	0.9854	1.0608	1.0772	1.0814	1.0357	1.0482	1.0450
V_{g5} (p.u.)	1.0782	1.0626	1.0723	1.0086	1.0215	1.0008	1.0172	1.0430	1.0493	1.0098	1.0208	1.0192
V_{g8}(p.u.)	1.0806	1.0808	1.0965	0.9967	1.0036	1.0056	1.0205	1.0434	1.0463	1.0092	1.0238	1.0180
V_{g11}(p.u.)	1.0996	1.0997	1.0981	1.0037	1.0092	1.0130	1.0995	1.0901	1.0847	1.0413	1.0436	1.0470
V_{g13}(p.u.)	1.0100	1.0999	1.0918	1.0125	1.0248	1.0142	1.0818	1.0933	1.0674	1.0271	1.0170	1.0324
T_{6-9}	1.0217	0.9948	1.0546	1.0438	1.0167	1.0217	0.9012	1.0026	0.9355	1.0683	1.0337	1.0001
T_{6-10}	1.0069	1.0402	1.0053	0.9542	0.9658	0.9893	0.9505	0.9390	1.0362	0.9004	0.9751	1.0057
T_{4-12}	0.9862	0.9887	0.9835	1.0055	1.0344	0.9944	0.9254	0.9733	0.9241	1.0179	1.0017	1.0164
T_{6-9}	1.0148	0.9931	1.0511	0.9937	0.9937	0.9898	0.9068	0.9137	0.9061	0.9604	0.9871	0.9610
Q_{c10}	0.0478	0.0437	0.0500	0.0463	0.0500	0.0500	0.0236	0.0490	0.0209	0.0017	0.0149	0.0023
Q_{c12}	0.0500	0.0496	0.0469	0.0500	0.0463	0.0361	0.0388	0.0340	0.0301	0.0130	0.0091	0.0010
Q_{c15}	0.0486	0.0500	0.0500	0.0468	0.0477	0.0500	0.0485	0.0383	0.0169	0.0248	0.0301	0.0309
Q_{c17}	0.0500	0.0483	0.0498	0.0182	0.0371	0.0137	0.0174	0.0270	0.0142	0.0168	0.0399	0.0357
Q_{c20}	0.0492	0.0500	0.0500	0.0493	0.0500	0.0500	0.0312	0.0226	0.0406	0.0499	0.0472	0.0486
Q_{c21}	0.0500	0.0500	0.0479	0.0489	0.0500	0.0500	0.0232	0.0388	0.0192	0.0499	0.0474	0.0458
Q_{c23}	0.0469	0.0494	0.0461	0.0500	0.0397	0.0482	0.0342	0.0266	0.0245	0.0479	0.0440	0.0368
Q_{c24}	0.0452	0.0499	0.0500	0.0500	0.0408	0.0396	0.0006	0.0312	0.0135	0.0496	0.0435	0.0355

continued on following page

Table 6. Continued

Techniques	Single objective									Multi-objective		
	Transmission loss minimization			Voltage deviation minimization			L-index minimization					
	BBO	CRO	HCRO	BBO	CRO	HCRO	BBO	CRO	HCRO	BBO	CRO	HCRO
Q_{c29}	0.0474	0.0453	0.0484	0.0439	0.0379	0.0377	0.0391	0.0250	0.0102	0.0107	0.0263	0.0006
Optimal location of SVC	19	21	19	26	26	29	23	23	21	22	20	18
Optimal setting of SVC (p.u.)	0.0890	0.0782	0.0793	0.0632	0.0899	0.0647	0.1030	0.0913	0.0836	0.0349	0.0733	0.0578
Optimal location of TCSC	9-11	10-17	10-17	18-19	21-22	18-19	15-18	15-18	18-19	21-22	21-22	24-25
Optimal setting of TCSC (p.u.)	0.0638	0.1106	0.0886	0.1367	0.1247	0.1097	0.1144	0.0938	0.0862	0.0937	0.0707	0.0856
TL (MW)	4.4783	4.4518	4.4237	5.5723	5.5006	5.4837	5.5712	5.5008	5.2937	5.1004	5.0245	4.9937
VD (p.u.)	2.3824	2.0645	2.1902	0.0884	0.0873	0.0869	2.1048	1.9976	2.1362	0.1178	0.1074	0.1039
L-index	0.1241	0.1238	0.1245	0.1402	0.1428	0.1450	0.1078	0.1071	0.1063	0.1197	0.1186	0.1173
CT (Sec.)	16.5837	8.1231	7.9882	16.6944	8.0642	7.8912	16.7322	8.2114	8.1134	16.6932	8.1346	7.9982

TCSC is a series FACTS device which can be used to get controlled power flow between the lines by changing the line reactance. Similarly, TCPS is also a series FACTS device which can be used to get controlled power flow between the lines by changing the phase shifter. This chapter aims at evaluating the optimal location of FACTS device and its effectiveness in minimization of the total cost of active power generation in a power system. A standard test system (IEEE 30 bus) is considered for simulation study to show the effectiveness of the proposed method over the other conventional methods as found during reference study. The result for the tested power system showed that all the techniques GWO, TLBO, BBO, KHA, CRO and HCRO can be used for finding out the optimal location and the best parameters of the SVC, TCSC and TCPSC easily, but the HCRO technique implemented here has superior features including high-quality solution, and good computation efficiency. Considering all the techniques discussed, it can be concluded that installation of TCSC device or TCPS device helps to improve power flow through the line and reduce the transmission loss at a reduced generation cost.

REFERENCES

Balamurugan, K., Muralisachithanandam, R., & Dharmalingam, V. (2015). Performance comparison of evolutionary programming and differential evolution approaches for social welfare maximization by placement of multi type FACTS devices in pool electricity market. *International Journal of Electrical Power & Energy Systems*, *67*, 517–528. doi:10.1016/j.ijepes.2014.12.007

Basu, M. (2008). Optimal power flow with FACTS devices using differential evolution. *International Journal of Electrical Power & Energy Systems*, *30*(2), 150–156. doi:10.1016/j.ijepes.2007.06.011

Bhattacharya, A., & Chattopadhyay, P. K. (2011). Application of biogeography based optimisation to solve different optimal power flow problems. *IET Generation, Transmission & Distribution*, *5*(1), 70–80. doi:10.1049/iet-gtd.2010.0237

Cai, L. J., & Erlich, L. (2003). Optimal choice and allocation of FACTS devices using genetic algorithms. *Proceedings on Twelfth Intelligent Systems Application to Power Systems Conference*, 1–6.

Duong, T. L., Gang, Y. J., & Truong, V. A. (2013). A new method for secured optimal power flow under normal and network contingencies via optimal location of TCSC. *International Journal of Electrical Power & Energy Systems*, *52*, 68–80. doi:10.1016/j.ijepes.2013.03.025

Dutta, S., Paul, S., & Roy, P. K. (2016). Optimal allocation of SVC and TCSC using quasi-oppositional chemical reaction optimization for solving multi-objective ORPD problem. *Journal of Electrical Systems and Information Technology*, *5*(1), 83–98. doi:10.1016/j.jesit.2016.12.007

Hooshmand, R. A., Morshed, M. J., & Parastegari, M. (2015). Congestion management by determining optimal location of series FACTS devices using hybrid bacterial foraging and Nelder–Mead algorithm. *Applied Soft Computing*, *28*, 57–68. doi:10.1016/j.asoc.2014.11.032

Kavitha, K., & Neela, R. (2017). Optimal allocation of multi-type FACTS devices and its effect in enhancing system security using BBO, WIPSO & PSO. *Journal of Electrical Systems and Information Technology*. doi:10.1016/j.jesit.2017.01.008

Kesselm, P., & Glavitsch, H. (1986). Estimating the voltage stability of power system. *IEEE Transactions on Power Delivery*, *1*(3), 346–354. doi:10.1109/TPWRD.1986.4308013

Lie, T. T., & Deng, W. (1997). Optimal flexible AC transmission systems (FACTS) devices allocation. *International Journal of Electrical Power & Energy Systems*, *19*(2), 125–134. doi:10.1016/S0142-0615(96)00036-1

Lima, G. M., Galiana, F. D., Kockar, I., & Munoz, J. (2003). Phase shifter placement in large scale system via mixed integer linear programming. *IEEE Transactions on Power Systems*, *18*(3), 1029–1034. doi:10.1109/TPWRS.2003.814858

Mahdad, B., & Srairi, K. (2016). Adaptive differential search algorithm for optimal location of distributed generation in the presence of SVC for power loss reduction in distribution system. *Engineering Science and Technology, an International Journal*, *19*, 1266–1282.

Mukherjee, A., & Mukherjee, V. (2016). Solution of optimal power flow with FACTS devices using a novel oppositional krill herd algorithm. *International Journal of Electrical Power & Energy Systems*, *78*, 700–714. doi:10.1016/j.ijepes.2015.12.001

Saravanan, M. S., Slochanal, M. R., Venkatesh, P., Stephen, J., & Abraham, P. (2007). Application of particle swarm optimization technique for optimal location of FACTS devices considering cost of installation and system loadability. *Electric Power Systems Research*, *77*(3-4), 276–283. doi:10.1016/j.epsr.2006.03.006

Sebaa, K., Bouhedda, M., Tlemçani, A., & Henini, N. (2014). Location and tuning of TCPSTs and SVCs based on optimal power flow and an improved cross-entropy approach. *International Journal of Electrical Power & Energy Systems*, *54*, 536–545. doi:10.1016/j.ijepes.2013.08.002

Sharma, A. K., (2006). Optimal number and location of TCSC and loadability enhancement in deregulated electricity markets using MINLP. *International Journal of Emerging Electric Power Systems*, *5*(1).

Shin, J. R., Kim, B. S., Park, J. B., & Lee, K. Y. (2007). A new optimal routing algorithm for loss minimization and voltage stability improvement in radial power systems. *IEEE Transactions on Power Systems*, *22*(2), 648–657. doi:10.1109/TPWRS.2006.889112

Singh, S. N., & David, A. K. (2001). Optimal location of FACTS devices for congestion management. *Electric Power Systems Research*, *58*(2), 71–79. doi:10.1016/S0378-7796(01)00087-6

Sundar, K. S., & Ravikumar, H. M. (2012). Selection of TCSC location for secured optimal power flow under normal and network contingencies. *International Journal of Electrical Power & Energy Systems*, *34*(1), 29–37. doi:10.1016/j.ijepes.2011.09.002

Wirmond, V. E., Fernandes, T. S. P., & Tortelli, O. L. (2011). TCPST allocation using optimal power flow and Genetic Algorithms. *International Journal of Electrical Power & Energy Systems*, *33*(4), 880–886. doi:10.1016/j.ijepes.2010.11.017

Zhang, W., & Liu, Y. (2008). Multi-objective reactive power and voltage control based on fuzzy optimization strategy and fuzzy adaptive particle swarm. *International Journal of Electrical Power & Energy Systems*, *30*(9), 525–532. doi:10.1016/j.ijepes.2008.04.005

Chapter 9
Optimal Location of STATCOM, SSSC, and UPFC Devices for Solving OPF and ORPD Problem Using Different Evolutionary Optimization Techniques

ABSTRACT

The development of FACTS devices based on the advance of semiconductor technology opened up new opportunities for controlling the power flow and extending the load ability of the power transmission network. Amongst the various FACTS devices, the UPFC is considered the most versatile FACTS device that can simultaneously control bus voltage and both active and reactive power flow through the transmission line. This chapter discusses the implementation of grey wolf optimization (GWO), teaching-learning-based optimization (TLBO), biogeography-based optimization (BBO), krill herd algorithm (KHA), chemical reaction optimization (CRO), and hybrid CRO (HCRO) approaches to find the optimal placement and parameter setting of unified power flow controller (UPFC) to achieve the optimal performance of optimal power flow (OPF) and optimal reactive power dispatch (ORPD) problems. Two test systems, namely IEEE 14-bus and IEEE 30 with valve-point non-linearity, are considered to demonstrate the effectiveness of proposed approaches.

DOI: 10.4018/978-1-5225-6971-8.ch009

INTRODUCTION

The electric power grid is the largest man-made machine in the world. It consists of synchronous generators, transformers, transmission lines, switches and relays, active/reactive components, and loads. Power system networks are complex systems that are nonlinear, non-stationary, and prone to disturbances and faults. Reinforcement of a power system can be accomplished by improving the voltage profile, increasing the transmission capacity and others. Nevertheless, some of these solutions may require a considerable investment that could be difficult to recover. Installation of FACTS devices in optimal position is one alternate to address some of those problems (Hingorani & Gyugyi, 2001).

Optimal reactive power dispatch (ORPD), a sub problem of optimal power flow (OPF), has a significant influence on the economic and secure operation of power systems. Optimal reactive power dispatch (ORPD) is an important tool for power system operators for both planning and reliable operation in the present day power systems. The important aspect of ORPD is to determine the optimal settings of control variables for minimizing transmission loss, improve the voltage profile and voltage stability, while satisfying various equality and inequality constraints. The ORPD problem is in general non-convex, non-linear and exists many local minima.

In the view of the above, the main objective of this chapter is to propose new, efficient optimization techniques to locate various FACTS devices in an optimal position in order to minimize transmission loss and voltage stability of short, medium and large-scale power systems. There are a number of FACTS devices available, such as, the static synchronous series compensator (SSSC) (Basu, 2008; Vijayakumar & Kumudinidevi, 2008) which can be used to generate and insert a series voltage, and it can be regulated to change the impedance of the transmission line. In this way, the power flow of transmission line, where the SSSC is connected is controlled. Similarly, Static synchronous compensator (STATCOM) is connected in parallel with the power system which can be controlled the power flow of the transmission line. UPFC is the most versatile FACTS device which can simultaneously controls the active power and the reactive power of the bus voltage at which it is connected (Gyugyi, 1992). It is used in the transmission system to reduce the flows in heavily loaded lines, resulting in increased load ability, low system loss and improved stability of transmission system by controlling the power flows in the network.

LITERATURE SURVEY

In the literature, the optimal location and setting of FACTS devices has retained the interest of worldwide researchers in power systems, where various methods and criteria are used in this field. A harmony search algorithm (HSA) was developed by Sirjani *et al.* (2012) for simultaneous minimization of total cost, the voltage stability index and voltage profile and power loss of IEEE 57-bus test system using shunt capacitors, SVC and static synchronous compensators (STATCOM). Mahdad *et al.* (2007) implemented a simple and dynamic methodology, based on reasoning-based rules from fuzzy logic theory for finding out the optimal position and setting of SVC in ORPD problem. The objective functions selected in this study were the real power losses and the voltage profile improvement. IEEE 30-bus test system was used in this paper to test the effectiveness of the proposed approach. Bhargava *et al.* (2006) proposed a simple and efficient algorithm for incorporating STATCOM losses in the load flow calculation. In this technique, both the STATCOM switching loss and the step-down transformer ohmic loss were considered and calculated iteratively. The proposed algorithm was tested on IEEE 30-bus and IEEE 57-bus power systems. The computational results showed that the proposed algorithm was found to possess good convergence characteristics. A multi-control functional model of SSSC in Newton power flow algorithm was presented by Zhang (2003). This multi-control functional model was used for steady state control of one of the following parameters such as the active power flow on the transmission line, the reactive power flow on the transmission line, the voltage at the bus or the reactance of the transmission line. The proposed method was used in IEEE 30-bus system, IEEE 118-bus system and IEEE 300-bus system to illustrate the feasibility of the SSSC model and the performance of the Newton power flow algorithm. Vinkovic and Mihalic (2008) presented a current-based model of a static synchronous series compensator (SSSC) for power-flow calculations by applying the Newton Raphson method. To check the feasibility, the proposed method was tested on the IEEE 57-bus system and compared to the existing injection SSSC load-flow model. Sedighizadeh *et al.* (2014) proposed a multi-objective optimization approach to find the optimal location and setting of TCSC and SVC devices in order to simultaneously minimize power loss, an increase of the stability margin and voltage profile improvement. Interior point method (IPM) and evolutionary particle swarm optimization (EPSO) based hybrid method was developed by Raju *et al.* (2009) to solve OPF in power system incorporating flexible AC transmission systems (FACTS) such as static synchronous series compensator (SSSC) for security enhancement in the power systems of IEEE 30-bus test . Hagh, *et al.* (2014) introduced the impact of STATCOM and SSSC on the predictability of power system in standard IEEE 14-bus, and IEEE 57-bus test systems. Panda proposed GA based solution (Panda, 2009) to solve the transient performance of a power system subjected to a severe disturbance by damping the multi-modal oscillations namely; local mode, inter-area mode and inter-plant mode using SSSC based controller. Zonkoly (2008) presented PSO approach for the optimal location and sizing of SSSC in the transmission system to reduce the transmission losses. Gholipour & Nosratabadi (2015) proposed a new seeker optimization algorithm (SOA) based on Pareto optimum method with high convergence speed for solving the nonlinear model to design of power system stabilizer (PSS) and SSSC coordinated controllers. Lai & Ma (1996) proposed a binary coded GA approach to solve the OPF problem with FACTS devices. In this work UPFC was used to regulate the power flow such that the real power losses were minimized. Several contingencies were considered to successfully test the proposed approach. A binary coded GA coupled with a power flow algorithm was used by Leung & Chung (2000) to minimize the quadratic fuel cost function with the inclusion of UPFC settings as additional control variables. A technique based on GA to solve OPF incorporating UPFC was reported by Ippolito *et al.* (2006) This technique provided the optimal number and locations of UPFC devices in an assigned power system network to maximize system capabilities with minimum transmission loss. Lai & Ma (1995) introduced EP to minimize the real power losses in power networks by regulating the power flow via optimal UPFC settings. The proposed approach was tested under various contingency scenarios. Venkatesh *et al.* (2004) made use of EP to solve the OPF problem with UPFC. They addressed the problem considering two fuzzy objective functions, minimum of the real power losses and the best voltage profile.

Motivation of the Present Work

The ongoing restructuring of power sector, together with continuously growing electricity demand, has led to the operation of power system networks closer to the transmission system thermal and stability limit. In recent years, there are emerging technologies available which can help to deal with problems of power system. One of such technologies is flexible AC Transmission system (FACTS).

Over the last two decades, many researchers performed a lot of research on ORPD. Various optimization techniques have been evolved to solve ORPD problem. These algorithms are generally divided into two categories: the classical mathematical optimization algorithms and intelligent optimization algorithms. The basic idea of classical algorithm is starting from an initial point, continuously improving the current solution through a certain orbit, and ultimately converging to the optimal solution. These algorithms include benders decomposition (Amjady & Ansari, 2014) linear programming (LP) (Urdaneta et al., 1999) quadratic programming (QP) (Grudinin, 1998), non-linear programming (NLP) (Lai & Ma, 1997) and mixed integer linear programming (MILP) (Kim et al., 1998) etc. Some of these techniques, however, have a good convergence but most of them suffer from local optimality. Since ORPD is multimodal and non-linear optimization problem and severely depends on the initial guess, the classical techniques are unable to produce the global optimal solution. To overcome this problem some intelligent optimization algorithms known as heuristic techniques are applied to solve ORPD problem. Some of the well popular optimization techniques are evolutionary programming (EP) (Venkatesh et al., 2003), genetic algorithm (GA) (Wei et al., 2006), simulated annealing (SA) (Keyan et al., 2006), tabu search (TS) (Lin et al., 2008), differential evolution (DE) (Shaheen et al., 2009), particle swarm optimization (PSO) (Zhao et al., 2005) and artificial bee colony (ABC) (Karaboga & Basturk, 2007) etc.

In this chapter, various optimization techniques discussed in earlier chapters are successfully implemented to find the optimal location of STATCOM, SSSC and UPFC device for solving OPF and ORPD problem. The effectiveness of the proposed algorithms is demonstrated by implementing them in standard system namely, IEEE 30-bus system and their performances are compared with PSO, DE and other optimization techniques recently published in the literature.

STATIC MODEL OF STATCOM, SSSC AND UPFC

Static models of the STATCOM, SSSC and UPFC devices have already been discussed in chapter 6**MATHEMATICAL PROBLEM FORMULATION**

Objective Function of ORPD

The conventional formulation of ORPD (Gayatri et al., 2018) problem determines the optimal setting of control variables such as generator terminal voltages, transformers tap setting, reactive power of shunt compensators, controllable voltage source of STATCOM (Rao & Rao 2017) and SSSC and there phase angle to minimize the transmission loss, improve the voltage profile and voltage stability of the power system while satisfying the operational constraints.

Minimization of Total Real Power Loss

It may be expressed by

$$f_1(x,y) = P_{loss} = \sum_{k=1}^{NTL} G_k [V_m^2 + V_n^2 - 2|V_m||V_n|\cos(\delta_m - \delta_n)] \tag{1}$$

where, $f_1(x,y)$ is the transmission loss minimization objective function; P_{loss} is the total power loss; G_k is the conductance of the kth line connected between them m^{th} and n^{th} buses; V_m, V_n are the voltage of the mth and n^{th} buses respectively; δ_m, δ_n are the phase angle of the mth and n^{th} bus voltages. x is the vector of dependent variable consisting of load voltages ($V_{l_1},...V_{l_{NL}}$), generators' reactive powers ($Q_{g_1},\cdots,Q_{g_{NG}}$), transmission lines' loadings ($S_{l_1},\cdots,S_{l_{NTL}}$), controllable voltage source of STATCOM ($E_{p_1},...,E_{p_n}$) and phase angle of STATCOM ($\delta_{p_1},...,\delta_{p_n}$), controllable voltage source of SSSC ($E_{S_1},...,E_{S_n}$) and phase angle of SSSC ($\delta_{S_1},...,\delta_{S_n}$); y is the vector of independent variables consisting of generators' voltages ($V_{g_1},\cdots,V_{g_{NG}}$), transformers' tap settings ($T_1,\cdots,T_{NT}$), reactive power injections ($Q_{i_1},..,Q_{i_{NC}}$) and voltage of the buses where STATCOMs are used ($V_{STATCOM_1},..,V_{STATCOM_n}$)

Therefore x & y can be expressed as

$$x = [V_{l_1},....,V_{l_{NL}},Q_{g_1},....Q_{g_{NG}},S_{l_1},.....,S_{l_{NTL}},E_{p_1},...,E_{p_n},\delta_{p_1},...,\delta_{p_n}] \tag{2}$$

$$y = [V_{g_1},..,V_{g_{NG}},T_1,..,T_{NT},Q_{i_1},..,Q_{i_{NC}},V_{STATCOM_1},..,V_{STATCOM_n}] \quad (3)$$

where, NG, NL are the number of generator and load buses; NTL, NT, NC are the number of transmission lines, regulating transformers and shunt compensators, respectively; V_{ref}, V_{Li} are the reference voltage of the load buses and load voltage of the ith bus, respectively.

Minimization of Voltage Deviation

Minimization of deviations of voltages from desired values is done since bus voltage is one of the most important security and service quality indices. The objective function of voltage profile improvement, i.e. voltage deviation minimization at load buses, may be expressed as:

$$f_2(x,y) = \min\left(\sum_{f=1}^{N_L} \left|V_{L_f} - V_{L_f}^{sp}\right|\right) \quad (4)$$

where, V_{L_f} is the voltage at the ith load bus; $V_{L_f}^{sp}$ is the desired voltage at the ith load bus, usually set to 1.0 $p.u.$

Minimization of L-Index

It is very important to maintain constantly acceptable bus voltage at each bus under normal operating conditions, when the system is being subjected to a disturbance. The non-optimized control variables may lead to a progressive and uncontrollable drop in voltage resulting in an eventual widespread voltage collapse. In this work, voltage stability enhancement is achieved through minimizing the voltage stability indicator L-*index*. The indicator values vary in the range between 0 (the no-load case) and 1 which corresponds to voltage collapse.

For multi-node system,

$$I_{bus} = Y_{bus}.V_{bus} \quad (5)$$

Segregating the nodes into two categories, that is, the load bus set ψ_L and generator bus set ψ_G can be written as

$$\begin{bmatrix} I^L \\ I^G \end{bmatrix} = \begin{bmatrix} Y_1 & Y_2 \\ Y_3 & Y_4 \end{bmatrix} \cdot \begin{bmatrix} V^L \\ V^G \end{bmatrix} \tag{6}$$

The transmission system itself is linear and allows a representation in terms of a hybrid matrix H:

$$\begin{bmatrix} V^L \\ I^G \end{bmatrix} = H. \begin{bmatrix} I^L \\ V^G \end{bmatrix} = \begin{bmatrix} Z^{LL} & F^{LG} \\ K^{GL} & Y^{GG} \end{bmatrix} \cdot \begin{bmatrix} I^L \\ V^G \end{bmatrix} \tag{7}$$

where V^L and I^L are vectors of voltages and currents at load buses; V^G and I^G are vectors of voltages and currents at generator buses; Z^{LL}, F^{LG}, K^{GL}, Y^{GG} are sub-matrices the H-matrix generated from Y_{bus} partial inversion, respectively. A voltage collapse can take place in systems or subsystems and can appear quite abruptly.

L-index L_j of the jth bus can be expressed through the following equation:

$$\begin{cases} L_j = \left| 1 - \sum_{i=1}^{N_{PV}} F_{ji} \dfrac{V_i}{V_j} \right|, \quad j = 1,2,\ldots,N_{PQ} \\ F_{ji} = -\left[Y_1\right]^{-1}\left[Y_2\right] \end{cases} \tag{8}$$

where, N_{PV} and N_{PQ} are number of PV and PQ bus, respectively. L-*index* is calculated for all the load buses. Thus, a global power system representative L describing the stability of the complete system is represented as follows:

$$L = \max(L_j), \; j = 1,2,\ldots,N_{PQ} \tag{9}$$

In the ORPD problem, incorrect tuning of continuous and discrete control variable settings may boost the value of L-*index,* that may reduce the voltage stability outskirt of the system. Let the maximum value of L-*index*, be denoted as $L_{\max}$. Therefore, to enhance the voltage stability and to move the system far from the voltage collapse margin, the following objective function can be calculated as:

$$f_3(x,y) = \min \ L_{\max} \tag{10}$$

OBJECTIVE FUNCTION OF OPF

Single Objective Function

Minimizing the Total Cost

The main objective of this work is to determine the optimal location and the optimal parameter setting of the UPFC (Dutta et al., 2016) in the power network to minimize the total cost which consists of the power generation cost and UPFC installation cost. Therefore, the objective of this work may be expressed as

$$C_T = C_{TG} + C_{UPFC} \tag{11}$$

The generation cost and UPFC installation cost may be formulated as follows:

Total Generation Cost

The total generation cost consists of active and reactive power production cost which maybe expressed as (Taher & Amooshahi, 2012):

$$C_{TG} = C_{act} + C_{rec} \tag{12}$$

$$C_{act} = \sum_{i=1}^{NG} a_i + b_i P_{gi} + c_i P_{gi}^{\ 2} \ (\$/\text{h}), \tag{13}$$

$$C_{rec} = \sum_{i=1}^{NG} a_i + b_i Q_{gi} + c_i Q_{gi}^{\ 2} \,(\$/\text{h}), \tag{14}$$

UPFC Installation Cost

The installation cost of UPFC may be expressed in terms of operating range of the UPFC in MVA and capital recovery factor and is given by (Taher & Amooshahi, 2012)

$$C_{UPFC} = \sum_{j=1}^{D} (F_{UPFCj} \times S_j \times 1000 \times \beta) \,/\, 8760 \,(\$/\text{h}), \tag{15}$$

$$F_{UPFCj} = 0.0003 S_j^{\ 2} \ - 0.2691 S_j + 188.22 \,(\$/\text{KVA}), \tag{16}$$

$$\beta = \frac{r(1+r)^n}{(1+r)^n - 1} \tag{17}$$

Improvement of Voltage Profile

Bus voltage is one of the most important security and service quality indices. Improving voltage profile can be obtained by minimizing the load bus voltage deviations from desired voltage, which is usually set to $1.0\,p.u.$ This objective function can be expressed as

$$f_v = \min\left(\sum_{i=1}^{NL} \left|V_i - V_i^{sp}\right|\right) \tag{18}$$

Minimization of System Power Losses

The minimization of system active power losses f_l is considered as another objective which may be expressed as follows

$$f_l = \min\left[\sum_{k=1}^{N_{TL}} G_k \left(\left|V_m^2 + V_n^2 - 2V_m V_n \cos\alpha_{mn}\right|\right)\right] \tag{19}$$

Multi-Objective Functions

The multi-objective function can be expressed in terms of active power loss and voltage deviation which may be formulated as

$$\text{Minimize } \Psi(f_v, f_l) \tag{20}$$

As the bus voltages are permitted to swing between lower and upper limits, a constant penalty factor is introduced and the problem is solved as a simple single objective optimization problem as follows:

$$Minimize \Psi = f_l + pf \times f_v \tag{21}$$

where, pf is penalty factor of load-bus voltage violation. With this penalty, any violation of bus voltage will be reflected in the fitness function. The choice of pf determines the accuracy and speed of convergence. There are several methods for choosing penalty factors which establish a family of intervals for every constraint. However, the choice of penalty terms depends on the complexity of the problem. Different values of pf are tried to find the suitability and after extensive experimentation, it is set as 0.01. With this penalty, HCRO attempts to minimize the real power losses while trying to maintain the load flow voltages close to 1.0 *p.u.*

Constraints

The OPF and ORPD with FACTS devices are subjected to the following constraints:

Equality Constraints

These constraints represent the load flow equations as given below:

$$\begin{cases} \sum_{i=1}^{NB}\left(P_{Gi} - P_{Li}\right) + \sum_{i=1}^{SSSC} P_{is} = \sum_{i=1}^{NB} \sum_{j=1}^{NB} \left|V_i\right|\left|V_j\right|\left|Y_{ij}\right| \cos\left(\theta_{ij} - \delta_{ij}\right) \\ \sum_{i=1}^{NB} Q_{Gi} - Q_{Li} + \sum_{i=1}^{SSSC} Q_{is} = -\sum_{i=1}^{NB} \sum_{j=1}^{NB} \left|V_i\right|\left|V_j\right|\left|Y_{ij}\right| \sin\left(\theta_{ij} - \delta_{ij}\right) \end{cases} \tag{22}$$

where P_{Li} is the active power demand of the ith bus; Q_{Gi}, Q_{Li} are the reactive power generation and demand of the ith bus, respectively; P_{is}, Q_{is} are the injected active and reactive powers of TCPS at the ith bus, respectively; Y_{ij} is the admittance of transmission line connected between the ith bus and the jth bus; θ_{ij} is the admittance angle of transmission line connected between the ith bus and the jth bus; NB is the number of buses; SSSC is the number of SSSCs.

Inequality Constraints

Some inequality constraints such as generator constraints, load bus constraints, transmission line constraints, transformer tap constraints, and shunt compensator constraints, have already been discussed in chapter 4.

STATCOM voltage constraints:

$$E_{p_i}^{\max} \leq E_p \leq E_{p_i}^{\min} \tag{23}$$

STATCOM phase angle constraints:

$$\delta_{p_i}^{\max} \leq \delta_p \leq \delta_{p_i}^{\min} \tag{24}$$

SSSC voltage constraints:

$$E_S^{\min} \leq E_S \leq E_S^{\max} \tag{25}$$

SSSC phase angle constraints:

$$0 \leq \delta_S \leq 2\pi \tag{26}$$

UPFC series source constraints:

$$E_{Si}^{\min} \leq E_{Si} \leq E_{Si}^{\max} \qquad i \in NUPFC \tag{27}$$

$$\delta_{Si}^{\min} \leq \delta_{Si} \leq \delta_{Si}^{\max} \qquad i \in NUPFC \tag{28}$$

UPFC shunt source constraints:

$$E_{Pi}^{\min} \leq E_{Pi} \leq E_{Pi}^{\max} \qquad i \in NUPFC \tag{29}$$

$$\delta_{Pi}^{\min} \leq \delta_{Pi} \leq \delta_{Pi}^{\max} \qquad i \in NUPFC$$

where, $\delta_{p_i}^{\max}, \delta_{p_i}^{\min}$ are the maximum and minimum phase angle limits of the ith STATCOM; $E_{p_i}^{\max}, E_{p_i}^{\min}$ are the maximum and minimum voltage limits of the ith STATCOM; $E_S^{\min}, E_S^{\max}$ are the maximum and minimum voltage limits of the SSSC; NG, NL, NTL, NT, NC are the number of generator bus, load bus, transmission line, regulating transformer and shunt compensator; respectively.

EVOLUTIONARY OPTIMIZATION TECHNIQUES APPLIED TO OPF AND ORPD INCORPORATING OPTIMAL LOCATION OF STATCOM, SSSC AND UPFC DEVICES

BBO Algorithm Applied to OPF and ORPD With Optimal Location of STATCOM, SSSC and UPFC Devices

The procedure for implementing the BBO algorithm in solving optimal STATCOM, SSSC and UPFC placement to resolve constrained OPF and ORPD problem can be summarized by the following steps:

Step 1: Initialize the population randomly in the search space, i.e. the initial positions of *SIVs* (independent variables such as generators' voltages, tap settings of regulating transformers, reactive power injections, power flow through the transmission lines where SSSC devices are placed and bus voltage of the STATCOM connected bus) of each habitat should be randomly selected while satisfying different operational constraints. Based on the population size, several feasible solution sets are generated.

Step 2: Perform STATCOM and SSSC based load flow using these *SIVs* and determine all dependent variables such as load voltages, active power of slack bus, generators' reactive powers, voltage and phase angle of STATCOM and SSSC devices.

Step 3: Evaluate the transmission loss, voltage deviation and L-index (i.e. *HSI*) for each habitat of the population set using

$$f_1(x,y) = P_{loss} = \sum_{k=1}^{NTL} G_k [V_m^2 + V_n^2 - 2|V_m||V_n|\cos(\delta_m - \delta_n)],$$

$$f_2(x,y) = \min\left(\sum_{f=1}^{N_L} \left|V_{L_f} - V_{L_f}^{sp}\right|\right) \text{and}\, L = \max(L_j),\ j = 1,2,\ldots,N_{PQ},$$

respectively for ORPD.

Similarly, evaluate the fuel cost, voltage profile, system power loss and multi-objective of active power loss and voltage deviation (i.e. *HSI*) for each habitat of the population set using $C_T = C_{TG} + C_{UPFC}$,

$$f_l = \min\left[\sum_{k=1}^{N_{TL}} G_k \left(\left|V_m^2 + V_n^2 - 2V_m V_n \cos\alpha_{mn}\right|\right)\right]$$

and $Minimize\Psi = f_l + pf \times f_v$ respectively for OPF.

Step 4: Sort the habitats from the best to the worst using *HSI* values.
Step 5: Identify few elite habitats based on the *HSI* values. These best habitat sets are kept as it is without any modification.
Step 6: Based on *HSI value, a*ssign number of species to each habitat.
Step 7: Non-elite habitats' independent variables (*SIVs*) are modified using hybrid migration operation as explained in an earlier chapter.
Step 8: Update species count probability of each habitat. Based on habitat modification probability, modify the independent variables (*SIVs*) of non-elite habitats by probabilistic mutation operation.
Step 9: Check whether the independent variables (rating of DG) violate the operating limits or not. If any independent variable is less than the minimum level, it is made equal to the minimum value and if it is greater than the maximum level it is made equal to the maximum level.
Step 10: Repeat Steps 2-9 for a predefined number of iterations.

TLBO Algorithm Applied to OPF and ORPD With the Optimal Location of STATCOM, SSSC and UPFC Devices

The main steps of the proposed TLBO (Nusair & Alomoush 2017) approach applied to ORPD problem are listed as follows.

Step 1: Depending upon the population size, several numbers of initial solutions are initialized randomly within their specified operating limits.
Step 2: The independent variables of ORPD problem (generators' voltages, tap settings of regulating transformers, reactive power injections, power flow through the transmission lines where SSSC are placed and bus voltage of the STATCOM connected bus) of each initial solution string are updated as follows:

$$OP_{r,s} = a_s + b_s - P_{rs}$$

where $r = 1,2,...,n_p; s = 1,2,...,n_d; P_{r,s}$ is the sth independent variables of the rth vector of the population; $OP_{r,s}$ is the sth independent variables of the rth vector of the opposite population; n_p is the population size and n_d is the number of independent variables.

Step 3: The dependent variables of ORPD problem (slack bus power, load bus voltages, generator reactive power outputs, transmission line loadings, voltage magnitude and phase angle of the voltage source of the STATCOM and SSSC devices) are evaluated using load flow analysis.

Step 4: The fitness values of the current population *(P)* and oppositional population *(OP)* sets are evaluated.

Step 5: n_p number of fittest individuals are selected from $\{P \cup OP\}$.

Step 6: The entire feasible solution sets are sorted from best to worst and few elite solutions are selected based on the fitness values.

Step 7: The independent variables (generators' voltages, tap settings of regulating transformers, reactive power injections, power flow through the transmission lines where STATCOM and SSSC devices are placed and power injections of STATCOM and SSSC devices) of each non-elite solution are modified by performing teaching and learning phase of TLBO algorithm.

Step 8: The dependent variables of ORPD (slack bus power, load bus voltages, generator reactive power outputs, transmission line loadings) are evaluated. Afterward, the fitness value of the population set is calculated.

Step 9: Based on a jumping rate j_r (i.e. jumping probability), the opposite population of the current population is evaluated and fitness values of the opposite population are calculated.

Step 10: n_p fittest individuals are selected from the current and the opposite population.

Step 11: Repeat steps (6-10) until the convergence criteria are reached.

HCRO Algorithm Applied to OPF and ORPD With Optimal Location of STATCOM, SSSC and UPFC Devices

The algorithmic steps of HCRO (Goswami et al., 2017) for solving STATCOM and SSSC based ORPD problem are enumerated as follows:

Step 1: The each independent control variable such as generators' voltages, tap settings of regulating transformers, reactive power injections, power flow through the transmission lines where SSSC devices are placed and bus voltage of STATCOM connected buses are randomly generated within the specified limit.

Step 2: Initialize the individuals of the population. The molecular structure of each molecule should be randomly initialized such that each candidate solution lies within the feasible operating region.

Step 2.1: Depending upon the number of SSSC or STATCOM, their positions are randomly generated as follows (see Box 1.)

Step 3: Run the Newton-Raphson load flow to find the slack bus power, load voltages, generators' reactive powers, voltage magnitude and phase angle of the source voltage of STATCOM and SSSC, optimal location of STATCOM and SSSC.

Step 4: Evaluate the fitness value. Based on the objective value, sort the molecules and few are kept as elite molecules.

Step 5: Assign PE same as fitness value to each molecule of the population set.

Step 6: Apply on-wall ineffective collision operations, decomposition operation, inter-molecular ineffective collision and synthesis collision operation on non elite molecules as described in Steps 4, 5, 6 and 7 of the section "HCRO Algorithm Applied to OPF and ORPD With TCSC and TPCS Devices" to modify the independent variables i.e. generators' voltages, tap settings of regulating transformers, reactive power injections, power flow through the transmission lines where SSSC are placed and bus voltage of STATCOM connected buses.

Step 7: To check the feasibility, Newton-Raphson based load flow is run and all the dependent variables of the ORPD problem are evaluated. If any dependent variable violates the operating limit, that solution set is replaced by new feasible solution.

Step 8: Fitness value for each newly generated feasible solution set is computed.

Step 9: If the termination criterion (maximum iteration limit) is satisfied then stop the iteration otherwise go to step 4 for next generation.

Box 1.

```
for i = 1 : N_P
   a = randperm(NTL)
    for j = 1 : NUPFC
       loc(j) = a(j)
    end
end
```

GWO Algorithm Applied to OPF and ORPD With Optimal Location of STATCOM, SSSC, and UPFC Devices

Step 1: Read line data, bus data, fuel cost co-efficient, load flow parameters, etc. of the power system and specify the upper and lower limits of each control variables of OPF like active power generation, generators' bus voltage, load bus voltages, reactive power generations, tap settings of the tap changing transformers, VAr injection of the shunt compensating devices and line flow through each transmission lines. Afterward, randomly select the independent control variables within their operating limits. Thereafter, Newton Raphson based load flow method is run to evaluate the slack bus power and the other dependent variables. Check whether the control variables are within the limits or not. If any of them violates the limits, the corresponding set is discarded and is initialized again. In this fashion, a feasible initial solution is generated. New several initial solutions set depending upon the population size are generated. A feasible solution set (control variables) represents the position of different GWO individuals.

Step 2: The input parameters such as a, A and C of GWO algorithm is initialized.

Step 3: Evaluate the fitness function of the initial solution set and elite solutions are identified.

Step 4: Find the updated best position of alpha, beta and delta wolves and independent variables of OPF are modified. Thereafter, Newton Raphson based load flow is run to obtain the dependent variables. The feasibility of the controlled variables is checked. If any one of them does not satisfy the inequality constraints, the particular set is replaced by the best feasible solution set.

Step 5: Evaluate the fitness value of the feasible population set.

Step 6: Check if the number of iteration is less than the maximum specified iteration or not.

Step 7: When the maximum iteration is reached display the fitness value and the updated final position of the wolves denoted by X(t+1) from the prey which is generally a circle around the prey.

Step 8: If the maximum iteration is not reached, iteration cycle is incremented.

Step 9: Go to Step 4 until a stopping criterion is satisfied.

KHA Algorithm Applied to OPF and ORPD With Optimal Location of STATCOM, SSSC, and UPFC Devices

The procedure for implementing the KHA algorithm in solving optimal power flow problem to get minimum cost may be summarized by the following steps:

Step 1: All independent variables such as all generators' active power except slack bus, generators' voltages, tap settings of regulating transformers, should be randomly selected between their operating limits

Step 2: Run the Newton-Raphson load to determine the slack bus power, load voltages, power flow through the transmission line, and check whether they satisfy the operating limits or not. If any of these parameters violate the operating limits; discard that population set and re-initialize the corresponding set. Depending upon the population size, several solutions are generated. Each feasible solution set represents the initial position of each krill individuals.

Step 3: Evaluate the fitness value and sort the solutions from best to worst.

Step 4: Select few best solutions for elitism.

Step 5: Evaluate the three motion index, namely, motion induced by other individual; foraging motion and random diffusion.

Step 6: The non-elite krill individuals' position is modified.

Step 7: Applying crossover and mutation to update the position of each non-elite krill individuals.

Step 8: Check whether the independent variables violate the operating limits or not. If any independent variable is less than the minimum level it is made equal to the minimum value and if it is greater than the maximum level it is made equal to maximum level.

Step 9: Run Newton-Raphson based load flow to determine the dependent variables and check whether they satisfy the system operating constraints or not. The infeasible solutions are replaced by the best feasible solutions.

Step 10: Go to Step 4 until the current iteration number reaches the pre specified maximum iteration number.

SIMULATION RESULTS AND DISCUSSIONS

The standard IEEE 30-bus system is used to evaluate the correctness and the relative performance of the different heuristic techniques method. This system consists of 6 generators, 4 regulating transformers, 9 shunt compensators and 41 transmission lines. The full system data are taken from (Bhargava et al., 2006) and transmission line data and the load data. The maximum and minimum voltage limits at all the buses are taken as 1.10 p.u and 0.95 p.u, respectively. The upper and lower tap settings limits of regulating transformers are taken as 1.10 *p.u.* and 0.9 *p.u.*, respectively. The upper and lower voltage limits of STATCOM are taken as 1.10 p.u and 0.9 p.u, respectively. The limits of the phase angle of STATCOM are taken as $-20^0 \leq \delta_p \leq 0^0$. The resistance and reactance of equivalent STATCOM converter is 0.01 *p.u.* and 0.1 *p.u.*, respectively. Similarly for SSSC the upper and lower voltage limits are taken as 0.10 p.u and 0.01 p.u, respectively. The limits of phase angle of SSSC are taken as $-20^0 \leq \delta_p \leq 20^0$. The resistance and reactance of equivalent SSSC converter is 0.01 *p.u.* and 0.1 *p.u.*, respectively. The upper and lower limits of the series and shunt voltage sources of the UPFC are taken as (0 to 0.2) *p.u.* and (0 to 1.1) *p.u.*, respectively. The shunt and series impedances of installed UPFC are taken as (0.01+j0.1) *p.u.* and (0.001+j0.2) *p.u.*, respectively.

ORPD Result of IEEE 30-Bus System

The performance of the different heuristics methods are demonstrated by applying it in conventional ORPD problem and ORPD with STATCOM and SSSC and its results are compared with those of other methods.

Case 1 (Transmission Loss Minimization)

The effectiveness of the HCRO method along with BBO, KHA, TLBO, GWO and CRO are initially verified by applying it in IEEE 30-bus system without any FACTS. The transmission loss and the optimal settings of control variables obtained by BBO, KHA, TLBO, GWO, CRO and HCRO algorithms are reported in the Table 1. The results show that the transmission loss found by the HCRO method is lower than the other algorithms.

In order to check the feasibility of the different method for a complicated network, it is applied to solve ORPD with STATCOM of the same test system. The simulation results of transmission loss, the controlled variables, the

Table 1. Comparison of simulation results obtained by different algorithms without FACTS

Real power loss minimization						
Algorithms	HCRO	CRO	GWO	TLBO	KHA	BBO
V_1 (*p.u.*)	1.1000	1.0998	1.0997	1.0989	1.0986	1.0985
V_2 (*p.u.*)	1.0941	1.0939	1.0918	1.0928	1.0923	1.0919
V_5 (*p.u.*)	1.0746	1.0743	1.0721	1.0724	1.0707	1.0707
V_8 (*p.u.*)	1.0764	1.0762	1.0733	1.0738	1.0740	1.0727
V_{11} (*p.u.*)	1.0998	1.0997	1.1000	1.0999	1.0990	1.0988
V_{13} (*p.u.*)	1.1000	1.0999	1.0984	1.0996	1.1000	1.0997
TC_{6-9}	1.0387	1.0380	1.0353	1.0328	1.0243	1.0215
TC_{6-10}	0.9000	0.9011	0.9022	0.9054	0.9085	0.9149
TC_{4-12}	0.9726	0.9727	0.9711	0.9736	0.9663	0.9727
TC_{27-28}	0.9628	0.9636	0.9592	0.9616	0.9588	0.9607
QC_{10} (*p.u.*)	0.0500	0.0499	0.0499	0.0493	0.0500	0.0494
QC_{12} (*p.u.*)	0.0500	0.0499	0.0498	0.0497	0.0496	0.0498
QC_{15} (*p.u.*)	0.0500	0.0499	0.0498	0.0496	0.0499	0.0495
QC_{17} (*p.u.*)	0.0500	0.0499	0.0496	0.0499	0.0499	0.0500
QC_{20} (*p.u.*)	0.0420	0.0422	0.0434	0.0415	0.0436	0.0423
QC_{21} (*p.u.*)	0.0500	0.0499	0.0496	0.0492	0.0499	0.0499
QC_{23} (*p.u.*)	0.0262	0.0263	0.0263	0.0265	0.0267	0.0269
QC_{24} (*p.u.*)	0.0500	0.0500	0.0500	0.0499	0.0500	0.0490
QC_{29} (*p.u.*)	0.0224	0.0228	0.0213	0.0220	0.0217	0.0226
SVD (*p.u.*)	2.0973	2.0888	2.0686	2.0611	2.1076	2.0564

optimal position of STATCOM and its voltage rating obtained by BBO, KHA, TLBO, GWO, CRO and HCRO are shown in the Table 2. The simulation results show that using STATCOM the transmission loss has substantially reduced for all the algorithms. Moreover, the results indicate that the HCRO algorithm gives more reduction in loss (4.4812 MW) compared to CRO (4.5297 MW), GWO (4.4887 MW), TLBO (4.4903 MW), KHA (4.4936 MW), BBO (4.5037 MW). . It is also observed that all parameter settings of STATCOM and SSSC device are within the specified limits.

In order to further check the feasibility of the different methods for another complicated network, it is applied to solve ORPD with SSSC of the same test system. The simulation results of transmission loss, the controlled variables, the optimal position of SSSC and its voltage rating obtained by BBO, KHA,

Table 2. Comparison of simulation results obtained by different algorithms with STATCOM

Real power loss minimization						
Algorithms	HCRO	CRO	GWO	TLBO	KHA	BBO
V_1 (*p.u.*)	1.0920	1.1000	1.0930	1.0906	1.0970	1.0741
V_2 (*p.u.*)	1.0822	1.0943	1.0850	1.0815	1.0866	1.0630
V_5 (*p.u.*)	1.0636	1.0748	1.0461	1.0634	1.0577	1.0452
V_8 (*p.u.*)	1.0668	1.0767	1.0515	1.0562	1.0677	1.0483
V_{11} (*p.u.*)	1.0998	1.0997	1.0689	1.0987	1.0626	1.0892
V_{13} (*p.u.*)	1.0993	1.0951	1.0787	1.0967	1.0920	1.0908
TC_{6-9}	1.0003	1.0393	1.0414	0.9504	1.0575	1.0263
TC_{6-10}	0.9227	0.9000	0.9108	1.0066	0.9182	0.9105
TC_{4-12}	0.9720	0.9725	1.0205	0.9668	1.0079	0.9876
TC_{27-28}	0.9553	0.9629	0.9806	0.9615	1.0098	0.9535
QC_{10} (*p.u.*)	0.0469	0.0500	0.0210	0.0346	0.0494	0.0464
QC_{12} (*p.u.*)	0.0482	0.0500	0.0373	0.0384	0.0128	0.0471
QC_{15} (*p.u.*)	0.0490	0.0500	0.0298	0.0345	0.0274	0.0280
QC_{17} (*p.u.*)	0.0489	0.0500	0.0324	0.0450	0.0483	0.0410
QC_{20} (*p.u.*)	0.0405	0.0419	0.0458	0.0483	0.0410	0.0433
QC_{21} (*p.u.*)	0.0495	0.0500	0.0438	0.0463	0.0355	0.0432
QC_{23} (*p.u.*)	0.0262	0.0261	0.0355	0.0430	0.0486	0.0312
QC_{24} (*p.u.*)	0.0496	0.0500	0.0342	0.0382	0.0475	0.0499
QC_{29} (*p.u.*)	0.0230	0.0223	0.0317	0.0250	0.0386	0.0196
Optimal location	23	30	29	29	30	24
STATCOM voltage (*p.u.*)	1.0347	1.0790	1.0544	1.0681	1.0882	1.0623
STATCOM angle (deg.)	-8.3613	-10.7025	-7.4278	-13.4091	-8.9007	-11.1320
SVD (*p.u.*)	1.7231	2.0869	2.1047	2.0036	1.3524	2.0347
TL (MW)	4.4812	4.5297	4.4887	4.4903	4.4936	4.5037

TLBO, GWO, CRO and HCRO are shown in the Table 3. The simulation results show that using SSSC the transmission loss has substantially reduced for all the algorithms. Moreover, the results indicate that the HCRO algorithm gives more reduction in loss (4.4832 MW) compared to other reported algorithms.

Table 3. Comparison of simulation results obtained by different algorithms with SSSC

Real power loss minimization						
Algorithms	**HCRO**	**CRO**	**GWO**	**TLBO**	**KHA**	**BBO**
V_1 (*p.u.*)	1.0863	1.0908	1.0803	1.0876	1.0767	1.0696
V_2 (*p.u.*)	1.0778	1.0865	1.0675	1.0764	1.0692	1.0613
V_5 (*p.u.*)	1.0520	1.0590	1.0438	1.0548	1.0474	1.0402
V_8 (*p.u.*)	1.0636	1.0613	1.0516	1.0604	1.0496	1.0393
V_{11} (*p.u.*)	1.0913	1.0980	1.0978	1.0976	1.0981	1.0966
V_{13} (*p.u.*)	1.0938	1.0937	1.0989	1.0996	1.0995	1.0989
TC_{6-9}	1.0134	0.9997	0.9711	1.0128	1.0007	0.9869
TC_{6-10}	0.9127	0.9648	0.9268	0.9018	0.9130	0.9162
TC_{4-12}	0.9841	0.9868	0.9551	0.9639	0.9645	0.9614
TC_{27-28}	0.9521	0.9623	0.9441	0.9542	0.9530	0.9430
QC_{10} (*p.u.*)	0.0459	0.0467	0.0486	0.0486	0.0497	0.0485
QC_{12} (*p.u.*)	0.0493	0.0499	0.0498	0.0499	0.0496	0.0479
QC_{15} (*p.u.*)	0.0472	0.0439	0.0484	0.0483	0.0481	0.0468
QC_{17} (*p.u.*)	0.0488	0.0498	0.0487	0.0498	0.0499	0.0488
QC_{20} (*p.u.*)	0.0390	0.0472	0.0420	0.0425	0.0418	0.0437
QC_{21} (*p.u.*)	0.0499	0.0449	0.0497	0.0480	0.0481	0.0477
QC_{23} (*p.u.*)	0.0372	0.0326	0.0262	0.0268	0.0288	0.0304
QC_{24} (*p.u.*)	0.0426	0.0482	0.0499	0.0487	0.0469	0.0479
QC_{29} (*p.u.*)	0.0229	0.0178	0.0266	0.0253	0.0286	0.0253
Optimal location	25-26	10-22	22-24	22-24	10-17	10-17
SSSC voltage (*p.u.*)	0.0891	0.0518	0.0647	0.0313	0.0802	0.0616
SSSC angle (deg.)	12.4804	-09.4518	-10.8243	11.9256	-11.7106	16.0034
SVD (*p.u.*)	1.6813	1.7234	1.8942	1.9139	1.7143	1.7092
TL (MW)	4.4832	4.5218	4.4893	4.4925	4.4971	4.5127

Case 2 (Voltage Deviation Minimization)

To assess the feasibility of the proposed approaches, they are applied to solve voltage deviation minimization objective. The results obtained by BBO, KHA, TLBO, GWO, CRO and HCRO for this objective function without FACTS devices are reported of Table 4 It is observed that voltage deviation obtained using HCRO is better as compared to that obtained by BBO, KHA, TLBO, GWO and CRO algorithms.

Table 4. Comparison of simulation results obtained by different algorithms without FACTS

Voltage deviation minimization						
Algorithms	**HCRO**	**CRO**	**GWO**	**TLBO**	**KHA**	**BBO**
V_1 (*p.u.*)	1.0081	1.0092	1.0193	1.0207	1.0220	1.0209
V_2 (*p.u.*)	1.0031	1.0050	1.0179	1.0192	1.0207	1.0187
V_5 (*p.u.*)	1.0198	1.0195	1.0175	1.0196	1.0210	1.0186
V_8 (*p.u.*)	1.0039	1.0031	1.0009	0.9994	0.9965	0.9986
V_{11} (*p.u.*)	1.0278	1.0390	1.0699	1.0429	1.0407	1.0318
V_{13} (*p.u.*)	1.0264	1.0144	0.9910	1.0023	1.0075	1.0176
$TC_{6\text{-}9}$	1.0420	1.0551	1.0920	1.0510	1.0455	1.0419
$TC_{6\text{-}10}$	0.9021	0.9001	0.9000	0.9111	0.9109	0.9016
$TC_{4\text{-}12}$	1.0076	0.9937	0.9424	0.9614	0.9721	0.9903
$TC_{27\text{-}28}$	0.9655	0.9663	0.9676	0.9580	0.9579	0.9589
QC_{10} (*p.u.*)	0.0500	0.0499	0.0485	0.0500	0.0470	0.0395
QC_{12} (*p.u.*)	0.0106	0.0424	0.0017	0.0011	0.0009	0.0005
QC_{15} (*p.u.*)	0.0500	0.0500	0.0500	0.0484	0.0500	0.0499
QC_{17} (*p.u.*)	0.0000	0.0000	0.0001	0.0004	0.0001	0.0019
QC_{20} (*p.u.*)	0.0500	0.0500	0.0500	0.0500	0.0499	0.0494
QC_{21} (*p.u.*)	0.0500	0.0500	0.0500	0.0500	0.0500	0.0497
QC_{23} (*p.u.*)	0.0500	0.0500	0.0500	0.0495	0.0497	0.0493
QC_{24} (*p.u.*)	0.0500	0.0500	0.0500	0.0499	0.0499	0.0465
QC_{29} (*p.u.*)	0.0248	0.0258	0.0275	0.0171	0.0191	0.0190
SVD (*p.u.*)	0.0891	0.0894	0.0920	0.0957	0.0984	0.1005
TL (MW)	5.8637	5.8067	5.5749	5.5627	5.5640	5.5206

In order to further improve the voltage deviation, STATCOM is optimally placed in the IEEE 30-bus system using the proposed algorithms. It is observed from the simulation results of Table 5 that voltage deviation is improved by incorporating STATCOM from 0.0891 *p.u.* to 0.0774 *p.u.* by HCRO, from 0.0894 *p.u.* to 0.0803 *p.u.* by CRO, from 0.0920 *p.u.* to 0.0814 *p.u.* by GWO, from 0.0957 *p.u.* to 0.0883 *p.u.* by TLBO, from 0.0984 *p.u.* to 0.0905 *p.u.* by KHA, from 0.1005 *p.u.* to 0.0917 *p.u.* by BBO.

In order to validate the effectiveness of the proposed methods, the SSSC based ORPD is used as another case study. The optimal settings of control variables (i.e. voltages of generator buses, tap settings of the regulating transformers, var injections of the shunt capacitors, optimal location of SSSC,

Table 5. Comparison of simulation results obtained by different algorithms with STATCOM

Voltage deviation minimization						
Algorithms	**HCRO**	**CRO**	**GWO**	**TLBO**	**KHA**	**BBO**
V_1 (*p.u.*)	1.0335	1.0293	1.0256	1.0213	1.0387	1.0216
V_2 (*p.u.*)	1.0316	1.0272	1.0255	1.0160	1.0416	1.0125
V_5 (*p.u.*)	1.0228	1.0220	1.0198	1.0191	1.0307	1.0206
V_8 (*p.u.*)	0.9996	0.9997	1.0031	0.9971	0.9848	0.9980
V_{11} (*p.u.*)	1.0160	0.9983	1.0217	1.0366	0.9941	1.0162
V_{13} (*p.u.*)	0.9888	1.0177	0.9960	1.0223	0.9947	1.0488
TC_{6-9}	0.9991	0.9773	1.0274	1.0399	0.9551	1.0271
TC_{6-10}	0.9250	0.9341	0.9058	0.9171	0.9454	0.9020
TC_{4-12}	0.9466	0.9808	0.9453	0.9998	0.9626	1.0486
TC_{27-28}	0.9451	0.9564	0.9463	0.9602	0.9418	0.9764
QC_{10} (*p.u.*)	0.0300	0.0284	0.0336	0.0499	0.0459	0.0450
QC_{12} (*p.u.*)	0.0185	0.0000	0.0086	0.0004	0.0432	0.0006
QC_{15} (*p.u.*)	0.0487	0.0425	0.0270	0.0498	0.0495	0.0393
QC_{17} (*p.u.*)	0.0008	0.0109	0.0002	0.0044	0.0001	0.0035
QC_{20} (*p.u.*)	0.0497	0.0454	0.0498	0.0497	0.0498	0.0497
QC_{21} (*p.u.*)	0.0495	0.0459	0.0483	0.0498	0.0495	0.0491
QC_{23} (*p.u.*)	0.0490	0.0437	0.0497	0.0496	0.0498	0.0488
QC_{24} (*p.u.*)	0.0488	0.0457	0.0499	0.0499	0.0496	0.0492
QC_{29} (*p.u.*)	0.0022	0.0209	0.0022	0.0213	0.0082	0.0495
Optimal location	24	22	29	29	24	13
STATCOM voltage (*p.u.*)	1.0643	1.0408	1.0718	1.0823	1.0137	1.0448
STATCOM angle (deg.)	-11.4326	-6.0648	-11.3709	-10.1356	-13.4602	-9.8113
SVD (*p.u.*)	0.0774	0.0803	0.0814	0.0883	0.0905	0.0917
TL (MW)	5.5232	6.1345	5.5746	5.5167	5.9981	5.5902

voltage magnitude and phase angle of SSSC), transmission loss and SVD that are obtained by the different methods are shown in the Table 6. The results clearly show that the voltage deviation has improved after incorporating SSSC. It may also be noticed in Table 6 that the objective function value (SVD) obtained by HCRO is better than those obtained by BBO, KHA, TLBO, GWO and CRO.

Table 6. Comparison of simulation results obtained by different algorithms with SSSC

Voltage deviation minimization						
Algorithms	**HCRO**	**CRO**	**GWO**	**TLBO**	**KHA**	**BBO**
V_1 (*p.u.*)	1.0092	1.0180	1.0178	1.0186	1.0297	1.0230
V_2 (*p.u.*)	1.0042	1.0147	1.0153	1.0108	1.0196	1.0219
V_5 (*p.u.*)	1.0248	1.0198	1.0263	1.0221	1.0195	1.0274
V_8 (*p.u.*)	0.9937	1.0034	0.9842	1.0004	0.9987	0.9947
V_{11} (*p.u.*)	1.0267	1.0112	1.0598	1.0022	1.0130	0.9820
V_{13} (*p.u.*)	1.0476	1.0156	1.0274	1.0359	1.0151	1.0149
$TC_{6\text{-}9}$	1.0322	1.0243	0.9871	1.0145	1.0188	0.9886
$TC_{6\text{-}10}$	0.9057	0.9015	1.0024	0.9001	0.9188	0.9000
$TC_{4\text{-}12}$	1.0472	0.9906	1.0153	1.0191	0.9903	0.9941
$TC_{27\text{-}28}$	0.9602	0.9561	0.9539	0.9512	0.9447	0.9524
QC_{10} (*p.u.*)	0.0492	0.0499	0.0452	0.0492	0.0443	0.0500
QC_{12} (*p.u.*)	0.0001	0.0055	0.0291	0.0001	0.0241	0.0337
QC_{15} (*p.u.*)	0.0500	0.0497	0.0495	0.0321	0.0484	0.0499
QC_{17} (*p.u.*)	0.0024	0.0009	0.0255	0.0025	0.0378	0.0006
QC_{20} (*p.u.*)	0.0500	0.0495	0.0500	0.0497	0.0458	0.0500
QC_{21} (*p.u.*)	0.0500	0.0499	0.0498	0.0497	0.0424	0.0499
QC_{23} (*p.u.*)	0.0500	0.0498	0.0498	0.0472	0.0380	0.0499
QC_{24} (*p.u.*)	0.0500	0.0496	0.0490	0.0492	0.0488	0.0500
QC_{29} (*p.u.*)	0.0240	0.0136	0.0209	0.0117	0.0021	0.0165
Optimal location	15-18	12-13	15-18	10-17	10-17	15-18
SSSC voltage (*p.u.*)	0.0209	0.0368	0.0543	0.0716	0.0861	0.0743
SSSC angle (deg.)	-6.0813	-13.3616	-9.1345	-12.1387	-8.1260	-9.0047
SVD (*p.u.*)	0.0786	0.0811	0.0820	0.0897	0.0918	0.0923
TL (MW)	4.8940	5.6188	5.8193	5.6702	5.4623	5.7219

Case 3 (Minimization of L-Index Voltage Stability)

To further investigate the efficiency of the different intelligent optimization algorithms, it is applied on the same IEEE 30-bus system to minimize voltage stability index. Tables 7, 8 and 9 show the optimal settings of control variables, optimal locations and optimal parameter setting of STATCOM and SSSC obtained by applying BBO, KHA, TLBO, GWO, CRO and HCRO techniques for normal, STATCOM based ORPD and SSSC based ORPD problems. It is observed from the simulation results that after installing STATCOM the voltage stability index is significantly reduced. It is also observed that the best voltage deviation is obtained using HCRO method for all the cases (i.e. without Facts, with STATCOM and with SSSC).

Robustness

In order to check the robustness, the simulations of normal ORPD for transmission loss, voltage deviation and voltage stability index minimization are carried out for 50 trials with the proposed BBO, KHA, TLBO, GWO, CRO and HCRO methods and their statistical results are compared with

Table 7. Comparison of simulation results obtained by different algorithms without FACTS

Voltage stability index minimization						
Algorithms	**HCRO**	**CRO**	**GWO**	**TLBO**	**KHA**	**BBO**
V_1 (*p.u.*)	1.0987	1.0998	1.0972	1.0910	1.0953	1.0947
V_2 (*p.u.*)	1.0968	1.0939	1.0997	1.0914	1.0799	1.0910
V_5 (*p.u.*)	1.0991	1.0743	1.0992	1.0925	1.0891	1.0800
V_8 (*p.u.*)	1.0990	1.0762	1.0998	1.0988	1.0942	1.0994
V_{11} (*p.u.*)	1.0982	1.0997	1.0996	1.0998	1.0760	1.0656
V_{13} (*p.u.*)	0.9506	1.0999	0.9521	0.9536	0.9595	0.9552
TC_{6-9}	0.9009	1.0380	0.9071	0.9048	0.9211	0.9374
TC_{6-10}	0.9030	0.9011	0.9032	0.9031	0.9105	0.9191
TC_{4-12}	1.0998	0.9727	1.0960	1.0956	1.0950	1.0837
TC_{27-28}	0.9012	0.9636	0.9064	0.9068	0.9015	0.9038
QC_{10} (*p.u.*)	0.0500	0.0499	0.0496	0.0488	0.0498	0.0351
QC_{12} (*p.u.*)	0.0004	0.0499	0.0005	0.0003	0.0145	0.0020
QC_{15} (*p.u.*)	0.0001	0.0499	0.0000	0.0002	0.0178	0.0043
QC_{17} (*p.u.*)	0.0500	0.0499	0.0484	0.0489	0.0389	0.0230
QC_{20} (*p.u.*)	0.0497	0.0422	0.0474	0.0442	0.0133	0.0357
QC_{21} (*p.u.*)	0.0496	0.0499	0.0457	0.0459	0.0337	0.0292
QC_{23} (*p.u.*)	0.0499	0.0263	0.0473	0.0494	0.0264	0.0375
QC_{24} (*p.u.*)	0.0487	0.0500	0.0497	0.0467	0.0494	0.0454
QC_{29} (*p.u.*)	0.0493	0.0228	0.0497	0.0493	0.0466	0.0454
SVD (*p.u.*)	1.8791	2.0888	1.8438	1.7982	1.5382	1.4695
TL (MW)	7.9377	4.5322	7.8093	7.8644	7.4118	7.0389
L- index	0.1148	0.1156	0.1240	0.1252	0.1259	0.1287

Table 8. Comparison of simulation results obtained by different algorithms with STATCOM

Voltage stability index minimization						
Algorithms	HCRO	CRO	GWO	TLBO	KHA	BBO
V_1 (*p.u.*)	1.1000	1.1000	1.0972	1.0996	1.0739	1.0578
V_2 (*p.u.*)	1.0994	1.0943	1.0998	1.1000	1.0944	1.0482
V_5 (*p.u.*)	1.0972	1.0748	1.0992	1.0993	1.0803	1.0821
V_8 (*p.u.*)	1.0985	1.0767	1.0983	1.0997	1.0923	1.0926
V_{11} (*p.u.*)	1.0994	1.0997	1.0993	1.0684	1.0785	1.0943
V_{13} (*p.u.*)	0.9592	1.0951	0.9529	0.9504	0.9538	0.9931
TC_{6-9}	0.9047	1.0393	0.9002	1.0979	0.9306	0.9787
TC_{6-10}	0.9006	0.9000	0.9028	1.0970	0.9632	0.9993
TC_{4-12}	1.0988	0.9725	1.0990	1.0992	1.0956	1.0755
TC_{27-28}	0.9021	0.9629	0.9006	0.9008	0.9023	0.9344
QC_{10} (*p.u.*)	0.0483	0.0500	0.0491	0.0212	0.0047	0.0334
QC_{12} (*p.u.*)	0.0012	0.0500	0.0000	0.0008	0.0496	0.0014
QC_{15} (*p.u.*)	0.0009	0.0500	0.0000	0.0000	0.0366	0.0287
QC_{17} (*p.u.*)	0.0487	0.0500	0.0500	0.0105	0.0363	0.0195
QC_{20} (*p.u.*)	0.0493	0.0419	0.0500	0.0003	0.0153	0.0286
QC_{21} (*p.u.*)	0.0481	0.0500	0.0499	0.0014	0.0444	0.0247
QC_{23} (*p.u.*)	0.0499	0.0261	0.0488	0.0000	0.0003	0.0228
QC_{24} (*p.u.*)	0.0483	0.0500	0.0487	0.0019	0.0427	0.0353
QC_{29} (*p.u.*)	0.0491	0.0223	0.0491	0.0499	0.0456	0.0465
Optimal location	14	30	14	23	14	14
STATCOM voltage (*p.u.*)	0.9863	1.0790	1.0349	0.9583	1.0207	1.0444
STATCOM angle (deg.)	-7.1782	-10.7025	-12.9929	-14.1845	9.0004	-17.1034
SVD (*p.u.*)	1.9183	2.0869	1.8941	1.3262	1.3311	1.4279
TL (MW)	7.6840	4.5297	7.9222	6.6915	7.4203	7.3116
L- index	0.1135	0.1142	0.1198	0.1207	0.1214	0.1240

those of GA-1 (Subburaj et al., 2007) and GA-2 (Devaraj & Roselyn, 2010). The statistical results reported in Table 10 shows that the best, worst and the average results obtained by the discussed methods are very near to each other and the variation is negligible for HCRO algorithm. It is observed that the worst, average, and best values of the fitness values achieved by HCRO are much better than those of other methods. These results clearly indicate the robustness of the solutions obtained by HCRO.

Table 9. Comparison of simulation results obtained by different algorithms with SSSC

Voltage stability index minimization						
Algorithms	HCRO	CRO	GWO	TLBO	KHA	BBO
V_1 (*p.u.*)	1.0947	1.0517	1.0781	1.0993	1.0739	1.0642
V_2 (*p.u.*)	1.0862	1.0841	1.0827	1.0311	1.0944	1.0944
V_5 (*p.u.*)	1.0616	1.0579	1.0069	1.0925	1.0803	1.0899
V_8 (*p.u.*)	1.0993	1.0978	1.0931	1.0926	1.0923	1.0955
V_{11} (*p.u.*)	1.0836	1.0662	1.0771	1.0986	1.0785	1.0938
V_{13} (*p.u.*)	0.9503	0.9541	0.9676	0.9533	0.9538	0.9747
TC_{6-9}	0.9026	0.9084	0.9003	0.9003	0.9306	0.9472
TC_{6-10}	0.9107	0.9129	0.9124	0.9124	0.9225	0.9050
TC_{4-12}	1.0977	1.0733	1.0939	1.0939	1.0956	1.0881
TC_{27-28}	0.9042	0.9069	0.9049	0.9049	0.9023	0.9065
QC_{10} (*p.u.*)	0.0415	0.0075	0.0424	0.0155	0.0047	0.0302
QC_{12} (*p.u.*)	0.0236	0.0044	0.0017	0.0339	0.0186	0.0230
QC_{15} (*p.u.*)	0.0045	0.0300	0.0042	0.0105	0.0366	0.0110
QC_{17} (*p.u.*)	0.0426	0.0001	0.0360	0.0135	0.0363	0.0161
QC_{20} (*p.u.*)	0.0465	0.0190	0.0158	0.0194	0.0153	0.0400
QC_{21} (*p.u.*)	0.0361	0.0036	0.0452	0.0379	0.0444	0.0421
QC_{23} (*p.u.*)	0.0466	0.0165	0.0198	0.0473	0.0003	0.0459
QC_{24} (*p.u.*)	0.0491	0.0462	0.0407	0.0314	0.0205	0.0488
QC_{29} (*p.u.*)	0.0428	0.0460	0.0410	0.0490	0.0329	0.0497
Optimal location	14-15	18-19	10-22	10-22	15-18	10-22
SSSC voltage (*p.u.*)	0.0261	0.0673	0.9878	0.0348	0.0076	0.0142
SSSC angle (deg.)	-13.4193	-9.0093	-9.9878	-14.1674	-12.1787	-10.1864
SVD (*p.u.*)	1.6370	1.1982	1.3733	1.4004	1.3221	1.6528
TL (MW)	7.7138	8.4008	8.6189	11.8324	7.6218	7.0763
L- index	0.1132	0.1144	0.1191	0.1203	0.1209	0.1233

OPF Result of IEEE 30-Bus System

Case 1 (Cost Minimization)

To analyze the performance of the different evolutionary methods under the overload condition, the simulations are conducted at 1.6 times the base load condition. In this case, two UPFC devices are optimally placed using HCRO, CRO, GWO, TLBO, KHA and BBO algorithms to solve security constraints

Table 10. Statistical comparison (50 trials) among various algorithms for IEEE 30-bus without FACTS

Real power loss minimization								
Techniques	**HCRO**	**CRO**	**GWO**	**TLBO**	**KHA**	**BBO**	**GA-1 (Subburaj, Sudha, Rajeswari, Ramar, and Ganesan, 2007)**	**GA-2 (Devaraj, and Roselyn, 2010)**
Best loss (MW)	4.5304	4.5322	4.5432	4.5401	4.5594	5.1170	4.5800	4.5550
Mean loss (MW)	4.5308	4.5413	4.5477	4.5516	4.5601	NA	NA	NA
Worst loss (MW)	4.5313	4.5476	4.5513	4.5595	4.5617	NA	NA	NA
Voltage deviation minimization								
Techniques	HCRO	CRO	GWO	TLBO	KHA	BBO	DE-1 (Abou, Ela, Abido, & Spea, 2011)	
Best Voltage deviation (MW)	0.0891	0.0849	0.0984	0.0913	0.0856	0.2424	0.0911	
Mean Voltage deviation	0.0894	0.0863	0.0997	0.0934	0.0872	NA	NA	
Worst Voltage deviation	0.0897	0.0898	0.1003	0.0988	0.0907	NA	NA	
Voltage stability index minimization								
Techniques	HCRO	CRO	GWO	TLBO	KHA	BBO	DE-1 "Abou, Ela, Abido, & Spea, (2011)"	
Best L index	0.1148	0.1156	0.1259	0.1252	0.1242	0.1397	0.1246	
Mean L index	0.1151	0.1163	0.1268	0.1254	0.1245	NA	NA	
Worst L index	0.1156	0.1172	0.1274	0.1258	0.1247	NA	NA	

OPF problem. The generator active power, the active and reactive power injected by UPFC devices are taken as the independent variables where as the load voltages, power flow through the transmission lines, voltage and phase angle of series and parallel voltage sources of UPFC devices are considered as dependent variables. It is worth mentioning that the comparison is carried out with the same system data and constraints. The results of this comparison are given in Table 11. Moreover, it is observed that the total cost obtained using HCRO is less compared to those obtained by CRO, GWO, TLBO, KHA and BBO.

Table 11. Simulation results obtained by different techniques for OPF incorporating UPFC (IEEE 30-bus system)

Techniques	HCRO	CRO	GWO	TLBO	KHA	BBO
P_{g1} (MW)	202.20	203.10	214.38	207.94	211.23	212.37
P_{g2} (MW)	94.16	100.30	99.18	99.79	101.39	100.82
P_{g5} (MW)	33.90	27.99	29.34	32.60	35.61	37.59
P_{g8} (MW)	71.53	76.41	75.24	78.13	70.47	68.43
P_{g11} (MW)	32.74	31.46	32.47	32.09	33.48	31.88
P_{g13} (MW)	36.05	31.94	30.83	31.57	30.85	34.56
Q_{g1} (MVAr)	-13.10	-43.85	35.41	39.47	-33.47	-42.31
Q_{g2} (MVAr)	55.91	22.58	21.03	-21.03	29.82	37.12
Q_{g5}(MVAr)	39.69	38.47	36.48	32.06	38.72	30.15
Q_{g8} (MVAr)	38.91	42.31	49.62	45.37	50.07	51.38
Q_{g11} (MVAr)	23.21	22.83	20.27	26.73	22.16	28.37
Q_{g13} (MVAr)	23.45	22.96	22.96	25.82	29.17	20.58
Optimal location of UPFC1	21-22	21-22	21-22	21-22	6-2	28-6
Optimal location of UPFC2	27-28	15-18	15-18	15-18	27-28	15-18
Series source voltage of UPFC1 (*p.u.*)	0.0145	-0.0349	0.0241	-0.0762	-0.0346	0.0159
Series source voltage of UPFC2 (*p.u.*)	-0.0995	0.0144	-0.0388	0.0167	-0.0870	-0.0421
Series source phase angle of UPFC1 (deg)	15.5928	25.2459	32.09	56.18	111.93	-22.47
Series source phase angle of UPFC2 (deg)	45.1001	16.8928	-23.38	-12.47	60.48	22.05
Shunt source voltage of UPFC1 (*p.u.*)	1.0092	0.9978	0.9872	1.0515	1.0234	0.9904
Shunt source voltage of UPFC2 (*p.u.*)	0.9774	1.0090	1.0645	1.0178	0.9947	1.0472
Shunt source phase angle of UPFC1 (deg)	14.8648	15.1464	13.4506	14.1623	07.9807	08.4723
Shunt source phase angle of UPFC2 (deg)	2.8432	15.8514	17.9013	11.0034	01.2348	-09.1583
Power loss (MW)	26.14	17.76	28.00	28.68	29.59	32.21
Active power cost ($/hr)	1547.84	1573.44	1570.81	1581.17	1584.75	1598.07
Reactive power cost ($/hr)	19.75	20.69	20.37	19.28	24.08	21.43
Cost of UPFC ($/hr)	16.92	16.92	16.83	17.42	17.94	17.16
Total cost ($/hr.)	1584.5	1611.0	1608.01	1617.86	1626.78	1636.7

NA means not available

Case 2 (Transmission Loss Minimization)

In this case, the HCRO, CRO, GWO, TLBO, KHA and BBO techniques are applied on IEEE 30 bus system to minimize the transmission loss. The optimal control variables settings corresponding to the best value of transmission loss without and with UPFC are given in Tables 12 and 13, respectively. The active power losses achieved by the proposed HCRO algorithm without and with UPFC are equal to 2.4174 MW and 1.9257 MW respectively, while they are equal to 2.7346 MW and 2.2583 MW by the CRO method, 2.6315 MW to 2.1846 MW using GWO, 2.7410 MW to 2.2946 MW using TLBO, 2.7676 MW to 2.3122 MW using KHA and 2.8070 MW to 2.3607 MW using BBO. As can be derived from the results, the HCRO algorithm with UPFC gives the best performance in comparison with the other methods.

Case 3 (Voltage Deviation Minimization)

Here, the proposed approaches are implemented to search for the minimization of voltage deviation. Initially, this problem is solved using HCRO, CRO, GWO, TLBO, KHA and BBO without incorporating any FACTS devices. Later, it is applied to UPFC based OPF problem. The optimal number of UPFC devices, their position and parameter settings of this case are adjusted in such a way that the system voltage profile is further reduced. The simulation results obtained using HCRO, CRO, GWO, TLBO, KHA and BBO are tabulated in Tables 14 and 15, respectively, for normal OPF and UPFC based OPF. It is clear from the simulation results that the voltage profile is greatly improved incorporating UPFC. It is reduced from 0.2033 *p.u.* to 0.1536 for HCRO, from 0.2478 *p.u.* to 0.1947 *p.u.* for CRO, from 0.2462 *p.u.* to 0.1926 *p.u.* using GWO, from 0.2629 *p.u.* to 0.2013 *p.u.* using TLBO, from 0.2718 *p.u.* to 0.2146 *p.u.* using KHA and from 0.2729 *p.u.* to 0.2238 *p.u.* using BBO.

Case 4 (Simultaneous Minimization of Transmission Loss and Voltage Deviation)

The best compromising solutions of transmission loss and voltage deviation along with optimal settings of control variables by HCRO, CRO, GWO, TLBO, KHA and BBO algorithms, without UPFC and with UPFC are listed in Tables 16 and 17, respectively. It is evident that the HCRO approach outperforms the other algorithms for both normal OPF and UPFC based

Table 12. Simulation results obtained by different techniques for Transmission loss minimization without UPFC (IEEE-30 bus system)

Techniques	Transmission loss minimization					
	HCRO	CRO	GWO	TLBO	KHA	BBO
P_{g1} (MW)	50.4275	50.0369	51.1193	53.5814	52.0189	51.1552
P_{g2} (MW)	48.9636	22.9428	32.7243	23.0187	45.8706	43.6925
P_{g5} (MW)	97.8007	97.2364	138.3096	141.1903	114.7431	139.2410
P_{g8} (MW)	68.6869	84.4384	60.0923	63.4256	59.4717	36.5037
P_{g11} (MW)	72.3743	59.7798	58.4147	53.7011	43.4193	49.5141
P_{g13} (MW)	0.1851	42.5503	16.2213	22.0739	41.4940	36.9505
V_{g1} (*p.u.*)	1.0893	1.0449	1.0117	1.0109	1.0594	1.0359
V_{g2} (*p.u.*)	1.0823	1.0352	1.0103	1.0081	1.0546	1.0347
V_{g5} (*p.u.*)	1.0711	1.0179	1.0057	1.0198	1.0425	1.0355
V_{g8} (*p.u.*)	1.0735	1.0315	1.0088	1.0091	1.0493	1.0273
V_{g11} (*p.u.*)	1.0998	1.0766	1.0810	1.0714	1.0652	1.0962
V_{g13} (*p.u.*)	1.0998	1.0549	1.0972	1.0404	1.0596	1.0972
Q_{C10} (*p.u.*)	0.0497	0.0379	0.0458	0.0279	0.0352	0.0499
Q_{C12} (*p.u.*)	0.0498	0.0458	0.0394	0.0488	0.0485	0.0488
Q_{C15} (*p.u.*)	0.0499	0.0467	0.0444	0.0465	0.0487	0.0485
Q_{C17} (*p.u.*)	0.0493	0.0470	0.0482	0.0485	0.0463	0.0484
Q_{C20} (*p.u.*)	0.0499	0.0490	0.0477	0.0466	0.0375	0.0497
Q_{C21} (*p.u.*)	0.0499	0.0397	0.0471	0.0483	0.0480	0.0484
Q_{C23} (*p.u.*)	0.0491	0.0408	0.0474	0.0362	0.0477	0.0488
Q_{C24} (*p.u.*)	0.0339	0.0395	0.0496	0.0356	0.0438	0.0477
Q_{C29} (*p.u.*)	0.0319	0.0453	0.0445	0.0440	0.0204	0.0422
T_{6-9}	1.0053	1.0438	0.9158	0.9995	1.0142	0.9689
T_{6-10}	0.9051	0.9255	0.9210	0.9290	0.9692	0.9029
T_{4-12}	0.9674	0.9718	0.9563	0.9865	1.0211	0.9607
T_{28-27}	0.9515	0.9388	0.9295	0.9457	0.9572	0.9307
Ploss (MW)	2.4174	2.7346	2.6315	2.7410	2.7676	2.8070
Voltage deviation	1.9299	0.7185	1.0626	0.2468	0.6704	1.3549

Table 13. Simulation results obtained by different techniques for transmission loss minimization incorporating UPFC (IEEE 30-bus system)

Techniques	Transmission loss minimization					
	HCRO	CRO	GWO	TLBO	KHA	BBO
P_{g1} (MW)	50.1001	50.1016	51.0616	50.0806	50.0714	50.2050
P_{g2} (MW)	55.6716	52.8788	28.6928	34.2374	23.5687	26.5733
P_{g5} (MW)	86.5105	75.2212	129.5936	130.8404	147.4729	112.9526
P_{g8} (MW)	50.7586	75.7485	57.8697	57.7542	53.6744	55.0657
P_{g11} (MW)	49.8927	49.5667	56.1828	54.4778	49.4738	63.4323
P_{g13} (MW)	63.2422	52.9915	33.0341	29.1542	32.3010	48.3818
V_{g1} (*p.u.*)	1.0714	1.0630	1.0685	1.0559	1.0724	1.0437
V_{g2} (*p.u.*)	1.0640	1.0488	1.0642	1.0498	1.0672	1.0379
V_{g5} (*p.u.*)	1.0446	1.0300	1.0627	1.0476	1.0694	1.0306
V_{g8} (*p.u.*)	1.0481	1.0407	1.0578	1.0453	1.0612	1.0368
V_{g11} (*p.u.*)	1.0763	1.0994	1.0990	1.0994	1.1000	1.0930
V_{g13} (*p.u.*)	1.0999	1.0998	1.0983	1.0988	1.0999	1.0926
Q_{C10} (*p.u.*)	0.0493	0.0500	0.0498	0.0497	0.0500	0.0488
Q_{C12} (*p.u.*)	0.0485	0.0498	0.0496	0.0495	0.0500	0.0498
Q_{C15} (*p.u.*)	0.0498	0.0499	0.0500	0.0499	0.0500	0.0473
Q_{C17} (*p.u.*)	0.0466	0.0499	0.0499	0.0498	0.0499	0.0486
Q_{C20} (*p.u.*)	0.0498	0.0486	0.0497	0.0495	0.0500	0.0476
Q_{C21} (*p.u.*)	0.0488	0.0497	0.0497	0.0496	0.0500	0.0471
Q_{C23} (*p.u.*)	0.0488	0.0500	0.0498	0.0497	0.0499	0.0488
Q_{C24} (*p.u.*)	0.0499	0.0496	0.0499	0.0498	0.0500	0.0473
Q_{C29} (*p.u.*)	0.0323	0.0321	0.0353	0.0382	0.0352	0.0338
$T_{6\text{-}9}$	0.9694	0.9620	0.9996	0.9843	1.0056	0.9631
$T_{6\text{-}10}$	0.9114	0.9007	0.9000	0.9000	0.9000	0.9180
$T_{4\text{-}12}$	0.9475	0.9394	0.9738	0.9660	0.9739	0.9541
$T_{28\text{-}27}$	0.9330	0.9219	0.9436	0.9386	0.9450	0.9184
Power loss (MW)	1.9257	2.2583	2.1846	2.2946	2.3122	2.3607
Voltage deviation (*p.u.*)	1.6921	1.6580	1.7006	1.5793	1.7593	1.4388

Table 14. Simulation results obtained by different techniques for Voltage deviation minimization without UPFC (IEEE-30 bus system)

Techniques	Voltage deviation minimization					
	HCRO	CRO	GWO	TLBO	KHA	BBO
P_{g1} (MW)	167.1554	288.2736	65.8777	113.2802	80.5590	158.4097
P_{g2} (MW)	38.6332	32.8346	62.2005	34.8827	67.4008	101.2536
P_{g5} (MW)	58.0018	16.4485	146.8859	96.7813	114.1736	24.4142
P_{g8} (MW)	63.4538	10.4706	38.0423	34.4182	45.8529	11.7306
P_{g11} (MW)	21.4778	13.6830	21.0107	27.6236	24.2442	49.2538
P_{g13} (MW)	14.2613	12.7842	25.5644	55.9216	27.9836	22.1845
V_{g1} (*p.u.*)	1.0602	1.0913	1.0551	1.0096	1.0359	1.0716
V_{g2} (*p.u.*)	1.0448	1.0688	1.0154	1.0473	0.9907	1.0246
V_{g5} (*p.u.*)	1.0159	1.0383	1.0309	0.9501	1.0245	1.0326
V_{g8} (*p.u.*)	1.0090	1.0154	1.0106	0.9956	0.9962	0.9924
V_{g11} (*p.u.*)	1.0109	1.0359	1.0441	1.0895	1.0789	1.0832
V_{g13} (*p.u.*)	1.0250	1.0242	1.0550	1.0458	1.0763	1.0462
Q_{C10} (*p.u.*)	0.0349	0.0500	0.0385	0.0428	0.0413	0.0286
Q_{C12} (*p.u.*)	0.0151	0.0004	0.0335	0.0089	0.0201	0.0489
Q_{C15} (*p.u.*)	0.0489	0.0468	0.0005	0.0270	0.0143	0.0134
Q_{C17} (*p.u.*)	0.0004	0.0005	0.0111	0.0319	0.0485	0.0314
Q_{C20} (*p.u.*)	0.0499	0.0499	0.0041	0.0172	0.0318	0.0198
Q_{C21} (*p.u.*)	0.0495	0.0498	0.0041	0.0326	0.0289	0.0295
Q_{C23} (*p.u.*)	0.0496	0.0499	0.0046	0.0447	0.0269	0.0452
Q_{C24} (*p.u.*)	0.0499	0.0496	0.0304	0.0481	0.0394	0.0253
Q_{C29} (*p.u.*)	0.0160	0.0027	0.0253	0.0324	0.0277	0.0170
T_{6-9}	0.9904	1.0611	0.9799	1.0014	1.0453	0.9349
T_{6-10}	0.9029	0.9012	0.9423	0.9575	0.9689	1.0180
T_{4-12}	0.9923	0.9994	0.9756	1.0264	0.9636	1.0368
T_{28-27}	0.9484	0.9497	0.9239	0.9155	0.9568	0.9039
Ploss (MW)	8.7333	8.7035	5.3315	8.6576	5.9641	12.9964
Voltage deviation	0.2033	0.2478	0.2462	0.2629	0.2718	0.2729

Table 15. Simulation results obtained by different techniques for Voltage deviation minimization incorporating UPFC (IEEE 30-bus system)

Techniques	Voltage deviation minimization					
	HCRO	**CRO**	**GWO**	**TLBO**	**KHA**	**BBO**
P_{g1} (MW)	87.9310	129.4311	125.8564	189.1546	115.4930	132.7429
P_{g2} (MW)	85.0326	100.1917	55.3282	73.8016	76.9495	93.6333
P_{g5} (MW)	51.0028	55.9123	76.1531	52.7045	66.7213	66.8907
P_{g8} (MW)	76.5758	39.8283	36.5283	10.0000	42.4827	21.7421
P_{g11} (MW)	47.4417	12.9221	53.2837	21.6142	33.9836	34.9946
P_{g13} (MW)	12.1586	23.9107	14.2243	19.3441	26.9840	13.3118
V_{g1} (*p.u.*)	1.0647	1.0814	1.0090	1.0345	1.0336	1.0224
V_{g2} (*p.u.*)	1.0236	1.0690	1.0096	1.0307	1.0031	1.0310
V_{g5} (*p.u.*)	0.9994	1.0220	1.0258	1.0054	1.0644	1.0400
V_{g8} (*p.u.*)	1.0024	1.0124	1.0069	1.0035	0.9982	0.9860
V_{g11} (*p.u.*)	1.0449	1.0366	1.0528	1.0978	1.0680	1.0011
V_{g13} (*p.u.*)	1.0344	1.0385	1.0583	1.0785	1.0496	1.0823
Q_{C10} (*p.u.*)	0.0457	0.0413	0.0169	0.0106	0.0366	0.0297
Q_{C12} (*p.u.*)	0.0009	0.0115	0.0443	0.0398	0.0097	0.0048
Q_{C15} (*p.u.*)	0.0459	0.0478	0.0079	0.0189	0.0442	0.0414
Q_{C17} (*p.u.*)	0.0018	0.0050	0.0100	0.0229	0.0433	0.0488
Q_{C20} (*p.u.*)	0.0491	0.0499	0.0448	0.0456	0.0361	0.0456
Q_{C21} (*p.u.*)	0.0447	0.0439	0.0029	0.0350	0.0366	0.0422
Q_{C23} (*p.u.*)	0.0499	0.0493	0.0248	0.0475	0.0099	0.0159
Q_{C24} (*p.u.*)	0.0491	0.0055	0.0470	0.0394	0.0390	0.0481
Q_{C29} (*p.u.*)	0.0409	0.0103	0.0043	0.0359	0.0008	0.0496
T_{6-9}	1.0059	1.0289	0.9670	1.0589	1.0645	0.9988
T_{6-10}	0.9181	0.9117	0.9119	0.9259	0.9236	0.9194
T_{4-12}	0.9885	1.0407	1.0264	1.0677	0.9603	1.0634
T_{28-27}	0.9561	0.9534	0.9160	0.9538	0.9071	0.9182
Power loss (MW)	5.8925	7.9462	7.1238	12.3678	8.3641	9.0654
Voltage deviation (*p.u.*)	0.1536	0.1947	0.1926	0.2013	0.2146	0.2238

Table 16. Simulation results obtained by different techniques for Simultaneous minimization of transmission loss and Voltage deviation without UPFC (IEEE-30 bus system)

Techniques	Simultaneous minimization of transmission loss and Voltage deviation					
	HCRO	CRO	GWO	TLBO	KHA	BBO
P_{g1} (MW)	39.4462	92.1529	50.4399	54.5363	62.7243	63.0540
P_{g2} (MW)	137.8412	25.5425	69.5753	76.7527	63.4116	54.4006
P_{g5} (MW)	85.2123	49.8485	100.9642	85.8150	72.9143	80.8172
P_{g8} (MW)	20.2006	69.2913	51.1119	64.2669	58.1136	68.9823
P_{g11} (MW)	24.0828	70.7889	52.4637	57.8216	51.5566	52.4687
P_{g13} (MW)	52.8576	52.0410	33.7394	19.3643	49.9023	38.8016
V_{g1} *(p.u.)*	1.0475	1.0625	1.0374	1.0603	1.0302	1.0194
V_{g2} *(p.u.)*	1.0434	1.0524	1.0570	1.0599	1.0218	1.0112
V_{g5} *(p.u.)*	1.0205	1.0111	1.0034	1.0045	1.0204	0.9938
V_{g8} *(p.u.)*	1.0226	1.0216	1.0216	1.0163	1.0027	0.9782
V_{g11} *(p.u.)*	1.0636	1.0638	1.0097	0.9764	1.0398	1.0550
V_{g13} *(p.u.)*	1.0454	1.0369	1.0482	1.0531	1.0608	1.0865
Q_{C10} *(p.u.)*	0.0314	0.0023	0.0390	0.0317	0.0225	0.0079
Q_{C12} *(p.u.)*	0.0201	0.0151	0.0061	0.0030	0.0196	0.0392
Q_{C15} *(p.u.)*	0.0406	0.0476	0.0232	0.0436	0.0080	0.0307
Q_{C17} *(p.u.)*	0.0156	0.0196	0.0330	0.0131	0.0287	0.0356
Q_{C20} *(p.u.)*	0.0469	0.0498	0.0449	0.0405	0.0077	0.0086
Q_{C21} *(p.u.)*	0.0426	0.0385	0.0478	0.0285	0.0391	0.0166
Q_{C23} *(p.u.)*	0.0451	0.0339	0.0226	0.0470	0.0388	0.0406
Q_{C24} *(p.u.)*	0.0439	0.0469	0.0436	0.0377	0.0477	0.0013
Q_{C29} *(p.u.)*	0.0429	0.0119	0.0370	0.0106	0.0092	0.0342
T_{6-9}	0.9535	0.9451	0.9817	0.9627	0.9425	0.9287
T_{6-10}	1.0376	1.0303	0.9211	0.9003	0.9374	1.0556
T_{4-12}	1.0316	1.0392	1.0220	1.0532	1.0505	0.9922
T_{28-27}	0.9599	0.9535	0.9692	0.9305	0.9229	0.9014
Ploss (MW)	5.3907	5.4151	4.0444	4.3068	4.3727	4.2744
Voltage deviation	0.2871	0.2931	0.2394	0.2213	0.2260	0.3393

Table 17. Simulation results obtained by different techniques for Simultaneous minimization of transmission loss and Voltage deviation incorporating UPFC (IEEE 30-bus system)

Techniques	Simultaneous minimization of transmission loss and Voltage deviation					
	HCRO	CRO	GWO	TLBO	KHA	BBO
P_{g1} (MW)	51.2482	52.2899	50.5210	53.8341	51.2032	57.5144
P_{g2} (MW)	32.1532	61.2124	93.4889	61.1636	53.9145	75.8413
P_{g5} (MW)	81.4394	86.4943	98.3232	96.5495	98.7636	114.0206
P_{g8} (MW)	37.7616	85.3869	46.4916	70.9430	35.6642	46.0677
P_{g11} (MW)	98.4263	41.4003	23.0941	42.6842	58.4282	21.7327
P_{g13} (MW)	57.6316	31.5036	46.2638	33.0717	60.2827	43.0859
V_{g1} (*p.u.*)	1.0379	1.0708	1.0551	1.0511	1.0226	1.0446
V_{g2} (*p.u.*)	1.0303	1.0668	1.0390	1.0230	1.0127	1.0374
V_{g5} (*p.u.*)	1.0102	1.0343	0.9979	0.9779	0.9884	1.0423
V_{g8} (*p.u.*)	1.0039	1.0358	1.0078	0.9999	0.9811	1.0258
V_{g11} (*p.u.*)	1.0293	1.0185	1.0663	1.0827	1.0713	1.0985
V_{g13} (*p.u.*)	1.0242	1.0373	1.0092	1.0225	1.0122	1.0372
Q_{C10} (*p.u.*)	0.0495	0.0443	0.0360	0.0036	0.0218	0.0322
Q_{C12} (*p.u.*)	0.0036	0.0269	0.0478	0.0307	0.0179	0.0155
Q_{C15} (*p.u.*)	0.0478	0.0275	0.0227	0.0366	0.0313	0.0473
Q_{C17} (*p.u.*)	0.0068	0.0274	0.0416	0.0458	0.0177	0.0294
Q_{C20} (*p.u.*)	0.0491	0.0494	0.0492	0.0448	0.0323	0.0228
Q_{C21} (*p.u.*)	0.0164	0.0398	0.0459	0.0453	0.0174	0.0242
Q_{C23} (*p.u.*)	0.0499	0.0499	0.0380	0.0485	0.0378	0.0227
Q_{C24} (*p.u.*)	0.0494	0.0248	0.0206	0.0463	0.0416	0.0223
Q_{C29} (*p.u.*)	0.0238	0.0234	0.0128	0.0163	0.0109	0.0499
T_{6-9}	0.9591	1.0092	1.0361	1.0216	0.9365	0.9671
T_{6-10}	0.9497	0.9489	0.9187	0.9265	0.9459	1.0169
T_{4-12}	0.9863	1.0342	0.9429	0.9789	0.9582	1.0216
T_{28-27}	0.9394	0.9729	0.9289	0.9405	0.9114	0.9684
Power loss (MW)	4.5307	4.0374	3.9326	3.9961	4.0064	4.0126
Voltage deviation (*p.u.*)	0.2136	0.2441	0.2288	0.2193	0.2248	0.2377

OPF. It is also observed that the voltage profiles and transmission loss are greatly improved by incorporating UPFC devices in optimal position using all the aforesaid approaches.

CONCLUSION

In this chapter HCRO, CRO, GWO, TLBO, KHA and BBO techniques, are used to find the optimal location of STATCOM and SSSC for solving optimal reactive power dispatch (ORPD) problem and optimal placement of UPFC for solving optimal power flow (OPF) problem for single and multi-objective problem. to evaluate the algorithms' performance, minimization the transmission loss, improve the voltage profile and voltage stability are considered as the objective function. It is observed that the optimally placed STATCOM and SSSC can reduce the transmission loss, voltage deviation and voltage stability index of a power system network effectively. Moreover, HCRO produces better solutions than so far best known results by any other method. Furthermore, from the statistical comparative results, it is found that the proposed algorithms are robust and suitable for sizing and locating STATCOM and SSSC devices in power system transmission network. Considering all these results of the study for ORPD problems with different characteristics, dimensions, and constraints it can be concluded that the proposed methods perform better, at least matching many of the previously reported methods.

REFERENCES

Abou, E. I., Ela, A. A., Abido, M. A., & Spea, S. R. (2011). Differential evolution algorithm for optimal reactive power dispatch. *Electric Power Systems Research*, *81*(2), 458–464. doi:10.1016/j.epsr.2010.10.005

Amjady, N., & Ansari, M. R. (2014). Non-convex security constrained optimal power flow by a new solution method composed of Benders decomposition and special ordered sets. *International Transactions on Electrical Energy Systems*, *24*(6), 842–857. doi:10.1002/etep.1742

Basu, M. (2008). Optimal power flow with FACTS devices using differential evolution. *International Journal of Electrical Power & Energy Systems, 30*(2), 150–156. doi:10.1016/j.ijepes.2007.06.011

Bhargava, A., Pant, V., & Das, B. (2006). An improved power flow analysis technique with STATCOM. *Proceedings of the International Conference on Power Electronics, Drives and Energy Systems*, 1-5. 10.1109/PEDES.2006.344231

Devaraj, D., & Roselyn, J. P. (2010). Genetic algorithm based reactive power dispatch for voltage stability improvement. *International Journal of Electrical Power & Energy Systems, 32*(10), 1151–1156. doi:10.1016/j.ijepes.2010.06.014

Dutta, S., Mukhopadhyay, P., Roy, P. K., & Nandi, D. (2016). Unified power flow controller based reactive power dispatch using oppositional krill herd algorithm. *International Journal of Electrical Power & Energy Systems, 80*, 10–25. doi:10.1016/j.ijepes.2016.01.032

Gayatri, M. T. L., Parimi, A. M., & Kumar, A. V. P. (2018). A review of reactive power compensation techniques in microgrids. *Renewable & Sustainable Energy Reviews, 81*(Part 1), 1030–1036. doi:10.1016/j.rser.2017.08.006

Gholipour, E., & Nosratabadi, S. M. (2015). A new coordination strategy of SSSC and PSS controllers in power system using SOA algorithm based on Pareto method. *International Journal of Electrical Power & Energy Systems, 67*, 462–471. doi:10.1016/j.ijepes.2014.12.020

Goswami, L., Dutta, S., & Roy, P. K. (2017). Optimal location of shunt compensating facts device for solving ORPD problem using hybrid chemical reaction optimization. *Proceedings of the Third International Conference on Science Technology Engineering & Management (ICONSTEM)*, 550 – 555. 10.1109/ICONSTEM.2017.8261384

Grudinin, N. (1998). Reactive power optimization using successive quadratic programming method. *IEEE Transactions on Power Systems, 13*(4), 1219–1225. doi:10.1109/59.736232

Gyugyi, L. (1992). A unified power flow control concept for flexible AC transmission systems. *IEE Proceedings C- Generation Transmission and Distribution, 139*(4), 323–31.

Hagh, M. T., Sharifian, M. B. B., & Galvani, S. (2014). Impact of SSSC and STATCOM on power system predictability. *International Journal of Electrical Power & Energy Systems*, *56*, 159–167. doi:10.1016/j.ijepes.2013.11.025

Hingorani, N. G., & Gyugyi, L. (2001). Understanding FACTS: Concepts and Technology of Flexible AC Transmission Systems. Wiley-IEEE Press.

Ippolito, L., Cortiglia, A. L., & Petrocelli, M. (2006). Optimal allocation of FACTS devices by using multi-objective optimal power flow and genetic algorithms. *International Journal of Emerging Electric Power Systems*, *7*(2), 1. doi:10.2202/1553-779X.1099

Karaboga, D., & Basturk, B. (2007). A powerful and efficient algorithm for numerical function optimization: Artificial bee colony (ABC) algorithm. *Journal of Global Optimization*, *39*(3), 459–471. doi:10.100710898-007-9149-x

Keyan, L., Wanxing, S., & Yunhua, L. (2006). Research on reactive power optimization based on adaptive genetic simulated annealing algorithm. *International conference on Power System Technology, PowerCon*, 1–6.

Kim, D. H., Lee, J. H., Hong, S. H., & Kim, S. R. (1998.). A mixed-integer programming approach for the linearized reactive power and voltage control-comparison with gradient projection approach. *Proceedings of International Conference on Energy Management and Power Delivery*, 1, 67–72. 10.1109/EMPD.1998.705422

Lai, L. L., & Ma, J. T. (1995). Power flow control in FACTS using evolutionary programming. *Proceedings of IEEE International Conference on Evolutionary Computation*, 109–113. 10.1109/ICEC.1995.489126

Lai, L. L., & Ma, J. T. (1996). Power flow control with UPFC using genetic algorithms. *Proceedings of International Conference on Intelligent Systems Applications to Power Systems*, 373–377. 10.1109/ISAP.1996.501101

Lai, L. L., & Ma, J. T. (1997). Application of evolutionary programming to reactive power planning-comparison with nonlinear programming approach. *IEEE Transactions on Power Systems*, *12*(1), 198–206. doi:10.1109/59.574940

Leung, H. C., & Chung, T. S., (2000). Optimal power flow with a versatile FACTS controller by genetic algorithm approach. *Proceedings of IEEE Power Engineering Society Winter Meeting*, 2806–2811.

Lin, W., Li, Y., Xu, X., & Li, M. (2008). Reactive power optimization in area power grid based on improved Tabu search algorithm. *Proceedings of the Third International Conference on Electric Utility Deregulation and Restructuring and Power Technologies*, 1472–1477.

Mahdad, B., Srairi, K., & Bouktir, T. (2007). Flexible Methodology Based in Fuzzy Logic Rules for Reactive Power Planning of Multiple Shunt FACTS Devices to Enhance System Loadability. Proceedings of Power Engineering Society General Meeting, 285-290.

Nusair, K. N., & Alomoush, M. I. (2017). Optimal reactive power dispatch using teaching learning based optimization algorithm with consideration of FACTS device STATCOM. *Proceedings of the 10th Jordanian International Electrical and Electronics Engineering Conference (JIEEEC), IEEE Conferences*, 1 – 12.

Panda, S. (2009). Multi-objective evolutionary algorithm for SSSC-based controller design. *Electric Power Systems Research*, *79*(6), 937–944. doi:10.1016/j.epsr.2008.12.004

Raju, C. P., Vaisakh, K., & Raju, S. S. (2009). An IPM-EPSO Based Hybrid Method for Security Enhancement Using SSSC. *International Journal of Recent Trends in Engineering*, *2*(5), 208–212.

Rao, V. S., & Rao, R. S. (2017). Optimal Placement of STATCOM using Two Stage Algorithm for Enhancing Power System Static Security. *Energy Procedia*, *117*, 575–582. doi:10.1016/j.egypro.2017.05.151

Sedighizadeh, M., Faramarzi, H., Mahmoodi, M. M., & Sarvi, M. (2014). Hybrid approach to FACTS devices allocation using multi-objective function with NSPSO and NSGA-II algorithms in Fuzzy framework. *International Journal of Electrical Power & Energy Systems*, *62*, 586–598. doi:10.1016/j.ijepes.2014.04.058

Shaheen, H. I., Rashed, G. I., & Cheng, S. J. (2009). Application of differential evolution algorithm for optimal location and parameters setting of UPFC considering power system security. *International Transactions on Electrical Energy Systems*, *19*(7), 911–932.

Sirjani, R., Mohamed, A., & Shareef, H. (2012). Optimal allocation of shunt Var compensators in power systems using a novel global harmony search algorithm. *International Journal of Electrical Power & Energy Systems*, *43*(1), 562–572. doi:10.1016/j.ijepes.2012.05.068

Subburaj, P., Sudha, N., Rajeswari, K., Ramar, K., & Ganesan, L. (2007). Optimum reactive power dispatch using genetic algorithm. *Academic Open Internet Journal*, *21*. Retrieved from http://www.acadjournal.com/2007/v21/part6/p6/

Taher, S. A., & Amooshahi, M. K. (2012). New approach for optimal UPFC placement using hybrid immune algorithm in electric power systems. *International Journal of Electrical Power & Energy Systems*, *43*(1), 899–909. doi:10.1016/j.ijepes.2012.05.064

Urdaneta, A. J., Gomez, J. F., Sorrentino, E., Flores, L., & Diaz, R. (1999). A hybrid genetic algorithm for optimal reactive power planning based upon successive linear programming. *IEEE Transactions on Power Systems*, *14*(4), 1292–1298. doi:10.1109/59.801887

Venkatesh, B., George, M. K., & Gooi, H. B. (2004). Fuzzy OPF incorporating UPFC. *IEE Proceedings. Generation, Transmission and Distribution*, *151*(5), 625–629. doi:10.1049/ip-gtd:20040611

Venkatesh, P., Gnanadas, R., & Padhy, N. P. (2003). Comparison and application of evolutionary programming techniques to combined economic emission dispatch with line flow constrained. *IEEE Transactions on Power Systems*, *18*(2), 688–697. doi:10.1109/TPWRS.2003.811008

Vijayakumar, K., & Kumudinidevi, R. P. (2008). A hybrid genetic algorithm for optimal power flow incorporating FACTS devices. *Asian Journal of Scientific Research*, *1*(4), 403–411. doi:10.3923/ajsr.2008.403.411

Vinkovic, A., & Mihalic, R. (2008). A current-based model of the static synchronous series (SSSC) compensator for Newton–Raphson power flow. *Electric Power Systems Research*, *78*(10), 1806–1813. doi:10.1016/j.epsr.2008.03.006

Wei, Y., Fang, L., Chung, C. Y., & Wong, K. P. (2006). A hybrid genetic algorithm-interior point method for optimal reactive power flow. *IEEE Transactions on Power Systems*, *21*(3), 1163–1169. doi:10.1109/TPWRS.2006.879262

Zhang, X. P. (2003). Advanced modeling of multicontrol functional static synchronous series compensator (SSSC) in Newton-Raphson power flow. *IEEE Transactions on Power Systems*, *18*(4), 1410–1416. doi:10.1109/TPWRS.2003.818690

Zhao, B., Guo, C. X., & Cao, Y. J. (2005). A multiagent-based particle swarm optimization approach for optimal reactive power dispatch. *IEEE Transactions on Power Systems*, *20*(2), 1070–1107. doi:10.1109/TPWRS.2005.846064

Zonkoly, A. E. (2008). Optimal sizing of SSSC controllers to minimize transmission loss and a novel model of SSSC to study transient response. *Electric Power Systems Research*, *78*(11), 1856–1864. doi:10.1016/j.epsr.2008.03.017

Conclusion

DISCUSSION

The objective of this chapter is to provide a coherent holistic view of all the chapters.

Chapter 1 develops and implements some evolutionary optimization techniques such as artificial bee colony (ABC), biogeography-based optimization (BBO), chemical reaction optimization (CRO), hybrid chemical reaction optimization (HCRO), teaching learning-based optimization (TLBO), Grew wolf Optimization (GWO) and Krill herd algorithm (KHA) techniques to solve economic load dispatch (ELD), optimal power flow (OPF) and optimal reactive power flow (ORPD) problems. All the optimization techniques are tested for different power system problems like OPF and ORPD without and with different FACTS device. Moreover, in order to judge the superiority, the results obtained using the proposed methods are compared with those of other algorithms reported in the literature.

Chapter 2 focuses on the theoretical background of economic load dispatch (ELD), optimal power flow (OPF) and optimal reactive power flow (ORPD) problems. Moreover, the limitations of the ELD and modified ELD solution are discussed. Furthermore, both single and multi-objective OPF and ORPD are presented in this chapter.

In Chapter 3, few stochastic search techniques named BBO, TLBO, KHA, GWO, CRO, and HCRO are proposed to solve the different types of non-linear ELD problems. Five different types of non-smooth and non-convex ELD problems with the nonlinearities of valve-point loading prohibited operating zones, ramp rate constraints and multiple fuels options are used to evaluate the performance of BBO, TLBO, KHA, GWO, CRO, and HCRO algorithms. To test the effectiveness, the proposed six solution methods are compared with some of the most recently published techniques available in the literature. Moreover, to better illustrate the robustness of the proposed methods, the statistical analysis is carried out on all the test systems. The simulation results show that the proposed BBO, TLBO, KHA, GWO, CRO, and HCRO are promising methods for solving various non-convex ELD problems. This clearly suggests that the proposed methods are promising and encouraging for further research in this direction.

In Chapter 4, BBO, TLBO, CRO, KHA, GWO and HCRO techniques have been proposed, developed and successfully implemented to solve OPF and ORPD problems considering the valve point discontinuities of the generating units. The proposed methods have been tested on the modified IEEE 30-bus test system for different single and multi-objective cases with the statistical comparison. The simulation results demonstrate the effectiveness and robustness of the proposed techniques to solve OPF and ORPD problem.

In Chapter 5, the operating principle of the major FACTS controllers is demonstrated. It is intended to be accompanied by the presentation material, which will also include details from actual FACTS installation for realizing different system objectives.

Chapter 6 describes how the conventional power flow equation could be modified to include multiple FACTS controllers such as TCSC, SVC, TCPS, STATCOM, SSSC, and UPFC. It is shown that the impacts of FACTS controllers on power flow can be accommodated for by adding new entries and modifying some existing entries in the linearized Jacobian equation of the underlying system with no FACTS controllers.

In Chapter 7, aforesaid techniques (i.e. BBO, KHA, TLBO, GWO, CRO and HCRO techniques) are developed and successfully implemented to solve OPF problems incorporating TCSC and TCPS devices considering the valve point discontinuities of the thermal generators and to solved FACTS based ORPD problem. In order to validate the results, the proposed approaches are tested on the modified IEEE 30-bus test system for different cases. The solution obtained by the above techniques was efficient, with satisfactory results and significant improvement of the overloads, beyond other benefits to the operation of the electrical system as load cuts and losses.

Chapter 8 aims at evaluating the optimal location of TCSC and TCPS device using the aforesaid techniques and their effectiveness is judge by minimizing the total cost of active power generation of a power system. A SVC is a shunt FACTS device which can be controlled the bus voltage for reducing the transmission loss, voltage deviation and voltage stability. A TCSC is a series FACTS device which can be used to get controlled power flow between the lines by changing the line reactance. Similarly, TCPS is also a series FACTS device which can be used to get controlled power flow between the lines by changing the phase shifter. The IEEE 30-bus system is considered for simulation study to show the effectiveness of the proposed methods over the other conventional methods as found during reference study. The result for the tested power system showed that all the techniques BBO, TLBO, KHA, GWO, CRO and HCRO can be used for finding out the optimal location and optimal settings of the TCSC and TCPS. It is also observed that the HCRO technique implemented here has superior features including high-quality solution, stable convergence characteristics, and good computation efficiency.

In Chapter 9, the optimal location of STATCOM and SSSC for solving ORPD problem is discussed. To evaluate the algorithms' performance, mminimization of the transmission loss, improve the voltage profile and voltage stability are considered as the objective function. The effectiveness of the proposed algorithms is demonstrated by implementing it in standard systems namely IEEE 30-bus systems and their performance is compared with other optimization techniques. The superiority of the proposed techniques is established by comparing the results with other meta-heuristic optimization techniques reported in the literature. Moreover, in this chapter, the effectiveness of the optimal installation of UPFC for optimizing various single objectives and one multi-objective of IEEE 30-bus is investigated. All the optimization techniques discussed earlier are successfully implemented for optimal tuning and location of UPFC device. Simulation results reveal that the proposed algorithms work satisfactorily for both the test systems at stresses condition of 125% and 175% of the base load. Furthermore, it is found that the proposed HCRO is superior to other algorithms in terms of solution quality and convergence speed.

About the Authors

Provas Kumar Roy was born in 1973 at Mejia, Bankura, West Bengal, India. He received the BE degree in Electrical Engineering from R.E. College, Durgapur, Burdwan, India in 1997; ME degree in Electrical Machine from Jadavpur University, Kolkata, India in 2001 and PhD from NIT Durgapur in 2011. Presently he is working as Professor in the department of Electrical Engineering, Kalyani Government Engineering College, Kalyani, India. He has published more than 125 research papers in international journals and conferences. His field of research interest includes Economic Load Dispatch, Optimal Power flow, FACTS, Unit Commitment, Radial Distribution System, State Estimation, Automatic Generation Control, Power System Stabilizer and Evolutionary computing techniques.

Susanta Dutta was born in 1976 at Bankura, West Bengal, India. He received the BE degree in Electrical Engineering from Dr. B.C. Roy Engineering College, Durgapur, Burdwan, West Bengal India in 2004; M. Tech. degree from NIT Durgapur, India in 2007. He completed his PhD from NIT Durgapur in 2015. Presently he is working as Associate Professor in the department of Electrical Engineering, Dr. B.C. Roy Engineering College, Durgapur, India. His field of research interest includes Economic Load Dispatch, Optimal Power flow, FACTS, Automatic Generation Control, Evolutionary computing techniques. He has published more than 25 research papers in international journals and conferences.

Index

www.ingramcontent.com/pod-product-compliance
Lightning Source LLC
LaVergne TN
LVHW061548290926
849162LV00004B/944